BACTERIAL INDICATORS/ HEALTH HAZARDS ASSOCIATED WITH WATER

A symposium
sponsored by ASTM
Committee D-19 on Water
AMERICAN SOCIETY FOR
TESTING AND MATERIALS
Chicago, Ill., 28-29 June 1976

ASTM SPECIAL TECHNICAL PUBLICATION 635
A. W. Hoadley, Georgia Institute of Technology,
and B. J. Dutka, Canada Centre for Inland Waters,
editors

List price $34.75
04-635000-16

AMERICAN SOCIETY FOR TESTING AND MATERIALS
1916 Race Street, Philadelphia, Pa. 19103

Library of Congress Catalog Card Number: 77-78026

NOTE

The Society is not responsible, as a body,
for the statements and opinions
advanced in this publication.

Printed in Tallahassee, Fla.
November 1977

Foreword

The International Symposium on Bacterial Indicators of Potential Health Hazards Associated With Water was held 28–29 June 1976 at Chicago, Ill. The American Society for Testing and Materials' Committee D-19 on Water sponsored the symposium, and A. W. Hoadley, Georgia Institute of Technology, served as symposium chairman. B. J. Dutka, Canada Centre for Inland Waters, and A. W. Hoadley served as editors of this publication.

The editors would like to thank R. Bordner, V. Cabelli, and L. Vlassoff for their help in organizing the symposium and chairing the sessions.

Related
ASTM Publications

Biological Monitoring of Water and Effluent Quality, STP 607 (1977), $24.25, 04-607000-16

Water Quality Parameters, STP 573 (1975), $29.50, 04-573000-16

Biological Methods for Assessment of Water Quality, STP 528 (1973), $16.25, 04-528000-16

A Note of Appreciation to Reviewers

This publication is made possible by the authors and, also, the unheralded efforts of the reviewers. This body of technical experts whose dedication, sacrifice of time and effort, and collective wisdom in reviewing the papers must be acknowledged. The quality level of ASTM publications is a direct function of their respected opinions. On behalf of ASTM we acknowledge their contribution with appreciation.

ASTM Committee on Publications

Editorial Staff

Contents

Introduction

In June 1976, the ASTM Section D19.01.04 on Microbiology of Subcommittee D.19.01 on Biological Monitoring of Committee D-19 on Water sponsored an international symposium on "Bacterial Indicators of Potential Health Hazards Associated with Water" in Chicago. The objectives of this symposium were to provide an opportunity for task groups of the Microbiology Section to report on their activities, views, and recommendations concerning the use of indicator groups in the estimation of water quality; introduce North American microbiologists to water quality standards applied in Europe and South Africa, their rationale, and methodology; identify research needs and provide a basis for establishing priorities for research and surveillance, and provide an authoritative source of background information to assist the individual or public agency establishing water quality objectives and standards, selecting bacterial indicators of potential health hazards, and interpreting water quality data.

The ASTM has established task groups which are reviewing and reevaluating concepts of indicator bacteria in water and the application of various indicator bacteria and bacterial pathogens in the evaluation of water quality for a variety of uses. This effort comes at a time when similar efforts are being carried out nationally and internationally through the World Health Organization and the International Organization for Standardization. This is a time also when concepts and applications of indicator bacteria are being challenged throughout the world. As a consequence many standards are being reexamined.

The reader will find in *Bacterial Indicators/Potential Health Hazards Associated with Water* an authoritative source of background information to assist the individual or public agency establishing water quality objectives and standards, selecting bacterial indicators of potential health hazards, and interpreting water quality data. North American, European, and South African views concerning the significance of indicator bacteria and pathogens in water are presented.

The book contains an introductory section dealing with historical perspectives of the application of indicator bacteria and the development of bacterial standards of water quality in different parts of the world. Of special interest are papers dealing with automated monitoring techniques and emerging concepts in the classification of environmental enteric strains. This section is followed by papers devoted to the classification, ecology, and significance of bacterial indicators and pathogens, including total coliforms, fecal coliforms, *Escherichia coli*, *Klebsiella*, vibrios, *Yersinia*, fecal streptococci, staphylococci, clostridia, *Pseudomonas aeruginosa*, and *Bifidobacterium*. Other chapters deal extensively with r-factors and salmonellae. In the concluding chapters, the use of these bacteria in the evaluation of water quality for water supply, shellfish, recreation and swimming pools are examined.

Through the medium of the ASTM symposium and this publication, we have attempted to bring the reader a current review on the traditional indicators used in water quality assessment in North America, Europe, and South Africa. We also have provided the reader with discussions on the classification, significance, and ecology of indicator pathogens of emerging interest such as *Yersinia*, staphylococci, *Bifidobacterium*, *Candida albicans*, vibrios, klebsiellae, and *Pseudomonas aeruginosa*. Other subjects to which we have tried to introduce the reader are the use of an automated biochemically based technique to monitor water quality and the potential significance of transferable drug resistance.

In retrospect, while there are always subjects which one would wish to add or develop further, we feel that our objectives have been largely satisfied. We feel that this volume does provide the reader with an insight into the background of bacteriological standards applied in North America, Europe, and South Africa. We also feel that we have provided a source of information on the significance and ecology of bacterial indicators of health hazards.

In our closing discussion we have attempted to summarize the conclusions reached by individual authors and identify some of the questions which remain. Perhaps these questions can be pursued in greater detail in a later edition.

We would like to acknowledge the support of the following companies in helping to make this symposium a success—Gelman Instrument Company and Sartorius Filters, Inc., which made possible the participation of the foreign speakers, Difco Laboratories, the Millipore Filter Corporation, Can-Lab, Fisher Scientific (Canada), and Med-Ox Chemicals Ltd.

A. W. Hoadley
School of Civil Engineering, Georgia Institute of Technology, Atlanta, Ga.; symposium chairman and coeditor.

B. J. Dutka
Microbiology Laboratories, Canada Centre for Inland Waters, Ontario, Canada; coeditor.

Yehuda Kott[1]

Current Concepts of Indicator Bacteria

REFERENCE: Kott, Yehuda, "**Current Concepts of Indicator Bacteria**," *Bacterial Indicators / Health Hazards Associated With Water, ASTM STP 635*, A. W. Hoadley and B. J. Dutka, Eds., American Society for Testing and Materials, 1977, pp. 3–13.

ABSTRACT: Pathogenic bacteria excreted in human feces are found in low counts in wastewater and are known to be more sensitive to environmental conditions than *Escherichia coli*. The latter, excreted both by humans and warm-blooded animals, are used as an indicator of fecal wastes which may contain the pathogen, but the validity of this practice is often questioned. In wells recharged with chlorinated lake water, regrowth of coliform bacteria reached 10^5 / 100 ml when the water was re-pumped 7 to 28 days later. Very often, the more sensitive *E. coli* die away and are replaced by other coliforms, so that counts do not in themselves indicate a change in water quality.

Recently, low numbers of *Samonella* have been detected with the aid of large sampling procedures. On the other hand, there has been little additional data on the efficiency of wastewater treatment plants. Survival varied from 0.003 up to 29 percent in a recent study. *S. typhimurium* survived in a sand column up to 44 days, while *S. typhi* did not last more than seven days. *Shigella flexneri*, introduced into an experimental oxidation pond, were recovered from the effluent but were killed by application of 8 mg / l chlorine after a short contact time.

The possibility of using coliphage as a means for identifying *E. coli* has been considered in an additional investigation. Coliphage are found in wastewater effluents in counts ranging from 10^4 to 10^5 / 100 ml. They are quite durable under adverse conditions and have been found to be highly resistant to chlorination. Enterovirus densities in wastewater vary between 100 and 2000/ 100 ml. Comparative studies have indicated that coliphage are found whenever enteroviruses are, and their ratio may vary from 1000 to one to as low as ten to one. It is believed that this recent technique employing indicator microorganisms will permit more rapid and precise evaluations of water quality and potential health hazards.

KEY WORDS: bacteria, chlorination, enteroviruses, indicator bacteria, pathogenic bacteria, recharge wells, wastewater, water

The rapid progress in the establishment of basic principles of bacteriology, on the one hand, and the current largescale migration from villages to metropolitan areas, on the other, have necessitated health safeguards in many areas including water quality.

[1]Professor, Environmental Engineering Laboratories, Technion-Israel Institute of Technology, Haifa, Israel.

The slow exchange of research information at the beginning of the century resulted in a diversity of pollution indicators and techniques for their measurements being adopted in different countries. The forementioned limitations caused the use of an indicator for water pollution.

The minimum requirement for an indicator is that it must be a biotype that is prevalent in sewage and excreted by humans or warm-blooded animals. In addition, the indicator should be present in greater abundance than pathogenic bacteria, incapable of proliferation—or at least not more capable than enteric bacteria, more resistant to various disinfectants than the pathogenic bacteria, and quantifiable by simple and rapid laboratory procedures.

Only a few of the numerous fecal microorganisms found in wastewater, and capable of contamination of drinking water, could be considered as pollution indicators. The less frequent pathogenic bacteria in wastewater such as *Salmonella, Shigella,* and *Vibrio* were ruled out because of difficulties of isolation, while attempts to use *Mycobacteria, Pasteurella, Entamoeba* cysts, *Leptospira*, and some others failed from the outset. Indicators most commonly used today are still the earliest ones considered—*Clostridium perfringens*, *Streptococcus faecalis*, the coliform bacteria (or more specifically, fecal coliforms), and *Escherichia coli.*

E. coli and Fecal Coliforms as the Most Commonly Accepted Indicators

Presently existing standard methods contain instructions on the procedures of water sampling and examination of coliforms, fecal coliforms, and *E. coli* counts, based on numerous studies (beginning with those by Escherich and others) [*1–3*].[2] The multiple-tube procedure is universally practicable, while membrane filtration is subject to limitations and unsuitable for turbid water. Fecal coliform counts necessitate selective media ana an elevated temperature. The IMVIC classification is cumbersome and cannot yield a clear-cut answer as to the origin of the isolated bacteria unless they are true ++–– *E. coli*. The uncertainty as to the best method for quantifying *E. coli* was illustrated by Kampelmacher [*4*], who compared nine media used for quantitative isolation of *E. coli* in water. Using six enrichment media, three confirmation media, and two incubation temperatures, he found that for combined detection of *E. coli* and the coli-aerogenes group the formate glutamate medium was the best. No decrease in *E. coli* counts was noticed using the most probable number (MPN) technique—frequently higher numbers were obtained as compared with the direct Eijkam 44c test [*4*]. It was also found that when *E. coli* was used in a comparative study of membrane filters, the density obtained was influenced markedly by the type of filter used [*5*]. Dutka [*6*] reported that for river water examined by three different procedures, counts were up to six times higher by one method than by the other two.

[2]The italic numbers in brackets refer to the list of references appended to this paper.

Coliform bacteria are found in practically all natural waters, and their populations should be markedly reduced by treatment, including disinfection. Chemical treatment very often includes disinfection. According to Burman, chlorine-damaged *E. coli* can be resuscitated at a Teepol medium on incubation of the samples for 4 h at 30°C with the temperature subsequently raised [*7*].

The foregoing studies suggest the question as to whether we should attempt to isolate and resuscitate coliform bacteria after treatment, or whether we should investigate the ability of damaged bacteria to develop in the human body. This point is very often overlooked. So far regrowth of coliform bacteria in good quality water after chlorination has received very little attention. Most studies dealing with regrowth mention the reproductive ability of surviving bacteria after wastewater treatment [*8,9*]. Table 1 shows that most of the coliform bacteria in secondary effluents examined were *E. coli*. With time, the latter decreased considerably, and other coliforms appeared in their place during ponding. While coliform densities decreased with time and chlorination, there was a significant change in the distribution of biotypes. This problem was examined in a winter recharge operation in Israel, where Lake Kinnereth water was chlorinated and pumped into wells

TABLE 1—*Coliform population in ponded secondary wastewater before and after chlorination.*[a]

Holding Time in Days	Total Coliform Count 100 ml	Percent of Total Coliforms			Remarks
		E. coli	Entero bacter-aerogenes	Other Coliforms	
Wastewater control	2.3×10^7	83	0	6	...
21	...	37	12	45	some colonies were not coliforms
56	7.3×10^3	8	29	58	
Wastewater before chlorination	4.6×10^7	...	...	...	...
Wastewater after chlorination	9.5×10^4	45	4	29	some colonies were not colonies
9	...	4	0	91	...
14	18×10^5	0	0	0	all isolated colonies were not coliforms[b]

[a]A 70 000 m^3 pond was used; secondary trickling filter effluents were pumped to the pond. In the chlorination experiments 20 mg/l of chlorine were applied for 2 h of contact before pumping.

[b]IMVIC test from isolated colonies grown on membrane filters.

with a coliform count of less than 2/100 ml; when the water was repumped after holding for 7 to 28 days, the count had increased to about 10^5/100 ml [*10,11*]. For this reason, the Israel drinking water standard (1974) states that when drinking water is pumped from a dual-purpose well,[3] it will have to be examined and found to contain no fecal coliforms, no fecal streptococcus, and no *Salmonella*; in addition, no more than 2 coliforms/100 ml will be allowed, and the water must be disinfected [*12*].

It is seen that, although coliform bacteria appear as the reference indicator in the standard, there are difficulties in isolation, identification, and interpretation of the results. However, it should be borne in mind that fecal coliform bacteria are not natural inhabitants of water and not necessarily attributable to fecal pollution. Recently published studies reported the presence of *Klebsiella* (some of whose biochemical reactions are identical with fecal coliforms) in floated timber and preliminarily attributed them to pollution of the water by other than fecal waste. In the presence of fecal coliforms in most water pollution studies, however, false positive results caused by *Klebsiella* are unlikely to occur.

Enterococci as Pollution Indicators

Fecal streptococci (more specifically, enterococci) are excreted regularly by human beings in lower abundance than *E. coli*. *Streptococcus faecalis* has been suggested as a pollution indicator by Allen [*13*]. A comprehensive study of stream pollution by fecal streptococci by Geldreich and Kenner revealed a *S. faecalis* biotype associated with vegetation and another associated with insects. *S. bovis* and *S. equinus* are not found in human feces, their sensitivity and rapid dieaway outside the animal gut would suggest that they can be found only close to a lower animal source. Fjerdingstad [*14*] suggested routine incubation at 10 and 45°C for identification of *S. faecalis* and at 10 and 50°C for *S. faecium*. Levin et al [*15*] demonstrated the superiority of the enterococcus method; on the average, recoveries were an improvement on the KF method by one order of magnitude.

From the scanty information presented, it can be seen that Enterococcus counts, whether by the MPN methods or by filtration, do not satisfy the criteria for pollution indicators and can serve at best as confirmation of the fecal nature of pollution in water in doubtful cases.

Current Trends

Improved scientific communication has accelerated the development of new methods, especially in the fields of membrane filtration, selective inhibitors, new media, and culture techniques. These, in turn, have been ap-

[3]In the winter, these wells are recharged with quality drinking water, and, in the summer, the water is pumped out.

plied in the development of methods for the isolation of the pathogens themselves.

Salmonella are carried and potentially spread to man by all types of animals, including those used for food. While their isolation from human clinical materials presents no problem, recovery of small populations from large volumes of water is a complex undertaking. The pad technique used first by Moore [*16*] proved to be a relatively satisfactory qualitative method. Since then, other concentration methods have been developed which have been employed to quantify *Salmonella* by the MPN technique, using tetrathionate broth or some other enrichment medium. Recoveries by the various methods have been compared for water, wastewater, treated wastewater, natural runoff, river water, and seawater [*17–21*]; however, quantitative data on *Salmonella* densities in raw wastewater, the extent of their reduction during treatment, and the final count in effluents are meager [*22*]. Grunnet in his comprehensive study [*23*] found *Salmonella* counts in raw wastes ranging from zero to 540 in 100 ml, and, with few exceptions, the count was usually less than ten in the final effluent. Further calculations, based on that study, are shown in Table 2, from which it may be seen that in raw wastes the ratio was one *Salmonella* to about 10^5 *E. coli*; in the effluent and after treatment, this ratio was generally one to 10^3 and one to 10^4. The percent survival of *Salmonella* after treatment fluctuates over a very wide range—from 0.003 to as high as 20 percent (arithmetic average 7.5 percent).

In another study [*24*], survival of *S. typhimurium* in a sand column was studied. The bacteria were suspended in tap water and passed through the column. Samples (wet) of the sand were taken and examined periodically. Figure 1 indicates that the organism count decreased continuously, faster in the first 25 days and more slowly later; after about 50 days, the organisms were undetectable. When the column was charged with lake water, *S. typhi-*

TABLE 2—Salmonella *survival in wastewater treatment plant.*[a]

Sample Serial No.	Ratio of *E. coli* to *Salmonella* Inlet	Outlet	Percent Survival
1	5.8×10^5	1.7×10^4	29.3
2	8.7×10^5	1.6×10^5	18.39
3	9.8×10^5	5.8×10^4	5.92
4	2.1×10^5	5.6×10^4	26.6
5	3.4×10^6	2.1×10^4	0.62
6	5.7×10^5	7.6×10^4	13.3
7	4.5×10^6	4.2×10^3	0.09
8	1.9×10^6	7.0×10^4	0.003
9	5.5×10^5	1.6×10^5	29
10	7.1×10^5	2.2×10^3	0.3
11	2.1×10^6	1.6×10^5	7.6
12	2.3×10^7	5.5×10^4	0.24
13	4.3×10^6	7.8×10^3	0.18

[a]All figures are based on calculated results taken from Table IX.2 of Grunnet's work (Ref *23*).

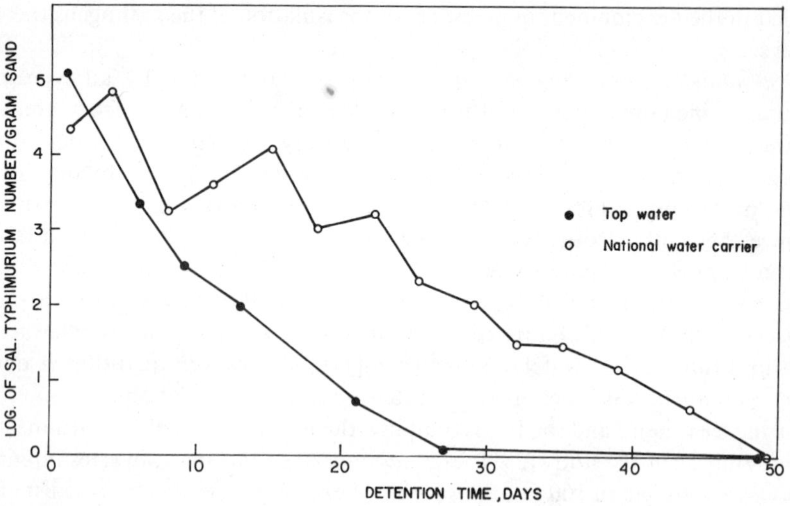

FIG. 1—*Survival of* S. typhimurium *bacteria in experimental column.*

murium suspension, algal cell debris, and other suspended matter, the *Salmonella* count could be expected to increase because of the organic material. The diagram indicates, however, that it decreased gradually and again approached zero after 50 days. When the experiment was repeated with *S. typhi* the bacteria disappeared after seven days, which indicates even higher sensitivity to the adverse conditions prevailing [*24,25*].

Shigella species are the etiological agents for acute diarrhea, most often transmitted by person-to-person contact. They are found only rarely in mammals and fish and, accordingly, were not included in recommendations for isolation from natural waters [*26*]. It is commonly accepted that *Shigella* can be transferred by wastewater or wastewater effluents, especially since about 7 percent of persons infected may be carriers for approximately a year. The need for reuse of treated wastewater in agriculture motivated a research project in which *Shigella flexneri*, type 2a was examined in a 70 litre experimental oxidation pond, with about 1000 bacteria introduced daily into raw wastewater for a seven day retention time. The effluents were examined by the MPN method in tubes containing liquid *Salmonella-Shigella* (SS) medium. After incubation of the tubes for 24 at 37°C, a loopful was streaked on SS agar plates with sulfathiazole. The colonies grown on that medium were identified as *Shigella flexneri* type 2a using a specific antiserum. Under steady operation, it was observed that of 21 effluent samples examined, only one was positive; it contained two *Shigella*/100 ml. When a dialysis tube containing 10^5 *Shigella*/100 ml was suspended in the experimental pond, not a single *Shigella* bacterium was isolated 48 h later (from 100-ml samples). In a similar experiment, the tube was suspended in effluent autoclaved at 121°C for 20 min and the *Shigella* level remained at 10^5/100 ml after four days; after 8 days, the density dropped to 7.9×10^3/100 ml. These results seem to indicate that in the absence of any other bacterial activity the *Shigella* survive for longer periods; in any case, they probably are limited in their ability to utilize organic matter found in the wastewater and die off in this environment. In still another series, one litre beakers were filled with the experimental pond fluid and 2.4×10^7 100 ml *Shigella* cells were introduced into each beaker. One beaker was exposed to sunshine (light intensity $16 \times 10^3 - 60 \times 10^3$ lx) and the other stored in darkness. The exposed bacteria disappeared within 24 h, while the unexposed ones survived at the level of about 1000 (per 100 ml) after 24 h [*27*]. Survival time in a sand column was much shorter than for *Salmonella*. The *Shigella* disappeared within 14 days [*25*]. In conclusion, the various results indicate that *Shigella flexneri*, type 2a are very sensitive to adverse environmental conditions, and those who advocate their exclusion as true water pollution bacteria are apparently right. Although 4000 confirmed shigellosis cases were officially recorded in Israel in 1975, it is believed that none of them were waterborne.

V. cholerae are regarded as highly pathogenic and are capable of causing epidemics. They are, however, extremely environment sensitive and often cannot survive under adverse conditions. According to Geldreich's [*28*] up-to-date historical review, their isolation under epidemic conditions is relatively easy. Studies of the epidemic (due to contaminated vegetables) which occurred in Jerusalem in 1970 [*29–31*], while dealing with various aspects of the findings, provided no answers to some very important questions including the fate of *V. cholerae* (El Tor) in oxidation ponds—a mode of treatment very common in rural areas which supply food to the cities. The bacteria, whose optimum growth pH is known to be about 8.0, could possibly start growing in this environment. On the other hand, Kott [*32*] demonstrated that the dieaway of the *V. cholerae* (El Tor) strain in oxidation ponds can be observed by the MPN technique. Although external conditions were favorable (pH, temperature, biochemical oxygen demand (BOD) below 35 mg/l) dieaway was rapid, and, from thousands of *Vibrio* in 100 ml applied to the raw wastewater, only 2.2/100 ml survived. Application of 8 mg/l of chlorine for ½ contact time yielded negative results.

In other experiments, these bacteria survived for three weeks in sterile wastewater and sterile secondary effluents; in a raw wastewater, they disappeared within 24 h.

Concepts of Indicators Measured at Real Time

An example of how recent techniques can be used to advantage in bacteriological research is Bachrach's [*33*] attempt to follow the release of $^{14}CO_2$ from ^{14}C-lactose by coliform bacteria incubated at 37°C; the method permitted detection of one of ten organisms within 6 h of incubation. Another approach permitting rapid results is based on the application of over 90 strains of *E. coli* bacteriophage filtered through a membrane filter (where, after exposure of the attached *E. coli* bacteria, other coliforms can be expected to develop) with a parallel test on an unexposed filter. The result, obtained by subtracting the colonies from the total count, yielded the *E. coli* count with over 90 percent accuracy within 18 to 24 h of incubation [*34*]. Thus, the method permits monitoring the proportion of *E. coli* bacteria in the total coliform population. It should be mentioned, however, that in the absence of similar data from other sources, it is not clear whether these 90 strains are the most prevalent or only specific to the region studied and, thus, incapable of universal application.

The use of *E. coli* bacteriophage as a tool for estimation of pollution is not new and was examined on several occasions in the past [*35–37*]. When introducing a new indicator, it should be borne in mind that superiority is obligatory. The advantage of the coliphage is the certainty of a final answer in less than 24 h. A MPN technique was developed for coliphage which permitted the counting of two organisms in a 100-ml sample [*38*], using membrane filters which accommodate large volumes of water [*39*]. Table 3 shows that the coliphage population in wastewater does not fluctuate. When coliphage were subjected to chlorination in oxidation ponds, their survival rate at pH 8.3 was significant; at pH 6.0, the decrease was more pronounced (as seen from Table 4), but the organisms were still present even at very high chlorine concentrations.

The high resistance of coliphage to adverse conditions and their inability to multiply in wastewater could make them a good indicator for enteroviruses. In principle, if this is proved, the analysis can be carried out in any laboratory (subject to a correlation being established between the coliphage and enteroviruses present in a sample), while a specialized laboratory and trained manpower are necessary for isolation of enteroviruses from wastewater. In addition, the isolation and incubation processes for enteroviruses are time consuming.

The various isolation methods are compared in Table 5, which is based on results from the Haifa trickling filter plant and shows the pattern of seasonal fluctuation. A number of research teams are currently attempting [*40–43*] to

TABLE 3—*Seasonal variation of coliphage in wastewater treatment plant.*[a]

Type of Wastewater	Season[b]			
	Winter[c]	Spring[c]	Summer[c]	Fall[c]
Trickling filter influent	1.8×10^6	7.6×10^6	8.1×10^6	3.8×10^6
Trickling filter effluent	9.8×10^5	4.6×10^6	2.5×10^5	1.8×10^5
Oxidation pond effluent	6.6×10^4	1.6×10^5	1.9×10^4	4.2×10^4

[a]Arithmetic mean of 4 to 12 samples in each season.
[b]Winter—December, January, Febraury; spring—March, April, May; summer—June, July, August, September; fall—October, November.
[c]Number of coliphage in 100 ml.

TABLE 4—*Percent survival of bacteriophage after chlorination in oxidation ponds effluent.*

	Chlorine Added mg/l[a]							
	20 pH		40 pH		60 pH		80 pH	
Bacteriophage Tested	8.3	6.0	8.3	6.0	8.3	6.0	8.3	6.0
Coliphages on *E. coli*, strain B	56	13.5	5.8	1.05	31	0.2	15	0.04
Coliphages on *E. coli*, strain K_{12}	52.1	19.1	17.6	0.5	28.9	0.3	13.1	0.01
f2	43.5	46.5	64	21.5	30.6	38	42.6	13
MS_2	25.5	21.2	22.6	9.5	7.4	2.1	10.6	3.4

[a]Contact time 1 h.

refine and simplify the available methods with a view to a standard technique.

In the meantime, there is reason to believe, on the basis of the present state of the art, that coliphage are at least as resistant to all natural adverse conditions, as well as to chemical and physical treatment as the enteroviruses

TABLE 5—*Number and ratio of enteric viruses in secondary effluents isolated by different methods.*

	Enteric Viruses in 100 ml			Ratio of Isolation	
Season[a]	Direct Inoculation[b]	Beef Extract[c]	Alginate	Direct Beef	Direct Alginate
Winter	400	150	20	2.6	20
	300	200	10	1.5	30
	400	70	12	5.7	33
	180	68	17	2.6	14
	200	80	12	2.5	16
	150	35	10	4.2	15
Spring	300	240	56	1.3	5
	720	200	41	3.6	18
	800	272	70	2.9	11
	560	240	65	2.3	9
Summer	1700	780	145	2.1	11
	1880	745	140	2.5	13
	1850	940	30	1.9	61
	2325	950	187	2.4	12
	975	560	20	1.7	48
	1800	1186	30	1.5	60
Fall	2350	640	35	3.6	67
	875	400	32	2.1	15
	825	315	55	2.6	15
	1175	430	5	2.7	235
	860	560	18	1.5	47

[a]Winter—December, January, February; spring—March, April, May; summer—June, July, August, September; fall—October, November.

[b]5 ml samples of secondary effluents were treated with high concentration of antibiotics incubated for 2 h at 37°C and inoculated on Boufallo green monkey (BGM) cell line.

[c]100 ml samples were treated with HCl to lower pH to 3.0 followed by filtration on membrane filter (Millipore) 0.45 micron pore size, elution of viral particles was done by filtering 3 percent beef extract through the membrane filter. The filtered beef extract was inoculated to BGM cell line.

TABLE 6—*Microorganisms population in a long-term holding reservoir of secondary effluents.*

Microorganisms Tested/100 ml	Holding Time in Days							
	Control[a]	8	15	29	34	40	47	73
Control	2.3×10^7	1.0×10^5	2.8×10^4	2.8×10^4	2.4×10^4	4.6×10^4	7.3×10^3	1.8×10^4
Fecal coliforms	1.0×10^5	1.0×10^5	2.8×10^3	3.6×10^3	2.4×10^3	2.7×10^3	3.0×10^2	2.4×10^3
Fecal streptococcus	1.1×10^6	1.0×10^4	5.8×10^3	2.0×10^3	5.1×10^3	1.3×10^3	7.8×10^2	5.0×10^2
E. coli strain B bacteriophage	7.8×10^6	1.4×10^6	7.9×10^5	7.9×10^5	3.3×10^6	7.0×10^5	1.3×10^5	1.3×10^4
Enterovirus	1.1×10^3	5.2×10^2	1.5×10^2	1.5×10^2	5.5×10^1	5.0×10^1	6.0×10^1	0

[a]Secondary effluents.

[44]. Table 6 shows a follow-up of the dieaway of enteropathogens in secondary effluents, with the decrease in human viruses much steeper than for coliphage. In natural waters (rivers, lakes, and springs) enteroviruses may range from 1000 to one to as low as ten to one. Thus, it may be concluded that future indicator standards should be based on ability to monitor a microorganism in the search for an immediate answer.

References

[1] *The Bacteriological Examination of Water Supplies*, Dept. of Health and Social Security, Welsh Office, Ministry of Housing, and Local Govern. Rep. No. 71 London 1969.
[2] PHS Drinking Water Standards 1962, U.S. Dept. of Health, Education and Welfare, Washington, D.C. 1962.
[3] International Standards for Drinking Water, 3rd ed., World Health Organization, Geneva, 1971.
[4] Kampelmacher, E. H., *Water Research*, Vol. 10, No. 4, April 1976, pp. 285–288.
[5] Hufham J. B., *Applied Microbiology*, Vol. 27, No. 4, April 1974, pp. 771–776.
[6] Dutka, B. J., *Journal of Environmental Health*, Vol. 36, No. 1, July/Aug. 1973, pp. 39–43.
[7] Burman, N. P., *Proceedings*, Society for Water Treatment Examination, Vol. 16, 1967, p. 40.
[8] Kott, Y. and Ben-Ari, H., *Water Research*, Vol. 1, 1967, pp. 451–459.
[9] Shuval, H. I. Cohen, J., and Kolodney, R., *Water Research*, Vol. 7, 1972, pp. 537–546.
[10] Eren, J., Reports, Mekoroth Water Co. Ltd. (Hebrew), 1967.
[11] Yitzhaki, J., *Israeli Journal of Medical Science*, Vol. 7, No. 9, Sept. 1971, p. 1103.
[12] Israel Drinking Water Standard, Jan. 17, 1974, Clause 14.
[13] Allen, L., Pierce, M., and Hazel Smith, *Journal of Hygiene*, Vol. 51, 1953, pp. 458–467.
[14] Fjerdingstad E., *Schweizerische Zeitschrift fur Hydrologie*, Vol. 32, 1970, pp. 429–438.
[15] Levin, M. A., Fisher, J. R., and Cabelli, V., *Journal of Applied Microbiology*, Vol. 30, No. 1, 1975, pp. 66–71.
[16] Moore, B. Perry, E. L., and Chard, S. T., *Journal of Hygiene*, Vol. 50, 1952, pp. 137–156.
[17] Greenberg, A. E. and Ongerth, J. H., *Journal of the American Water Works Association*, Vol. 58, No. 9, Sept. 1966, pp. 1145–1150.
[18] Geldreich, E. E., Best, L. C., Kenner, B. A., and Van Donsel, D. J., *Journal of the Water Pollution Control Federation*, Vol. 40, 1968, p. 1861.
[19] Dutka, B. J., Collins, P. G., Bell, J. B., and Vanderpost, J. M., Manuscript Report KR-70-1, Division of Public Health Engineering, Dept. of National Health and Welfare, Bacteriological Laboratories, Kingston, Ontario, 1970.
[20] Seligman, R. and Reitler, R., *Journal of the American Water Works Association*, Vol. 57, 1965, pp. 1572.
[21] Levin, M. A., Fisher, J. R., and Cabelli, V., *Journal of Applied Microbiology*, Vol. 28, No. 3, Sept. 1974, pp. 515–517.
[22] Butler, C. E. and Ludovici, P. P., *Journal of the Water Pollution Control Federation*, Vol. 41, No. 5, May 1969, pp. 738–744.
[23] Grunnet, K., *Salmonella in Sewage and Receiving Wastes*, Fadls Forlag, Copenhagen, 1975.
[24] Yitzhaki, J. "The Behaviour of Enterobacteriaceae in Recharging to Aquifer," D.SC. thesis, Technion, Israel Institute of Technology, Haifa, Israel, Sept. 1974.
[25] Kott, Y. unpublished data, 1971.
[26] "Analytical Control," *U.S. Environmental Protection Agency Newsletter*, July 1975.
[27] Slijkhus, H., Betzer, N., and Kott, Y., *Proceedings*, Microbiology Society of Israel, Dec. 1975.
[28] Geldreich, E. E. in *Water Pollution Microbiology*, R. Mitchell, Ed., Wiley-Interscience, New York, 1972, pp. 207–241.
[29] Gerichter, C. B., Sechter, I., Cohen, J., and Davis, M., *Israel Journal of Medical Science*, Vol. 9, No. 8, 1973 pp. 980–985.
[30] Gerichter, C. B., Sechter, I., and Cohen D. *Israel Journal of Medical Science*, Vol. 8, 1972, pp. 531.

[*31*] Sechter, I. Gerichter, C. B., and Cohen D., *Applied Microbiology*, Vol. 29, June 1975, pp. 814–818.
[*32*] Kott, Y. and Betzer, N., *Israel Journal of Medical Science*, Vol. 8, No. 12, 1972, pp. 1912 1916.
[*33*] Bachrach, U. and Bachrach, Z., *Applied Microbiology*, Vol. 28, No. 2, Aug. 1974, pp. 169–171.
[*34*] Buras, N. and Kott Y. in *Advances in Water Pollution Research*, S. H. Jenkins, Ed., Pergamon Press, 1972, pp. 73–81.
[*35*] Kott, Y. *Public Institute of Marine Science*, Vol. 11, 1966, pp. 1–6.
[*36*] Kott, Y. and Ben-Ari, H., *International Oceanographic Medicine*, Vol. 9, 1968, pp. 207.
[*37*] Suner, J. and Pinol, J. in *Advances in Water Pollution Research*, Vol. 3, J. Paz Maroto and J. Josa, Eds., Water Pollution Control Federation, Washington, D.C. 1967, pp. 105.
[*38*] Kott, Y., *Applied Microbiology*, Vol. 14, No. 2, March 1966, pp. 141–144.
[*39*] Kott, Y., "Indicators of Enteric Viruses in Waste and Other Waters," Annual Report, R80-1950, USEPA, Technion Research and Development Foundation, Haifa, Israel, 1975.
[*40*] Buras, N., *Water Research*, Vol. 10, 1976, pp. 295–298.
[*41*] Berg, G., Dahling, D. R., and Berman, D., *Applied Microbiology*, Vol. 22, No. 4, 1971, pp. 668.
[*42*] Jakobowski, W., Hill, W. F., and Clark N. A., *Applied Microbiology*, Vol. 30, No. 1, 1975, pp. 58–65.
[*43*] Scarpino, P. V. in *Water and Water Pollution Handbook*, Vol. 2, L. L. Ciaccio, Ed., 1971, pp. 639–761.
[*44*] Kott, Y. Roze, N. Sperber, S., and Betzer, N., *Water Research*, Vol. 8, 1974, pp. 165 171.

L. J. McCabe[1]

Epidemiological Consideration in the Application of Indicator Bacteria in North America

REFERENCE: McCabe, L. J., "**Epidemiological Consideration in the Application of Indicator Bacteria in North America,**" *Bacterial Indicators/Health Hazards Associated With Water, ASTM STP 635*, A. W. Hoadley and B. J. Dutka, Eds., American Society for Testing and Materials, 1977, pp. 15–22.

ABSTRACT: The microbiological needs of the epidemiologist for the investigation of waterborne disease outbreaks are reviewed. The density of indicators needs to be determined during outbreaks, but the isolation of pathogens from the water provides more definitive data. It is suggested that the measurement of chlorine residual is more useful than bacteriological monitoring of drinking water supplies.

KEY WORDS: bacteria, water, microbiology, waterborne disease

To control the spread of waterborne disease, information on water quality must be available in time to influence decisions concerning the use of the water. If real-time continuous monitoring techniques were available for all of the etiological agents of waterborne disease, they should be employed. Microbiologists have not developed such techniques, and reliance must be placed on historical data concerning the concentrations of easy to measure bacteria that may be indicative of the occurrence of pathogens. Water is considered safe to drink because coliform bacteria were found very infrequently in the finished water during the past year. Recreation is allowed in our lakes, rivers, and at beaches because in the past the coliform density was usually not too great. Waterborne disease still occurs and epidemiological investigation of these outbreaks requires more than the density of indicator bacteria. The density of the pathogen in the water is necessary to evaluate the health hazard from the water contact.

Only the etiological agent of guinea worm disease, *Dracunculus medinensis*, must come from drinking water. All other waterborne diseases are usually transmitted directly from person to person but also are transmitted

[1]Chief, Water Quality Division, Health Effects Research Laboratory, U. S. Environmental Protection Agency, Cincinnati, Ohio 45268.

indirectly by some environmental vehicle. Because of the number of consumers of a water supply, a single contamination can result in many cases of disease. Epidemiological investigations can be aided greatly by isolation of the causative agent from drinking water. Concentrations of indicator organisms can be helpful, but they are not definitive. Often the sampling locations have been so poorly picked that even coliform data are not available for the section of the distribution network implicated in the outbreak. The isolation of coliform bacteria from a community water supply during an outbreak investigation, particularly if it is a small water supply, cannot be considered as proving that the outbreak was waterborne. A random sample of small community water supplies might find one in seven with coliforms, and, in some parts of the country, the ratio would be lower than one in two for individual home water supplies. Fourteen percent of water supplies serving less than 5000 population failed to meet drinking water standards for coliform density [*1*].[2]

What Should an Indicator Indicate?

Seldom have epidemiological investigations been conducted relating indicators to health risks. The studies of community supplies by Petersen and Hines [*2*] are singular in their efforts to study the endemic occurrence of disease related to drinking water quality. Table 1 shows the dose-response data developed in this study. In the community with the lowest quality drinking water, 100 percent of the tubes (sets of five, 10-ml tubes) contained coliforms, and 17 percent of the exposed population reported gastrointestinal illness during the three summer months. This was approximately twice the background rate of illness reported by the three communities which provided higher quality drinking water in which none to 2½ percent of the tubes contained coliforms. The attack rate for gastrointestinal illness during the summer months among residents of more than two years was half that of shorter-time residents in the communities providing poor quality water, whereas the rates among long- and short-term residents of the communities

TABLE 1—*Drinking water quality and gastrointestinal illness.*[a]

Community	Percent of Water Sample Portions Positive for Coliforms	Overall Attack Rate 3 Summer Months
A	100.0	17.1
B	90.0	14.0
C	72.5	11.9
D	0.0	9.9
E	2.5	9.1
F	1.7	7.5

[a]From Ref *2*.

[2]The italic numbers in brackets refer to the list of references appended to this paper.

TABLE 2—*Drinking water quality and gastrointestinal illness.*[a]

Period of Exposure	Attack Rate Three Summer Months	
	Three Communities, Low-Quality Water, %	Three Communities, High-Quality Water, %
Less than two years residence	21.6	9.4
More than two years residence	10.8	8.5

[a]From Ref *2*.

providing satisfactory finished waters differed little (Table 2). Furthermore, the rate for long-term residents of the communities providing water of poor quality were only slightly higher than those among users of high-quality waters.

In Stevenson's Public Health Service (PHS) [*3*] and Cabelli's Environmental Protection Agency (EPA) [*4*] studies, the relationships of indicators to disease have been sought for recreational exposure. The protocol of three PHS studies called for a comparison of illness rates of swimmers at two beaches where it was expected there would be a marked microbiological difference between the beaches. The differences expected from historical records were not noted in the summers the studies were conducted. The study of the Chicago beaches by Stevenson could not make a beach to beach comparison and had to use a comparison of three high-coliform density days with three low-coliform days (Table 3). This three-day comparison at one of the beaches indicated a significant difference between all the illnesses reported when the most probable number (MPN) of coliforms on high days was 2300/100 ml compared with 43 coliforms/100 ml on the three low days. The EPA study protocol allowed for a comparison of illness rates between swimmers and nonswimmers on each day of the study. A significant increase of gastrointestinal illness of swimmers over nonswimmers was noted at 1000 coliforms/100 ml. Table 4 shows the comparison of two New York beaches, giving excess of swimmers over nonswimmers for gastrointestinal illness. Results are not as yet available, but studies are in progress to evaluate the effects of exposure to aerosolized wastewater and a comparison with indicator bacterial densities.

Most indicators are justified on the premise that they indicate fecal dis-

TABLE 3—*Chicago bathing water quality and recorded illness (all types).*[a]

Day of Swimming	Coliform MPN Per 100 ml	Percent Illness During the Week Following Swimming
North Beach		
3 high days	730	9.9
3 low days	31	8.7
South Beach		
3 high days	2300	12.2
3 low days	43	8.5

[a]From Ref *3*.

TABLE 4—*New York recreational water quality and additional gastrointestinal illness (following swimming).*

Location of Swimming	Coliform Count per 100 ml	Percent Illness During the Week Following Swimming (S − NS)[a]
Coney Island	983	4.8
Rockaways	39.8	3.5

[a]S −NS = Swimmers less nonswimmers: 7.2 − 2.4 = 4.8; 8.1 − 4.6 = 3.5.

charges of man or animals or both. The differentiation between contamination by man and other animals may have value in epidemiological investigations but can usually be accomplished by sanitary surveys. Monitoring for fecal discharges of all animals is necessary as indicated by the recent outbreak of giardiasis at Camas, Washington. The apparent reservoir of the parasites was beavers because *Giardia* were isolated from a beaver trapped on the watershed [*5*].

Ten pathogens have been associated with outbreaks of disease attributable to drinking water supplies in the United States in the past 29 years [*6–9*]. About half of these outbreaks were a gastrointestinal illness of an undetermined etiology. Many outbreaks come to official notice long after they occur, and the opportunity for determining their etiology has been lost. Often the microbiological workup is incomplete. For example, in 1974, an investigation of the gastroenteritis outbreak at Big Sky, Montana, revealed that *Yersinia enterocolitica* were present in the water supply, but rectal swab cultures were not examined for this particular species [*10*]. Frequently it is assumed that outbreaks of an unknown etiology are caused by viruses, but failure to identify the infectious agent could be related to poor or incomplete bacteriology.

The most frequent specific cause of reported waterborne disease is infectious hepatitis (Table 5). For the noncommunity water supplies, the problem uncovered in investigating the outbreaks was contamination of the source—usually poorly constructed wells. For the community water supplies, the contamination was related to cross connections or back siphonage [*11*]. The indicators now used may not be adequate to demonstrate the absence of hazard from infectious hepatitis. Bacteriological water quality monitoring has not been used effectively in cross-connection prevention programs. If a single high-count sample is found, the usual procedure is to resample and not to investigate further if contamination is not found the second time. The single contaminated sample may have indicated an intermittently operating cross connection, but this possibility is often overlooked or ignored. Experience from large well-operated water supplies indicates that consistently good water samples can be obtained, and the occasional high count should be a call for action, not just resampling.

Waterborne giardiasis has been recognized infrequently in the past, but,

TABLE 5—*Waterborne disease outbreaks distribution by etiology 1946–1974.*

Disease	Community Water Systems, %	Other Water Systems, %
Gastroenteritis	52.6	47.5
Infectious Hepatitis	16.3	13.7
Shigellosis	9.6	10.2
Chemical Poisoning	5.9	4.0
Giardiasis	5.2	2.5
Typhoid	4.4	15.8
Salmonellosis	4.4	2.8
Amebiasis	0.7	1.2
Poliomyelitis	0.7	0.0
Enteropathogenic *E. coli*	0.0	1.2
Tularemia	0.0	0.6
Leptospirosis	0.0	0.3
Number of Outbreaks	135	322

in 1974, a quarter of the reported outbreaks were this disease. It is doubtful that the coliform count is an adequate indicator of the hazard associated with *Giardia* in unfiltered water supplies because of the greater resistance of the protozoan cysts to disinfection.

Bacterial indicators may be useful for assessing the safety of small water supplies where it is desired to establish the adequacy of the protection of the sources, such as wells. Even in the case of properly constructed wells, total plate counts may be more informative than coliform counts to indicate nutritive conditions.

The usefulness of indicator bacteria for quality control in community water systems is subject to question. There is no possibility for immediate continuous monitoring for bacterial indicators. The use of chlorine residual measurements as a substitution was proposed for the new EPA drinking water standards in the 14 March 1975 *Federal Register*. When the "Interim Primary Drinking Water Regulations" were published on 24 December 1975, the provision to allow complete substitution of chlorine residual for coliform testing in the smaller water systems had been changed to substitution for a maximum of 75 percent of the monthly microbiological samples required. Four chlorine residual determinations are required in substitution for each microbiological determination omitted with at least daily determinations of chlorine residual.

Recognition of the usefulness of chlorine residual stems from two needs of drinking water quality control: one is the need for an immediate measurement and the other for a measure of the quality of all the water produced. The use of a free chlorine residual measurement will provide a real-time continuous monitoring. The loss of the chlorine residual can indicate unsafe conditions and treatment modified by means of feed-back controls. The chlorine residual test has been used successfully to locate an operating cross connection between a pressure sewer main that was connected to a water main in Tampa, Florida [*12*]. Indicator bacteria would have provided confirmatory data only after a delay of several days.

Chlorine residual testing could have been required in addition to the required microbiological monitoring, but, to promote the use of the more effective monitoring, it was decided to eliminate the more costly microbiological testing. Savings effected can be used for other water quality testing. Responsibility for first line quality control must rest with the waterworks operator. The present system of collecting limited samples at small water supplies and sending them to the state laboratory for testing introduces a mystical aspect to quality control. The operator is not involved personally and reports come back a week or more later that may indicate deficiencies in the past. When the operator determines a chlorine residual and does not find a pink color (DPD test), he is the only one to know there is a problem, and he must make a decision to protect the public's health. The responsibility for an acute problem is where it should be and at the only place where necessary corrective action can be taken. The publication of the causes of waterborne outbreaks must be intensified to reinforce the responsibility of the waterworks operator in the prevention of disease. This will require the reporting and investigation of outbreaks, and this activity can be supported with EPA's water supply program grants. The state laboratory microbiologist should no longer be overburdened with routine bacteriology samples and should be able to participate in outbreak investigations to determine etiological agents and their source.

A substantial reason for abandoning the coliform test is the record of its use in the past for routine drinking water quality surveillance. Nationally it has been found that 85 percent of water systems were not collected at the rate prescribed in the bacteriological surveillance criteria [*1*]. Samples were collected at less than half of the prescribed frequency at 69 percent of the supplies surveyed. The importance of maintaining a chlorine residual, if coliform organisms are to be eliminated from the distribution system, was also demonstrated in this survey. Although the practice of chlorination caused a dramatic decline in the frequency of the samples containing coliforms, these organisms are not eliminated merely by claiming that chlorination is practiced. Unless a chlorine residual was maintained in the distribution system, a significant portion of the samples from the distribution system contained coliforms. The coliforms are eliminated nearly completely in systems providing at least a trace of chlorine residual. In facing the reality of these data, the substitution of chlorine residual for the coliform test was developed.

If microbiological measurements should continue on treated supplies, it would seem best to consider an indicator that would be more resistant to the disinfectant used. This would provide more useful data than can be provided by the coliform. Research at the University of Illinois has progressed enough with wastewater to suggest that acid-fast bacilli and yeasts should be considered in the evaluation of the safety of drinking water. As now understood, the cultural characteristics of these more resistant organisms would require a much longer time in incubator for the test procedure. The coliform test does not provide a real-time indication of water quality, and longer test procedures would be less immediate, but, if a test organism was used that was more resistant to the disinfectant than pathogens, bacteria, and virus, the historic elimination of a health hazard would be known. When the chlorine residual substitution for bacteriological testing is used by a water system and a sample is found without a chlorine residual, then a bacteriological sample shall be taken from that point in the distribution system. Corrective action can be taken immediately to bring the chlorine residual up to standard, but the bacteriological sample provides an opportunity to determine what may have gone wrong in the distribution system. This should provide more interesting microbiological work than routine testing.

Recreational Water Quality

A more resilient bacterial indicator than coliforms would seem necessary for swimming pools. Bacteria more indicative of the nature of the hazard would be more appropriate. The danger is not from fecal matter but from other human discharges. Disease that have been associated with pools have been caused by viruses and mycobacterium.

A more interesting problem is presented by natural bathing areas. The quality of these areas can be influenced by both point and nonpoint waste discharges. Considering the amount of water-based recreation that occurs in or on polluted waters, the lack of associated disease is an anomaly.

There have been no systematic tabulations of waterborne disease outbreaks associated with recreational waters, but a casual review of the literature would indicate that the occurrence of this type of outbreak is rare. The diseases known to have occurred in outbreaks have been typhoid fever [*14–15*], infectious hepatitis [*16*], leptospirosis [*17*], amebic meningoencephalitis [*18*], and shigellosis [*19*]. Investigations of these outbreaks seldom included the enumeration of indicator organisms. Byran et al [*16*] indicated that the lake water associated with the hepatitis outbreak, when sampled during the year of the outbreak, frequently revealed gross contamination wih coliform organisms. The Rosenberg investigation [*19*] of the shigellosis outbreak on the Mississippi River below Dubuque, Iowa, was thorough and provided a relationship between concentration of the indicator organism and illness. The mean fecal coliform count was 17 500 organisms / 100 ml, and 13 percent of the persons swimming reported illness. The illness rate was 18 percent among persons who reported having water in their mouths.

Summary

If indicator bacteria are to be more useful to epidemiologists, the concentration of the indicator must be measured in the water implicated in a disease outbreak; pathogens must be isolated from the water during investigations of outbreaks to document the presence of the agent involved and thus provide definitive associations. Indicators may be useful in developing water sources but seem to be of limited value in monitoring water supplies to prevent waterborne disease. Chlorine residual measurement provides a more useful tool to control drinking water quality.

References

[1] McCabe, L. J., Symons, J. M., Lee, R. D., and Robeck, G. G., *Journal of the American Water Works Association*, 1970, Vol. 62, No. 11, pp. 670–687.

[2] Petersen, N. J. and Hines, V. D., *American Journal of Hygiene*, 1960, Vol. 71, pp. 314–320.

[3] Stevenson, A. H., *American Journal of Public Health*, 1953, Vol. 43, No. 5, pp. 529–538.

[4] Cabelli, V. J., Levin, M. A., Dufour, A. P., and McCabe, L. J., "The Development of Criteria for Recreational Waters." Gameson, Ed., *Discharge of Sewage from Sea Outfalls*, Pergamon Press, New York, pp. 63–73.

[5] Thomas, Allen, "Beavers Blamed for Illness," *The Columbian Newspaper*, Vancouver, Wash., 21 May 1976.

[6] Craun, G. F. and McCabe, L. J., *Journal of the American Water Works Association*, Vol. 65, No. 1, 1973, pp. 74–84.

[7] *Journal of the American Water Works Association*, Vol. 67, No. 2, 1975, pp. 95–98.

[8] Hughes, J. M., Merson, M. H., Craun, G. F. and McCabe, L. J., *The Journal of Infectious Diseases*, Vol. 132, No. 3, 1975, pp. 336–339.

[9] Horwitz, M. A., Hughes, J. M., and Craun, G. F., "Outbreaks of Waterborne Disease in the United States, 1974," *The Journal of Infectious Diseases*, Vol. 133, No. 5, 1976, pp. 588–593.

[10] *Morbidity and Mortality Weekly Reports*, Vol. 26, No. 16, April 19, 1975, pp. 141–142.

[11] McCabe, L. J., "Significance of the Virus Problem," *Proceedings*, American Water Works Association Technology Conference, 3–4 Dec. 1973.

[12] Communication Hillsborough County Health Department to the State Division of Health, 27 Nov. 1973.

[13] Engelbrecht, R. S., Foster, D. H., Greening, E. O., and Lee, S. H., "New Microbial Indicators of Wastewater Chlorination Efficiency," EPA Report 670/2-73-082, Environmental Protection Agency, Cincinnati, Ohio, Feb. 1974.

[14] "Typhoid at Covington State Park, Louisana," CDC Salmonella Surveillance, Communicable Disease Center, 18 Nov. 1963.

[15] "Typhoid Fever-Alabama," *Morbidity and Mortality Weekly Reports*, Vol. 21, No. 32, 12 Aug. 1972, p. 280.

[16] Bryan, J. A., Lehmann, J. D., Setiady, T. F., and Hatch, M. H., *American Journal of Epidemiology*, Vol. 99, No. 2, 1974, pp. 145–154.

[17] "Human Leptospirosis-Tennessee," *CDC Veterinary Public Health Notes*, Center for Disease Control, Jan. 1976.

[18] Butt, C. G., *New England Journal of Medicine*, Vol. 274, 1966, pp. 1173–1176.

[19] Rosenberg, M. L., Hazlet, K. K., and Schaefer, J., "Shigellosis from Swimming," presented American Public Health Association Meeting, 19 Nov. 1975.

H. Leclerc,[1] D. A. Mossel,[2] P. A. Trinel,[1] and F. Gavini[1]

Microbiological Monitoring— A New Test for Fecal Contamination

REFERENCE: Leclerc, H., Mossel, D. A., Trinel, P. A., and Gavini, F., "**Microbiological Monitoring—A New Test for Fecal Contamination,**" *Bacterial Indicators/Health Hazards Associated With Water, ASTM STP 635*, A. W. Hoadley and B. J. Dutka, Eds., American Society for Testing and Materials, 1977, pp. 21–31.

ABSTRACT: Bacteriological monitoring of drinking water has not changed essentially during the last decennium. *Escherichia coli* and fecal coliforms are still traditionally used as index organisms, although methods are slightly different in various countries.

Investigations carried out by the authors have demonstrated that conventional methods can still be criticized in two respects. The spontaneous dying off of the *Salmonella* in water often occurs at a considerably lower rate than that of *E. coli*, which can lead to falsely negative results. In addition, investigations on human enteric flora composition, using newer methodology, have demonstrated that coli-aerogenes bacteria other than *E. coli* occur in relatively high concentrations. Two approaches will allow the elimination of these deficiencies of conventional monitoring of drinking water. They are using the entire group of Enterobacteriaceae as index organisms or considerably increasing the sensitivity of testing for *E. coli*.

For the latter purpose, an instrumental, continuous method has been developed. Water samples are first passed through a hollow fiber ultrafiltration device. Thereupon enrichment is carried out by culturing in lactose broth at 41°C. After stripping off the carbon dioxide (CO_2) formed by fermentation, the liquid is mixed with a glutamate solution buffered at pH 3.9, the CO_2 formed set is freed by acidification, and then absorbed in a phenolphtalein buffer which is ultimately examined colorometrically. The sensitivity of this method is 1 colony forming unit (cfu) of *E. coli*/100 ml or better. It has the additional advantage of allowing much more representative sampling of a water supply than customarily carried out.

KEY WORDS: bacteria, water, automation, coliform bacteria, taxonomy, resistance

Tests for fecal contamination of water are relevant when they comply with the following criteria [*1–3*].[3] They are specificity—the origin of the organisms should be almost exclusively fecal, sensitivity—the more that bacteria occur in the intestinal contents of man and animal, the better are the

[1]Professor, doctor, and doctor, respectively, Institut Pasteur, Unite Inserm 146, Domaine du Certia-369, rue Jules Guesde, 59650 Villeneuve D'Ascq, France.

[2]Professor, Medical Food Microbiology, University of Utrecht, Biltstraat 172, Utrecht, The Netherlands.

[3]The italic numbers in brackets refer to the list of references appended to this paper.

chances to isolate them from contaminated water, and resistance—bacterial recovery depends greatly on their survival in water. Among the test bacteria, the coliform group is usually chosen because it is considered as being the most significant. However, this point of view merits some further consideration.

Coliform bacteria are defined as those Enterobacteriaceae that ferment lactose with the production of acid and gas in less than 48 h at a temperature ranging from 30 to 37°C. However this definition is not fully adequate for the following reasons.

1. By the use of lactose containing media lactose, negative strains are ignored; these bacteria have the same habitat and at least the same significance as the lactose positive types.

2. Gas formation from lactose depends on the formation of an adaptative enzyme, formic hydrogen lyase which is affected by temperature. The activity of this enzyme is quite often reduced and sometimes entirely suppressed under conditions that do not favor survival of coliforms in water. This phenomenon can have adverse consequences when coliforms are sought by the use of liquid media. When membrane-filtration techniques are used, coliforms are assessed irrespective of gas formation. This discrepancy makes a new and consistent definition of coliform bacteria compulsory.

3. The most suitable incubation temperature for the enumeration of coliform bacteria has never been accurately defined. In the United States, 35°C has been suggested [*4*]. In Europe, investigations by Buttiaux et al [*5*] and Henriksen [*6*] tend to show that the most suitable temperature should be 30°C.

4. In Europe, there is a long standing tradition to distinguish between total coliforms and *Escherichia coli* [*7*], because the latter species was considered more significant. Tests for *E. coli* depend on Eijkman's elevated temperature test which is also the basis of methods used in the United States for the detection of fecal coliforms. The definition of *E. coli* is based on taxonomical criteria. The designation fecal coliforms should be used whenever detection methods rely on a simplified approach, such as the elevated temperature test.

5. Extensive research by Schubert [*8*] in the German Federal Republic and ourselves [*9–11*] demonstrated the frequent occurrence of *Aeromonas* in drinking water. In an examination of 9036 water samples, we were able to isolate these bacteria from 30 percent of the samples showing a positive presumptive test for coliforms. *Aeromonas* bacteria are saprophytic organisms found in water and cold-blooded animals, and their presence does not seem to indicate any particular health hazards. It seems, therefore, essential to distinguish them from the coliforms [*12*]; this is being neglected by several methods and especially by the standard total coliform most probable number (MPN) tests advocated in the United States [*4*].

Types of Enterobacteriaceae Occurring in Human Feces

It is possible to remedy some of these deficiencies, whereas others appear to us as being more fundamental. These last ones pertain to the actual significance of the coliform test.

The microbiology of the intestinal system has been studied in great detail [*13,14*]. Yet, the frequency of occurrence of species other than *E. coli* is virtually unknown. Quite often the predominance of *E. coli* prevents other coliforms, that are less numerous, from being detected. In this field, reliable information can be derived only from quantitative analyses of stools. For this purpose, we studied the distribution of coliform bacteria in human feces using two sets of culture media; namely, one for *E. coli* [*5*] and the other for species such as *Klebsiella*, *Citrobacter*, *Levinea*, and *Enterobacter* [*15*]. The data in Tables 1, 2, and 3 summarize the results and particularly the frequency of the various groups of Enterobacteriaceae. These figures are quite different from those found in previous studies, particularly with regard to the genus *Citrobacter* which had been considered to occur only in very low numbers in human fecal flora [*1,14*].

Similarly, information is lacking on the distribution in the biosphere of the various types of coliform bacteria [*16*]. Up to now, the classification of the Enterobacteriaceae has been almost wholly based on data of a medical nature. Additional information should be collected from environmental sampling and particularly from the examination of water and soil. In the course of a numerical taxonomical study [*17*] dealing with 111 strains of

TABLE 1—*Human fecal flora: qualitative survey on 30 adults.*

	Species or Genus	Number of Positive Samples / 30	Presence, %	Frequency
Aerobic bacteria gram-negative				
	Escherichia coli	30	100	constant
	Citrobacter-Levinea	20	66	high
	Klebsiella	15	50	medium
	Enterobacter	3	10	rare
Aerobic bacteria gram-positive				
	Staphylococcus	15	50	medium
	Enterococcus	30	100	constant
	Bacillus	28	93	constant
Anaerobic bacteria gram-negative				
	Lactobacillus	30	100	constant
	Bacteroides	30	100	constant
Anaerobic bacteria gram-positive				
	Clostridium	23	76	high
Yeasts	. . .	20	66	high
Moulds	. . .	16	53	medium

TABLE 2—*Human fecal flora: quantitative survey (30 adults).*

	Species	Average cfu/g[a]	Number of Samples Where the Species is Found / 30
Total bacteria	. . .	1.5×10^{11}	24
Total aerobic bacteria	. . .	7×10^{8}	30
Aerobic bacteria gram-negative			
	E. coli	4×10^{8}	30
	Citrobacter-Levinea	1×10^{6}	20
	Klebsiella	5×10^{4}	14
	Enterobacter	1×10^{5}	3
Aerobic bacteria gram-positive			
	Enterococcus	2×10^{8}	30
	Staphylococcus	8×10^{6}	15
	Bacillus	3×10^{4}	28
Anaerobic bacteria gram-negative			
	Bacteroides	1×10^{10}	30
	Lactobacillus	1×10^{9}	30
Anaerobic bacteria gram-positive			
	Clostridium	4×10^{6}	23
Yeasts	. . .	5×10^{4}	20
Moulds	. . .	4×10^{4}	16

[a]cfu = colony forming units.

Enterobacteriaceae related to the genus *Citrobacter*, we have been able to identify six classes. Besides species that have been well described such as *Citrobacter freundii* hydrogen sulfide + (H_2S +) and H_2S– [*18*], *Levinea malonatica*, and *Levinea amalonatica*, numerical analysis revealed three new

TABLE 3—*Adult human fecal flora: overall quantitative survey.*

Type of Bacterial Flora/Genera or Species	Average, cfu/g
Primary	
Bacteroides	10^{10}
Lactobacillus	10^{9}
Escherichia coli	10^{8}
Enterococcus	10^{8}
Secondary	
Citrobacter-Levinea	10^{5} to 10^{6}
Klebsiella	10^{4} to 10^{5}
Enterobacter	10^{4} to 10^{5}
Clostridium	10^{5} to 10^{6}
Staphylococcus	10^{5} to 10^{6}
Bacillus	10^{4} to 10^{5}
Yeasts	10^{4} to 10^{5}
Moulds	10^{4} to 10^{5}
Rare	
Proteus-Providencia	10^{2} to 10^{3}
Pseudomonas aeruginosa	10^{2} to 10^{3}

classes: C, D, and F. The classes C and F represent strains originating from unpolluted soils or potable tap water. All are psychotrophic showing visible growth in two to three days at +4°C. The classes A, B, D, and E comprise strains that all grow at 41°C and often at 44.5°C and originate from human feces or from water samples that are highly contaminated, such as sewage and surface water. There is a striking relation between the strain characters and the biotope from which they originate and which might have exerted a selective influence (Table 4). In this respect, capacity of growth at an elevated temperature seems to be especially important.

A taxonomic study of Enterobacteriaceae belonging to, or related to, the genus *Enterobacter* revealed the same correlation between the origin of the strains and their phenotypical properties, especially temperature tolerance (Table 5). However, the genus *Hafnia* showed a certain heterogeneity with regard to growth temperatures as well as origin (Table 5).

Lack of sufficient data on the origin of Enterobacteriaceae encountered in water is the cause of the difficulties in attributing significance to the isolation of given types. Further ecological and taxonomical studies will allow better assessing of the value of these tests. Geldreich's fundamental studies [*19,20*] constitute one step towards achieving this aim, though yet having to be compared with taxonomical studies with the purpose of revealing their intrinsic value.

Fate of Various Enterobacteriaceae in Water

The resistance of bacteria to the environment determines their suitability as index organisms. Suitable organisms must survive as well as the pathogenic types they are meant to indicate [*21,22*]. Coliforms indeed do not serve this purpose too well.

As the data determined by us and summarized in Table 6 demonstrate, *Salmonella* survives 19 weeks in contaminated surface water, whereas *E. coli* only survives two weeks. On the contrary, *Streptococci* group D survive six weeks. The latter's resistance is therefore rather like that of non*E. coli* coliforms such as *Klebsiella*, *Enterobacter*, and *Citrobacter* (Table 6).

Remedying Existing Deficiencies

In the light of the ecological considerations presented in the previous two paragraphs, two routes can be followed, in principle, to make proper use of index organisms for the assessment of the sanitary condition of water.

The first approach is to use the entire group of Enterobacteriaceae and, thus, include (*a*) lactose positive coli-aerogenes bacteria that are or are not *E. coli* (Tables 1 to 3) and (*b*) the lactose negative types, including *Salmonella*, *Shigella*, and *Proteus* [*12,23*]. Alternatively, one can increase the sensitivity of the mode of detection of *E. coli* used and, thus, compensate for its very much lower extra-enteric survival rate (Table 7).

TABLE 4—*H_2S-Enterobacteriaceae related to the genus* Citrobacter.

	Class						
	A						
	A_1	A_2	B	C	D	E	F
Number of Isolates	8	10	15	20	10	14	28
Origin							
F	+	+	+	–	–	+	–
Su	+	+	+	–	+	+	–
Se	–	+	+	–	+	+	–
So	–	–	–	+	–	–	+
D	–	–	–	+	–	–	+
Growth +4°C, %	0	0	40	100	30	0	90
Growth +41°C, %	100	100	100	0	100	100	0
Growth +44°5C, %	75	80	67	0	70	80	0
Species	*Levinea malonatica*	*Levinea amalonatica*	*Citrobacter freundii* (biotype H_2S–)	new taxon	new taxon	*Enterobacter cloacae* (acetoine –)	new taxon

NOTE—F = fecal strains, Se = sewage strains, So = unpolluted soils, Su = surface water strains, and D = drinking water strains.

TABLE 5—*Enterobacteriaceae belonging to or related to the genus* Enterobacter.

	Class					
	G	H_1	H_2	H_3	I_1	I_2
Number of Isolates	63	11	20	28	19	12
Origin						
F	+	–	–	–	–	+
Su	+	–	–	–	+	+
Se	+	–	–	–	+	=
So	–	+	+	+	–	–
D	–	+	+	+	+	–
Growth +4°C, %	9	91	95	86	100	100
Growth +41°C, %	100	0	0	0	42	75
Growth +44°5C, %	46	0	0	0	0	0
Species	*Enterobacter cloacae*	new taxon	new taxon	new taxon	*Hafnia alvei* (biotype lactose –)	*Hafnia alvei* (biotype lactose +)

NOTE—F = fecal strains, Se = sewage strains, So = unpolluted soils, Su = surface water strains, and D = drinking water strains.

TABLE 6—*Survival of* E. coli *compared with the survival of other coliforms and of group D streptococci in contaminated well water stored at +4° C.*

Times (days)	*E. coli*/100 ml	Other Coliforms/100 ml	Streptococci, Group D/100 ml
1	120	1400	116
16	50	660	50
33	5	140	10
50	0	50	2
78	0	7	1
87	0	1	0

We have studied the latter solution, after a most sensitive, automated mehod of detection of *E. coli* had become available. The results of this preliminary study will be presented in the second part of this paper. One of the additional considerations prompting this study was a further deficiency of customarily used methodology for the bacteriological monitoring of water; namely, the low degree of sampling. It seems senseless to increase the sensitivity of methods of analysis without even wondering how well a 1000-ml sample of water represents the quality of some 10^5 m^3 of water, as daily distributed by urban networks. The automatic device studied continually draws samples and, thus, remedies the last mentioned deficiency of current methods for the assessment of the bacteriological condition of water.

New Test for Fecal Contamination and Its Automated Use

The new methodology is based on the following two principles. The choice of a sensitive and specific criterion of fecal contamination—glutamic acid decarboxylase (GAD) biosynthesized by *E. coli*, and use of an automatic device (Technicon Autoanalyser) that constantly draws samples of water to be analyzed and secures its culturing under conditions that result in the selection of *E. coli* and subsequent GAD activity.

In order to apply a test with high specificity, we carried out a survey on the occurrence of this enzyme in various bacterial species encountered in water [*24-26*]. The data in Table 8 show its presence in *E. coli* as well as in *Shigella*, *Proteus rettgeri*, *P. hauseri*, *Providencia*, and *Clostridium*. The figures in Table 9 illustrate its quantitative distribution. *Shigella* is only exceptionally isolated from water, and there is, hence, no possibility whatsoever of any interference of these bacteria with the detection of *E. coli*. The bacteria of the Proteus-Providencia group are more likely to be found in polluted waters. However, our experience has shown that they are always associated with *E. coli*—the latter being much more numerous. *Clostridium* species are unable to thrive under the very aerobic culture conditions of the suggested analytical system and, hence, will neither interfere.

In order to assess the sensitivity of the suggested procedure [*25*], we selected 240 *E. coli* strains and examined them for GAD activity. The results presented in Table 10 show that this is a constant attribute of the species, unrelated to the origin of the strain. Consequently the GAD procedure can

TABLE 7—*Survival of* Salmonella *and index organisms for fecal contamination in tap water (at + 20° C in the dark); the sample was naturally contaminated with fecal coliforms and streptococci, but was inoculated with* Salmonella.

Time (weeks)	*E. coli* /100 ml	Total Coliforms /100 ml	Group D Streptococci /100 ml	*Salmonella*/100 ml Determined at 43°	37°	Percentage of various *Salmonella* serotypes: *S. infantis* 43°	*S. infantis* 37°	*S. para B* 43°	*S. para B* 37°	*S. brandenburg* 43°	*S. brandenburg* 37°	*S. enteriditis* 43°	*S. enteriditis* 37°
0	10^6	10^6	10^5	24 000	24 000	68.7	68.7	6.2	3.1	18.7	21.9	6.2	6.2
1	10^4	10^4	10^2	24 000	24 000	79.3	80	6.9	4	10.3	12	3.4	4
2	8	10	10	24 000	24 000	75	86.4	4.2	0	8.3	4.5	12	9.1
3	0	0	10	21 000	21 000	69.6	39.1	13	30.4	17.4	30.4	0	0
4	0	0	1000	15 000	15 000	8.3	8.7	83.3	69.6	8.3	21.7	0	0
5	0	0	10	15 000	15 000	63.6	42.8	18.2	4.8	18.2	52.4	0	0
6	0	0	6	15 000	15 000	41.7	50	0	4.2	41.7	41.7	16.7	4.2
7	0	0	0	15 000	15 000	57.9	42.8	0	9.5	42.1	42.8	0	4.8
8	0	0	0	9 300	9 300	20	10.5	20	0	60	89.5	0	0
9	0	0	0	9 300	9 300	35.3	5	5.9	5	52.9	85	5.9	5
10	0	0	0	4 300	24 000	6.7	0	33.3	27.3	60	72.7	0	0
11	0	0	0	11 000	2 300	0	3.2	46.1	25.8	53.8	71	0	0
12	0	0	0	2 400	4 300	0	12.5	12.5	18.7	87.5	68.7	0	0
13	0	0	0	430	2 400	0	0	16.7	5.5	83.3	94.4	0	0
14	0	0	0	930	930	0	0	0	0	100	100	0	0
15	0	0	0	91	150	SR	0	SR	0	SR	100	SR	0
16	0	0	0	62	30	SR	SR	SR	SR	SR	SR	SR	SR
17	0	0	0	0	3	SR	SR	SR	SR	SR	SR	SR	SR
18	0	0	0	0	3	SR	SR	SR	SR	SR	SR	SR	SR
19	0	0	0	0	3	SR	SR	SR	SR	SR	SR	SR	SR
20	0	0	0	0	0	0	0	0	0	0	0	0	0

NOTE—SR = *Salmonella* species that cannot be serologically classified.

TABLE 8—*Occurrence of GAD in 230 bacterial cultures.*

Genus or Species	Number of Strains Examined	Negative	Positive
Shigella	10	1	9
Alcalescens	5	1	4
Escherichia coli	10		10
Salmonella	10	10	. . .
Arizona	1	1	. . .
Citrobacter	10	10	. . .
Klebsiella	10	10	. . .
Enterobacter	10	10	. . .
Hafnia	10	10	. . .
Serratia	10	10	. . .
Proteus morganii	5	5	. . .
Proteus rettgeri	5	1	4
Proteus hauseri	5	. . .	5
Providencia	5	. . .	5
Aeromonas	22	22	. . .
Vibrio	10	10	. . .
Pseudomonas	12	12	. . .
Acinetobacter lwoffi	5	5	. . .
Acinetobacter calcoaceticus	5	5	. . .
Brucella	6	6	. . .
Yerslnla pseudotuberculosis	3	3	
Streptococcus of Lance-field's group D	15	15	. . .
Micrococcus	11	11	. . .
Staphylococcus	5	5	. . .
Bacillus	10	10	. . .
Clostridium	10	5	5

be considered a specific and sensitive test for fecal contamination. The procedure has been discussed previously [*27–29*].

Figure 1 shows the construction and mode of operation of this device. Water samples to be examined are withdrawn continuously by a peristaltic pump and mixed in a given proportion with lactose phosphate broth. This

TABLE 9—*Quantitative study of GAD activity in enteric bacteria.*

Genus or species	Number	Results
Shigella dysenteriae	3	++
Shigella flexneri	3	++
Shigella boydii	1	++
Shigella sonnei	3	++
Alcalescens	5	−,+,++
Escherichia coli	10	++
Hafnia	10	+,+,++
Proteus mirabilis	2	+++
Proteus vulgaris	3	+++
Proteus morganii	5	−
Proteus rettgeri	5	−,+++
Providencia	5	+++
Clostridium perfringens	5	−,+,++
Clostridium bifermentans	5	−,+,++
Clostridium saccharobutyricum	5	−

TABLE 10—*GAD activity in* E. coli *strains of various origin.*

Origin	Number of Strains, GAD +	+	++	+++	++++
Human feces (enteropathogenic *E. coli*, and nonpathogenic *E. coli*)	60	23	36	1	0
Drinking water (well, spring, tap water)	99	29	36	34	0
Foods (milk, meat . . .)	80	13	55	10	2

mixture is provided with sterile air bubbles and pumped for 12 h through glass incubation coils and immersed in a water bath at 41°C. These culture conditions are semiselective, favoring the growth of fecal coliforms and especially of *E. coli.* In this circuit, consecutive samples are being separated by means of a bactericidal solution (phthalate HCL-buffer 0.047 mol pH 2.4) that prevents the successive samples from being contaminated. As shown in Table 11, this solution will kill water bacteria within 1 min at 41°C. Diffusion of this solution in the culture medium passing through the various incubation coils remains without effect, because the bactericidal activity of the solution is nullified by the pH level prevailing in the culture fluid [*27,28*].

When this fluid is leaving the incubation coils, the air bubbles are removed from the liquid in order to eliminate the gaseous fermentation products, and the liquid is subsequently diluted by means of sterile carbon dioxide (CO_2) free water and an acetate buffer, 0.1 mol, pH 3.4, at 45°C. Thereupon, the fluid is mixed with a GAD solution buffered with 0.1 mol acetate, pH 3.9.

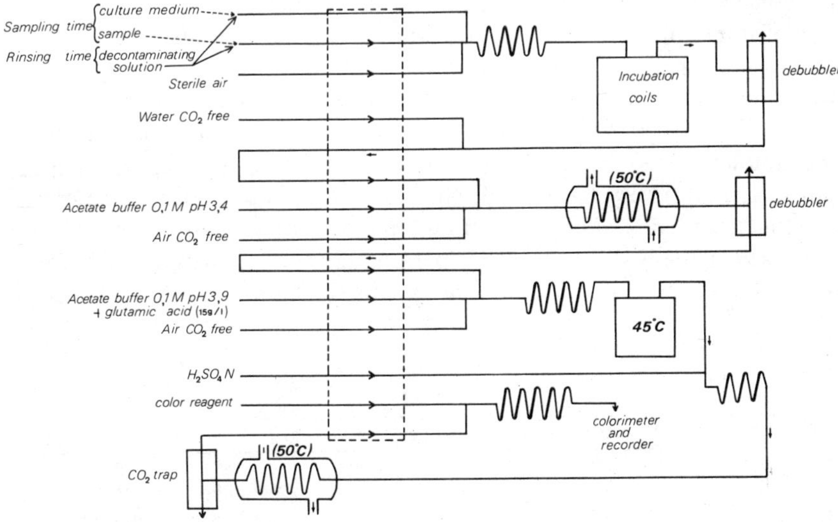

FIG. 1—*Apparatus for the continuous detection of* E. coli *in water; assembly drawing.*

TABLE 11—*Demonstration of the bactericidal action at 41°C of 0.047 mol, phthalate-HCl buffer of pH 2.4 on various bacterial species.*

Genus or Species	Culture After 24 h Incubation at 37°C	
	Control	Exposed, 1 min
E. coli	+	–
Klebsiella	+	–
Citrobacter	+	–
Enterobacter	+	–
Aeromonas	+	–
Pseudomonas aeruginosa	+	–
Staphylococcus	+	–
Streptococcus	+	–
Bacillus	+	–

The chosen conditions of pH, temperature, and GAD concentration are optimal for GAD activity, leading to the formation of γ-amino-butyric acid and CO_2. After a 25 min reaction period, the CO_2 is set free by the addition of sulfuric acid and recovered in the gaseous phase (air). This causes ultimately a proportional discoloration of a buffered phenolphtalein reagent.

As explained earlier, the aim of the use of the GAD method was to enable detection of very low levels of *E. coli*, that is, below 1/100 ml. For this purpose, we coupled the apparatus with a device that concentrates bacteria by means of hollow fiber ultrafiltration [*30*]. This apparatus allows quantitative recovery of *E. coli* from water, irrespective of filtration rate or initital bacterial density (Table 12). It is connected with the automatic analyzer as shown by Fig. 2 and enables the continuous concentration of bacteria in water. The pump PP1 draws samples and forwards them in the ultrafiltration apparatus where they are concentrated at a flow rate D_1. A fraction of the concentrate is drawn by the automatic apparatus at a flow rate D_2. The loss of liquid in the circuit, due to D_1 and D_2, is compensated by the supply of a new sample at a flow rate $D_0 = D_1 + D_2$. Consequently, there is a concentration gradient being formed with time, which tends to reach a limit that equals $C_0 \times (D_1 + D_2)D_2$ where C_0 = the bacterial concentration (cfu/ml) in the sample, D_1 = filtration rate (ml/min), and D_2 = concentration sampling rate (ml/min). Every 12 min, the whole device analyzes 100 ml of water with high accuracy. *E. coli* concentrations equal to or less than 1 *E. coli*/100 ml have been actually detected. The apparatus allows examination of 12 litres of water in 24 h, with a sensitivity that equals the most sophisticated conven-

TABLE 12—*Concentration of* E. coli *by means of Biomed MHFC-10-1 apparatus, containing SHF-36 fibers.*

Test No.	Filtration rate, l/h	*E. coli* in 200 ml	*E. coli* recovery, %
24	1.2	230	99
25	0.64	176	95
15	0.98	183	100
30	0.88	622×10^7	100

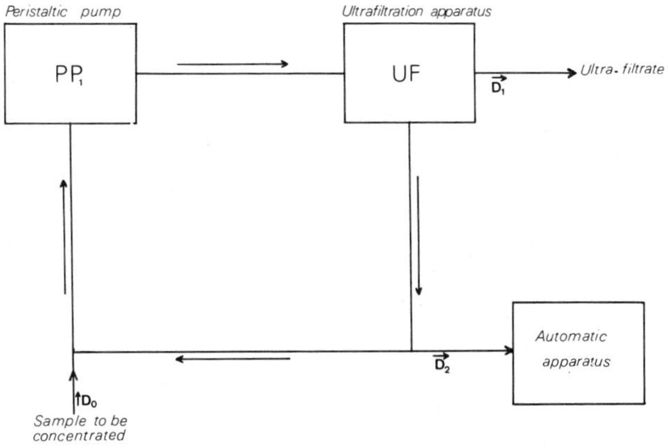

FIG. 2—*Continuous concentration of* E. coli *by means of ultrafiltration; assembly drawing with apparatus for automatic bacteriological analysis.*

tional method and within 12 to 13 h. It can be used in treatment plants and tanks as well as at crucial points of the distribution net of drinking water. Investigations are in progress to assess its efficiency in comparison with that of conventional methods under practical conditions.

References

[*1*] Buttiaux, R., *Journal of Applied Bacteriology*, Vol. 22, 1959, pp. 153–158.

[*2*] Buttiaux, R., "Revue des Fermentations et des Industries alimentaires." *Bruxelles*, Vol. 19, 1964, pp. 11–17.

[*3*] Buttiaux, R., and Mossel, D. A. A., *Journal of Applied Bacteriology*, Vol. 24, 1961, pp. 353–364.

[*4*] *Standard Methods for Examination of Water and Wastewater*, 13th Ed., American Public Health Association, Washington, D. C., 1971.

[*5*] Buttiaux, R., Samaille, J., and Pierens, Y., *Annals of the Institut Pasteur Lille*, Vol. 7, 1956, p. 137.

[*6*] Henriksen, S. D. A., *Acta Pathologica Et Microbiologica Scandinavica*, Vol. 37, 1955, p. 267.

[*7*] Kampelmacher, E. H., Leussink, A. B., and Van Noorle Jansen, L. M., *Water Research*, Vol. 10, 1976, pp. 285–288.

[8] Schubert, R. H. W., *Archiv für hygiene und Bakteriologie*, Vol. 150, 1967, pp. 689-708.

[*9*] Leclerc, H., *Bulletin De L'Association Des Diplomes De Microbiologie De La Faculte De Pharmacie De Nancy*, Vol. 88, 1962, pp. 12–20.

[*10*] Leclerc, H., *Annals of the Institut Pasteur Lille*, Vol. 14, 1963, pp. 49–110.

[*11*] Leclerc, H., and Buttiaux, R., *Annals of the Institut Pasteur*, Vol. 103, 1962, pp. 97–100.

[*12*] Mossel, D. A. A., *Journal of the Association of Official Analytical Chemists*, Vol. 50, 1967, pp. 91–104.

[*13*] Bettelheim, K. A., Ismail, N., Shinebaum, R., Shooter, R. A., Moorhouse, E., and Farrell, W., *Journal of Hygiene*, Vol. 76, 1976, p. 403.

[*14*] Drasar, B. S. and Hill, M. J., *Human Intestinal Flora*, London, Academic Press, London, England, 1974.

[*15*] Fung, D. Y. C. and Miller, R. D., *Applied Microbiology*, Vol. 25, 1973, pp. 793–799.

[16] Habs, H. and Muller, H., *Archiv für Hygiene und Bakteriologie*, Vol. 144, 1960, pp. 1-6.

[*17*] Gavini, F., Lefebvre, B., and Leclerc, H., *Annals of Microbiology (Institut Pasteur)*, Vol. 127A, 1976, pp. 275–295.

[*18*] Ewing, W. H., and Davis, B. R., *International Journal of Systematic Bacteriology*, Vol. 22, 1972, pp. 12–18.
[*19*] Geldreich, E. E., Bordner, R. H., Huff, C. B., Clark, H. F., and Kabler, P. W., *Journal of the Water Pollution Control Federation*, Vol. 34, 1962, pp. 295–301.
[*20*] Geldreich, E. E., Huff, C. B., Bordner, R. H., Kabler, P. W., and Clark, H. F., *Journal of Applied Bacteriology*, Vol. 25, 1962, pp. 87–93.
[*21*] Selenka, F. and Kissling, E., *Archiv für Hygiene und Bakteriologie*, Vol. 148, 1964. pp. 516–526.
[*22*] Selenka, F., *Archiv für Hygiene und Bakteriologie*, Vol. 150, 1967, pp. 660–681.
[*23*] Mossel, D. A. A., "Microbiological quality assurance of water in relation to food hygiene," Rend. Cont. Sup. Sanita., Roma, 1976, to be published.
[*24*] Leclerc, H., *Annals of the Institut Pasteur*, Vol. 112, 1967, pp. 713–731.
[*25*] Leclerc, H., and Catsaras, M., *Bulletin De L'Association Des Diplomes De Microbiologie De La Faculte De Pharmacie De Nancy*, Vol. 107, 1967, pp. 38–49.
[*26*] Leclerc, H., and Catsaras, M., *Annals of the Institut Pasteur*, Vol. 114, 1968, pp. 421–424.
[*27*] Leclerc, H., *Annals of the Institut Pasteur Lille*, Vol. 17, 1966, pp. 21–32.
[*28*] Leclerc, H., *Annals of the Institut Pasteur Lille*, Vol. 17, 1966, pp. 203–207.
[29] Trinel, P. A. and Leclerc, H., *Water Research*, Vol. 6, 1972, pp. 1445–458.
[30] Trinel, P. A. and Leclerc, H., *Annals of Microbiology (Institut Pasteur)*, Vol. 127B, 1976. pp. 201–212.

J. J. Farmer III[1] *and D. J. Brenner*[1]

Concept of a Bacterial Species: Importance to Writers of Microbiological Standards for Water

REFERENCE: Farmer, J. J. III and Brenner, D. J., "**Concept of a Bacterial Species: Importance to Writers of Microbiological Standards for Water,**" *Bacterial Indicators/Health Hazards Associated With Water, ASTM STP 635,* A. W. Hoadley and B. J. Dutka, Eds., American Society for Testing and Materials, 1977, pp. 37–47.

ABSTRACT: A bacterial species is operationally difficult to define. There is considerable phenotypic variation in some species, such as *Escherichia coli*; thus, the limits of the species are difficult to determine. Other species such as *Edwardsiella tarda* and *Serratia marcescens* are "tighter" because the strains which comprise these species show little phenotypic variation. The latter species are easier to define biochemically.

Newer taxonomic techniques have been helpful in defining bacterial species. Often, the results have agreed with those obtained by conventional methods. However, species such as *Enterobacter agglomerans* have been shown by DNA-DNA hybridization to be composed of over twelve different hybridization groups which could be considered "new species."

Many different bacterial species are washed from the soil into water. Some of these are soil bacteria; others make up the gut flora of insects, invertebrates, and lower vertebrates. Often, these environmental species are closely related to species known to be pathogenic for man. Sometimes environmental isolates can be differentiated from the human pathogens by simple tests; *S. liquefaciens* and *S. rubidaea* are phenotypically different from the human pathogen *S. marcescens* and can be recognized easily. Environmental strains of *Klebsiella*, for which the taxonomy has not been completely defined, cannot yet be differentiated from the human pathogens. Similarly, in *Enterobacter*, *Citrobacter*, *Proteus*, *Vibrio*, *Pseudomonas*, *Aeromonas*, and other groups, there are taxonomic gaps where species lines are not completely clear. Many isolates that have been classified in these genera may not be pathogenic for man. Consequently, to ensure that standards will have relevance in public health, writers of microbiological standards for water should proceed very carefully in their treatment of indicator organisms. Many species in water are probably not pathogenic for humans but are closely related to species which are pathogenic. Even with a complete battery of biochemical and serological tests, isolates cannot always be assigned to their correct species. We see many isolates which probably belong to new species which have not been defined.

KEY WORDS: bacteria, water, species concept, microbiology, fecal coliform, lactose

[1]Chief, Enteric Section, and director, respectively, National Laboratory for Enteric Bacteriophage Typing, Center for Disease Control, U. S. Public Health Service, Atlanta, Ga. 30333.

It is important that writers of microbiological standards for water consider the concept of a bacterial species. There are as many definitions for a bacterial species as there are people who have pondered the subject [*1–3*].[2] To say that a bacterial isolate is *Escherichia coli* is a rather profound statement. It means that the isolate belongs to the species *E. coli* and to no other species. It is not a *Pseudomonas aeruginosa*, *Candida albicans*, *Aeromonas hydrophila*, or *Citrobacter freundii*. Most bacteriologists have some concept of a bacterial species, but it is very difficult to define precisely.

The concept of a bacterial species is important in the discussion of indicator bacteria in water. During the past 20 years, the Enteric Section at the Center for Disease Control (CDC) has characterized (in great detail) over 20 000 strains of Enterobacteriaceae and their relatives. We have learned much about the "human pathogens" and something about their close relatives. Some of the principles that have emerged follow.

1. There are many different species of bacteria in the aquatic environment; many of these come from soil and the feces of lower animals.
2. Many of these species have close relatives which are well known as human pathogens.
3. An extremely sensitive technique, such as DNA-DNA hybridization, can show that these environmental isolates belong to species different from the human pathogens.
4. A simple battery of tests (and sometimes even a complete one) often leads to the conclusion that "the environmental isolate is the same species as the human pathogen." However, in reality, they are different species.

In this paper, we will illustrate these points. Until more is known about the fine taxonomic differences between human pathogens and their environmental relatives, those who write standards or propose guidelines relating to the health hazards of water for humans should exercise great caution.

Traditional View of a Bacterial Species

Until recently, most species have been defined on the basis of biochemical tests and a few other characteristics, such as Gram's stain, presence or absence of spores, cell morpholoy, and attachment of flagella [4,2,5]. Figure 1 shows a graphic representation of four species, based on phenotypic properties. This represents the amount of variation when hundreds of strains are tested for biochemical reactions such as IMViC, carbohydrate fermentation, and similar tests [6]. *Edwardsiella tarda* is a very "tight" species biochemically. Table 1 shows that *E. tarda* strains are extremely uniform in their reactions for the eleven tests shown in the table [*2*]. When the number of tests is expanded to 53, only the test for L-arabinose fermentation is variable (9 percent positive, 91 percent negative). The other tests are 98 percent positive or 98 percent negative for the strains. Since the reactions for *E. tarda* are constant, it is easy to arrive at a species definition—*E. tarda* is the bacterial

[2]The italic numbers in brackets refer to the list of references appended to this paper.

Biochemical Variation

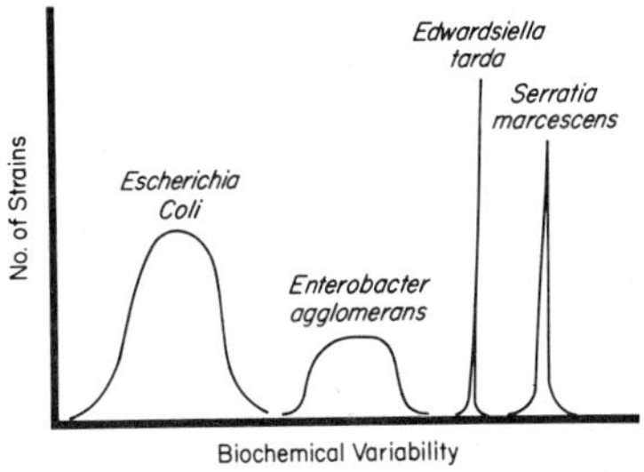

FIG. 1—*Representation of four species based on the amount of phenotypic variation found in a collection of several hundred strains. (The figure is based on actual data, but the data are not presented quantitatively.)*

species composed of strains which are indole$^+$, MR$^+$, VP$^-$, citrate$^-$, lysine$^+$, arginine$^-$, ornithine$^+$, lactose$^-$, and so on for all 53 tests.

A "wild type" strain of *E. tarda* would have the properties expected from data such as those in Table 1. Occasionally, a strain is encountered that has one (or occasionally more) characteristic which is different from the wild type. When the characteristic is negative and the wild type character is positive, the difference in phenotype is usually caused by a mutation at a locus within the gene which specifies the phenotypic character. Indole-negative strains are rare (less than 1 percent of all strains), and this altered phenotype is easily explained by a mutation in the gene which codes for the enzyme which transforms tryptophan to indole. A strain which cannot pro-

TABLE 1—*Biochemical variability of four species.*

	Percent Positive[a] for Strains of			
Test	*Escherichia coli*	*Enterobacter agglomerans*	*Edwardsiella tarda*	*Serratia marcescens*
Indole production	98.6	18.7	99.0	0.2
Methyl Red test	99.9	44.8	100	17.7
Voges Proskauer test	0	67.9	0	100
Citrate	0.2	66.6	0	98.6
Lysine decarboxylase	88.7	0	100	99.6
Arginine dihydrolase	17.6	0	0	1.3
Ornithine decarboxylase	64.2	0	99.7	99.5
Fermentation of				
Lactose	90.8	40.5	0	2.2
Raffinose	50.9	24.8	0	1.7
Salicin	40.0	63.6	0	95.1
Cellobiose	2.4	56.2	0	20.8

[a]Positive reaction within 48 h; data from Ref *2*.

duce indole could be designated as *E. tarda* indole$^-$. Thus, almost all strains of *E. tarda* that have been studied are wild type strains or are slightly atypical because of a probable gene mutation that either prevents production of an enzyme or results in formation of an inactive enzyme.

Serratia marcescens is another tight species (Figure 1), although it is more variable than *E. tarda* (Table 1) in the phenotypic characters usually studied [2]. Both *E. tarda* and *S. marcescens* are ideal species, and it is easy to coin an operational definition for them on the basis of simple biochemical tests.

E. coli and *Enterobacter agglomerans* are "phenotypically variable species" [7,8] in contrast to the phenotypically tight species *E. tarda* and *S. marcescens* (Table 1, Fig. 1). A species definition is very hard to write for *E. coli* and *E. agglomerans*. There are strains of *E. coli* which are very active metabolically and are positive on many of the biochemical tests; however, there are also strains of *E. coli* which are inactive biochemically. The latter strains are usually nonmotile, lactose$^-$, and gas$^-$. They were once called the Alkalescens-Dispar group but are now considered to be biochemically inactive strains of *E. coli*. Isolate No. 64–75 is an example of an inactive *E. coli*; it was positive on only eleven of 43 biochemical tests. In contrast, Isolate No. 4493-75 is an active strain of *E. coli* and was positive on 28 of 44 tests. Both of these isolates are members of the species *E. coli*, even though they vary greatly in the number of positive tests. Table 1 shows that 90.8 percent of *E. coli* strains at CDC fermented lactose. This variability in the species is important to the water microbiologist because lactose$^-$ isolates would not be members of the "coliform group" even though they are members of the species *E. coli* [7]. Fortunately, most strains of *E. coli* found in feces from humans are active biochemically and ferment lactose. If this were not true, the coliform concept probably would not have gained popularity.

What is simple operational definition of *E. coli*? Betty Davis of CDC's Enteric Section has studied and characterized more strains of *E. coli* than anyone else in this country and, perhaps, the world. We asked her to write a definition of *E. coli* for this paper. After considering the matter for several days, she said that it simply was not possible to write a good operational definition of this species because of its variations with respect to biochemical reactions, antigens, antimicrobial susceptibilities, and other properties. Many unusual strains are sent to CDC for identification. Even after doing 50 biochemical tests and serological typing, we often still cannot answer the question: Is the isolate a member of the species *E. coli*? We often do identifications with the aid of computer analysis. Even though the computer does about 2000 calculations in comparing the unknown with each of the known species, we often get a final printout such as probability that the isolate is *E. coli*: 0.86; probability that the isolate is *Citrobacter freundii*: 0.14. This means that even the computer cannot tell an *E. coli* from its close relatives in other species. It has become a standard joke that people come to the Enteric Section "knowing what an *E. coli* is" but leave "not knowing what an *E. coli* is." Figure 1 also shows that the newly defined *E. agglomerans* is an even more variable species than *E. coli* in its phenotypic characters. Species like *E. coli* and *E. agglomerans* are taxonomic problems because of their variability. Fortunately, many strains of *E. coli* are easy to identify, and it is usually true that isolates that look like *E. coli* on eosine methylene blue agar or similar lactose containing are really *E. coli*. If this were not true, the coliform concept would be too inaccurate to be of practical value [7]. However, critics of the coliform concept are quick to point up that many other bacterial species are indistinguishable from *E. coli* on these media and would be "false positives."

Changing Views of a Bacteria Species

The thinking summarized in the previous section dominated bacterial taxonomy until the 1960s. About this time, new taxonomic approaches became popular, and several of these methods have been used to compare the same groups of strains. The newer techniques [9] include determination of guanine-cytosine ratios, DNA-DNA hybridization, DNA-RNA hybridization, Adansonian analysis of phenotypic properties, cell wall analysis, gas chromatography, electrophoresis of cell proteins and enzymes, antibiograms, bacteriophage susceptibility, bacteriocin production, genetic mapping, and genetic exchange by transformation, transduction, or conjugation. These newer techniques usually confirmed the previous classification. However, in some instances the results indicated that a revised classification was needed [*5,10–13*].

DNA-DNA hybridization is one of the most powerful of the newer methods [*10*]. The advantage of DNA hybridization is that it determines strain similarity by comparing all the genes, whether or not they are functional in contrast to phenotypic characterization which examines the expression of only a fraction of the total genome. If two strains belong to the same species, then most of their DNA base sequences will be highly related, because DNA base sequences within a given species are conserved during evolution. In contrast, if two strains belong to different species, then their DNA sequences will not be as closely related because of nucleotide changes during long-term evolution (otherwise, they would still be the same species).

DNA hybridization has been used to analyze the classification used in most bacterial groups, and it has provided an unbiased definition of a bacterial species. We mentioned that the previous taxonomic concept of *E. coli* was that of a species composed of strains with considerable phenotypic variation; however an alternate hypothesis is also possible: *E. coli* is composed of many different species, but when they are lumped together, they appear to be a "single highly variable species." This is a circular argument, and we could not tell which hypothesis was correct. However, DNA hybridization showed that *E. coli* isolates from many different sources and with different phenotypic patterns were highly related and thus could be considered as a true species in the genetic sense [*11*]. Thus, the hypothesis—"*E.*

coli is a true species made up of phenotypically variable strains"—is correct. To the surprise of many clinical microbiologists, DNA hybridization showed that many strains of *Shigella* were as closely related (in their DNA base sequences) to *E. coli* as they were to other strains of *Shigella* [*10*]. The inescapable conclusion was that these species of *Shigella* really belonged to the same species as *E. coli* and that they could be considered as "pathogenic, serotypic, or biochemical varieties of this species as it is defined genetically." Although clinical microbiologists accept this concept of a species in the Shigella-E. coli group in principle, they have not used it in practical classification.

E. agglomerans has been defined as a "single species with considerable phenotypic variation among its strains" [*8*] (Fig. 1 and Table 1). DNA hybridization was used to test this definition. The results indicated *E. agglomerans* is composed of a dozen or more different DNA hybridization groups, most of which could be thought of as true bacterial species [*10*]. Unfortunately, it may take years to find simple phenotypic tests which correlate with the data from DNA hybridization. So at present we should think of *E. agglomerans* as a term that includes many different species which cannot be separated by simple tests. A better term would probably be the "Herbicola-Lathyri group," which does not represent these bacteria as a true species. This term would be similar to the "paracolon group," once used in enteric bacteriology as a convenience to signify bacteria which fermented lactose rapidly and were thus unlikely to be members of the pathogenic groups Shigella or Salmonella [*2*]. Today, the strains which were called "paracolon" 25 years ago are assigned to their proper genus and species on the basis of simple phenotypic characters. All this makes interesting reading to the microbial taxonomist, but what does it have to do with indicator bacteria in water? We hope to show that it is relevant and that the writer of microbiological standards for water quality should consider some of these fine taxonomic points.

Human Pathogens Which Have Close Relatives in the Environment

Over the years, the concept has justifiably evolved that strains of *Klebsiella pneumoniae* are generally pathogenic for man. The Public Health Service regulations on interstate quarantine define all species and serotypes of *Klebsiella* to be etiologic agents (an etiologic agent "means a viable microorganism or its toxins which cause, or may cause, human disease"). Recently, a number of investigators have studied *Klebsiella* in the hospital environment and natural waters [*14–17*]. The implication has been that these strains are "pathogenic *Klebsiella*." However, recent taxonomic studies have shown that there is a need for taxonomic changes in the genus *Klebsiella*. The species *K. rhinoscleromatis* and *K. ozaenae* are highly related to *K. pneumoniae* by DNA hybridization. These data suggest that they should perhaps be lowered to subspecies status [*10*]. Similarly, Jain, Radsak, and Mannheim recently showed by DNA hybridization that members of a Klebsiella group, previously known as the "Oxytocum group of *Klebsiella pneumoniae*" (these are indole$^+$ and gelatin$^+$) were not closely related to *K. pneumoniae* [*13*]. They proposed that the Oxytocum group be removed from the genus *Klebsiella* to a new genus which they did not specify. Strains of *Klebsiella* isolated from natural sources, such as soil or water, have been called *K. pneumoniae* on the basis of a few biochemical tests. The definition of *Klebsiella* and *K. pneumoniae* will probably undergo radical change as soon as more data become available. Some strains that CDC's Enteric Section has identified in the past as *K. pneumoniae* probably do not belong to this species. Thus, the assumption that they are pathogenic for man may not have a scientific basis. Klebsiella-like bacteria are easily isolated from soil, water, plants, and insects [*3*] (J. J. Farmer III, unpublished) and whether these bacteria are pathogenic for man is not known. Until the taxonomy of *Klebsiella* has been clarified and each species has been studied for pathogenicity and public health importance, the writer of guidelines for *Klebsiella* in the environment will be faced with a difficult task. If a reference laboratory cannot identify *Klebsiella* with 50 biochemical tests, computer analysis, and complete capsular typing, then how can an identification be based on a few phenotypic tests [*14*].

A similar situation exists in the genus *Serratia* where *S. marcescens* is by far the most important human pathogen in the genus. Over the last ten years, the Enteric Section has studied over 50 outbreaks of infection due to *Serratia*, and each one has been caused by *S. marcescens*. However, other species of *Serratia* are found in the environment [*18*]. Two other *Serratia* species, *S. liquefaciens* and *S. rubidaea*, have been described. Further, several additional species probably exist but have not been described. *S. marcescens* is found in the environment, but so are the other *Serratia* species. Thus, it would be wrong to equate all *Serratia* to the human pathogen *S. marcesens*. Similar patterns appear to exist in *Enterobacter*, *Yersinia*, *Vibrio*, *Pseudomonas*, *Aeromonas*, and *Citrobacter*. Each of these groups contains one or more recognized human pathogens, but also there are probably very closely related species which are found in the aquatic environment. Even with a large battery of biochemical tests and computer analysis, we find it difficult to assign some of these environmental strains to an existing species. Preliminary data from many laboratories suggest that these environmental isolates do not fit existing species and that they will be given species status as soon as the taxonomy has been refined. The importance of these new environmental species to human health is unknown.

Bacteria Species in Water

It is well known that soil washes into natural bodies of water, particularly after heavy rains. Many different kinds of bacteria also will be carried into water, including soil bacteria and bacteria from the the feces of animals, such as tiny invertebrates, insects, birds, and other "lower animals." In addition, water will often have its own bacterial flora, contributed by its resident species of animals. Bacterial flora from literally hundreds of different species of animals enters natural waters each week. The normal flora of man has been thoroughly studied, and the results fill many volumes [*19*]. The stool flora of man alone may comprise over a hundred different species of bacteria. Although less well studied, other animals have their own characteristic bacterial flora [*20,21*]. Many of the bacterial species which form the normal flora of man probably evolved from bacterial species which form the normal flora of lower animals. In the cases that have been carefully studied, it is easy to tell the difference between closely related species with a sensitive technique, such as DNA-hybridization [*10*]. Often, however, it is not easy to tell the difference between closely related species with the simple tests most applicable to routine identification. To illustrate this point, fecal samples were taken from insects, birds, and mammals (J. J. Farmer III, unpublished). These gut isolates were then identified with the same tests that have been used to identify human pathogens. Only 20 percent of the isolates could be assigned with certainty to a known species of Enterobacteriaceae, 30 percent could be assigned to a genus (but the isolates did not fit any of the previously described species), and 50 percent could not even be assigned to a genus. Most of these isolates are probably new species which just have not been encountered frequently by medical bacteriologists, who have dominated the taxonomy of enteric bacteria. These identifications were based on results from 45 biochemical tests. When a short battery of tests was used (15 or fewer tests), the isolates fit into named species. Thus, a superficial identification resulted in an incorrect identification. These and similar data [*2*] show that it is frequently difficult to assign an isolate to its correct genus and species. Even with 45 tests, it is sometimes difficult; with a few tests, it is impossible.

Discussion

We have tried to show the importance of taxonomy to writers of microbiological standards for water. The concept of a bacterial species is very important in this field and must be understood; otherwise, standards could be written which have little relevance to public health.

The taxonomist who is a purist will object to a term such as the coliform group. However, this is an operational term which has been precisely defined in the 13th edition of *Standard Methods* [*7*]: "The coliform group comprises all of the aerobic and facultative anaerobic, gram-negative, nonspore-forming, rod-shaped bacteria which ferment lactose with gas formation within 48 hr at 35 C." The taxonomist would point up that this says nothing about the species *E. coli* on which the coliform concept depends heavily. In fact, any species of Enterobacteriaceae (because they can acquire plasmids which enable them to ferment lactose) and many species of *Aeromonas* and *Vibrio* are coliforms according to this definition. Similarly, many strains which are true *E. coli* are not coliform, because they have lost one of the enzymes for the transport or catabolism of lactose, or have lost the enzyme formate hydrogen lyase, which is needed for gas production. Even with these shortcomings, the definition of the coliform group has been widely accepted throughout the years. This is because coliform is a very simple operational definition; a coliform can be easily tested for in the field or in a moderately equipped laboratory. An operational definition for *E. coli* could be made, but it would take three or four times as long to process each colony to be sure it fit the definition and was really a member of the species *E. coli*. The term coliform is much easier to test for than the species *E. coli* in terms of microbiological standards. Similarly, water microbiologists have used other operational definitions, such as "fecal streptococcus" and "enterococcus" and have wisely stayed away from taxonomic consideration of the species concept in *Streptococcus*, a very complex group [*4*]. We would advise those who consider future microbiological standards for water to avoid definitions based on bacterial species and continue to use operational terms such as coliform group and fecal streptococci, which have a precise meaning but beautifully avoid all taxonomic arguments. It would be easy to coin terms such as salmonella form, aeruginosa form, pseudomonas form, vibrio form, clolera form, and klebsiella form. Standards written in terms of actual species or genera should be avoided if possible. As we have pointed up, even the so-called experts cannot always agree on the precise definition or significance of these genera or species.

Standards can be written in terms of actual species, but this adds a further degree of sophistication and requires many additional laboratory tests. For example, a fecal coliform can be defined on the basis of the colony's appearance on a membrane filter after 24 h incubation at 44.5°C. To actually identify the colony as belonging to the species *E. coli* would require additional tests such as oxidase reaction, indole production, citrate utilization, methyl red and Voges-Proshauer Tests, and Gram's stain. Typical members of the species *E. coli* are gram , oxidase , idole$^+$, citrate$^-$, MR$^+$, and VP$^-$. However, there are exceptions to these generalizations. We have seen isolates of *E. coli* which are indole$^-$, citrate$^+$ or MR$^-$. Similarly, other species of bacteria found in water can appear to be *E. coli* when this limited number of tests is done. The lower number of false positives and false negatives for a standard based on the species *E. coli* could be easily outweighed by increased costs and delays in obtaining final results. We would also emphasize that *E. coli* is not a simple species to define, and any definition will have many exceptions. A good definition of *E. coli* has been proposed by others participating in this symposium; the only question is whether this is a practical concept for routine use.

Once terms have been coined and guidelines written, then the logical step would be to determine if they are meaningful in public health. For example, klebsiella forms found in rivers, lakes, or bathing water may have no significance because they belong to species which are not pathogenic for man. The human pathogens may be limited to the single species *K. pneumoniae* (including *K. ozaenae* and *K. rhinoscleromatis*). A reference laboratory can be of assistance in determining if the environmental isolates are really the same species as the known human pathogens. When the complex taxonomy has been worked out and correlated with the role of each species in human disease, then the guidelines for microbiological quality of water can be revised to include this new information. Until this time, the writers of standards may want to avoid all genus and species terms and write their guidelines in terms which can be precisely defined and easily tested. The terms coliform, fecal coliform, fecal streptococcus and enterococcus are not perfect, but they have stood the most important tests—the test of time and usefulness. New terms and standards should be patterned after these examples.

Acknowledgments

We thank our colleagues in the Enteric Section for their helpful discussions and for providing much of the data we have discussed.

References

[*1*] Cowan, S. T., *Journal of General Microbiology*, Vol. 61, 1970, pp. 145–154.

[*2*] Edwards, P. R. and Ewing, W. H., *Identification of Enterobacteriaceae*, 3rd ed., Burgess Publishing Co., Minneapolis, Minn., 1972.

[*3*] Seidler, R. J., Knittel, M. D., and Brown, C., *Applied Microbiology*, Vol. 29, 1975, pp. 819–825.

[*4*] *Bergey's Manual of Determinative Bacteriology*, 8th ed., R. E. Buchanan and N. E. Gibbons, eds., The Williams and Wilkins Co., Baltimore, Md., 1974.

[*5*] Stanier, R. Y., Palleroni, N. J., and Doudoroff, M., *Journal of General Microbiology*, Vol. 43, 1966, pp. 159–271.

[*6*] Ewing, W. H. and Davis, B. R., *Media and Tests for Differentiation of Enterobacteriaceae*, Center for Disease Control, Atlanta, Ga., 1970.

[7] *Standard Methods for the Examination of Water and Wastewater*, 13th Edition, American Public Health Association, Washington, D. C., 1971.

[*8*] Ewing, W. H. and Fife, M. A., *Biochemical Characterization of* Enterobacter agglomerans, Center for Disease Control, Atlanta, Ga., 1972.

[*9*] Marmur, J., Falkow, S., and Mandel, M., *Annual Review of Microbiology*, Vol. 17, 1963, pp. 329–372.

[*10*] Brenner, D. J., *Public Health Laboratory*, Vol. 34, 1976, pp. 48–55.

[*11*] Brenner, D. J., Fanning, G. R., Skerman, F. J., and Falkow, S., *Journal of Bacteriology*, Vol. 109, 1972, pp. 953–965.

[*12*] Juni, E., *Journal of Bacteriology*, Vol. 112, 1972, pp. 917–931.

[*13*] Radsak, K. J. and Mannheim, W., *International Journal of Systematic Bacteriology*, Vol. 24, 1974, pp. 402–407.

[*14*] Campbell, L. M., Roth, I. L., and Klein, R. D., *Applied and Environmental Microbiology*, Vol. 31, 1976, pp. 213–215.

[*15*] Davis, T. J. and Matsen, J. M., *Journal of Infectious Diseases*, Vol. 130, 1974, pp. 402–405.

[*16*] Duncan, D. W. and Razzell, W. E., *Applied Microbiology*, Vol. 24, 1972, pp. 933–938.

[*17*] Fallon, R. J., *Journal of Clinical Pathology*, Vol. 26, 1973, pp. 523–528.

[*18*] Farmer, J. J. III, Silva, F., and Williams, D. R., *Applied Microbiology*, Vol. 25, 1973, pp. 151–152.

[*19*] Rosebury, T., *Microorganisms Indigenous to Man*, McGraw Hill, New York, N. Y., 1961.
[*20*] Kloos, W. E., Zimmerman, R. J., and Smith, R. F., *Applied and Environmental Microbiology*, Vol. 31, 1976, pp. 53–59.
[*21*] Saphir, D. A. and Carter, G. R., *Journal of Clinical Microbiology*, Vol. 3, 1976, pp. 344–349.

A. P. Dufour[1]

Escherichia coli: The Fecal Coliform

REFERENCE: Dufour, A. P., ***Escherichia coli:*** **The Fecal Coliform,"***Bacterial Indicators/Health Hazards Associated With Water, ASTM STP 635*, A. W. Hoadley and B. J. Dutka, Eds., American Society for Testing and Materials, 1977, pp. 48–58.

ABSTRACT: The use of *Escherichia coli* as an indicator of fecal pollution and the significance of its presence in surface waters was reviewed. A minidefinition which identifies this organism at least 95 percent of the time was proposed. Membrane filtration and most probable number methods for the enumeration of *E. coli* were discussed.

KEY WORDS: bacteria, water, coliform bacteria, fecal pollution indicator, fecal coliform, enumeration techniques

For many years, the coliform group was the mainstay of the sanitarian's tools for detecting the presence of fecal contamination in aquatic environments. The broad general characteristics which define this group have allowed it to be one of the most useful of bacterial indicators and at the same time have been responsible for its displacement as an indicator of fecal contamination. Coliforms are defined as including all of the gram-negative, nonspore forming facultatively anaerobic bacilli which ferment lactose with the production of gas within 48 h at a temperature of 35° C [*1*].[2] This utilitarian definition, the main facets of which were adopted circa 1917 [*2*], accounts for the easy, unambiguous methodology used to identify and enumerate coliforms. The simplicity of the definition was also its main drawback, since it includes multiple genera such as the *Escherichia*, *Klebsiella*, *Enterobacter*, and *Citrobacter*—some of which are seldom associated with fecal contamination. One of the earliest efforts to limit the assay to those coliforms related to feces was carried out by Eijkman in 1904 [*3*]. He used gas production from glucose at an elevated incubation temperature to detect coliforms associated with fecal pollution. Modifications of his medium by others [*4–6*] has led to the one used today to detect those coliforms labelled as fecal coliforms. This indicator group is defined as having all the charac-

[1]Microbiologist, Marine Field Station of the Health Effects Research Laboratory (Cincinnati, Ohio), U.S. Environmental Protection Agency, West Kingston, R. I. 02892.

[2]The italic numbers in brackets refer to the list of references appended to this paper.

teristics of coliforms; in addition, it is able to ferment lactose with the production of gas in 24 h at an incubation temperature of 44.5°C [*1*]. The improved specificity of the fecal coliform index relative to that of the coliform index led to its acceptance and subsequent widespread use. Although the fecal coliform test was a great improvement in methodology for detecting fecal contamination in aquatic environments, it soon became apparent that it too had a shortcoming. Various situations arose where a positive test for fecal coliforms was observed even though no evidence for human or warm-blooded animal fecal pollution could be found. These observations were usually made on samples taken from waters that received industrial effluents containing high concentrations of carbohydrate materials [*7–9*]. These aberrations were invariably due to *Klebsiella* species. Thus, once again, the heterogeneity of an indicator group was found to affect the specificity of the index.

The fecal coliform test selects mainly for the genera *Escherichia* and *Klebsiella* with occasional positive reactions being given by other genera. This lack of specificity was one of the determining factors which prompted the call for a reexamination of the definition for fecal coliforms or, more properly, the definition of a coliform indicator consistently and specifically associated with fecal contamination.

Description of Indicator Species

The proposed definition for an indicator of the presence of fecal pollution is a coliform capable of fermenting lactose with the production of acid and gas at 44.5°C within 48 h, which does not produce cytochrome oxidase, which produces indole from tryptophane, which is incapable of utilizing sodium citrate as a sole source of carbon, which is incapable of producing acetyl methyl carbinol, which is incapable of hydrolyzing urea, and which gives a positive result with the methyl red test.

If incubation at 44.5°C is excluded from this proposed defintion, it becomes at once obvious that the biochemical characteristics describe, in part, the species *E. coli*, which is well defined in *Bergey's Manual of Determinative Bacteriology* [*10*]. The elevated temperature characteristic in the definition, which does not appear in *Bergey's Manual*, is included because the ability to grow and ferment lactose at an elevated temperature has been shown over the years to be highly characteristic of *E. coli*.

The abbreviated nature of this proposed definition is also obvious; however, it should be kept in mind that it was designed to accomplish a very limited goal. That goal is to be able to identify the indicator bacterium at least 95 percent of the time with a limited number of tests.

The justification for the definition is quite simple. *E. coli* is the only coliform that is an undoubted inhabitant of the gastrointestinal tract. *Klebsiella*, *Citrobacter*, and *Enterobacter* have also been isolated from human fecal samples but in small numbers [*2,11*]. Actually, there is a paucity of quantification data for these genera, especially *Klebsiella*. In order to examine the distribution and densities of these genera, our laboratory examined 28 fecal samples with a membrane filter procedure [*12*]. Table 1 shows the distribution and density of coliforms isolated from the 28 fecal samples. Overall, 96.8 percent of the coliforms detected were *E. coli*, 1.5 percent were *Klebsiella*, and 1.7 percent were in the Enterobacter-Citrobacter group. These percentages are not too dissimilar from those described by Prescott, Winslow, and McCrady [*2*] who summarized the data of many workers and concluded that 95 percent of the coliforms in feces are *E. coli*, the majority of which are indole formers, and the remaining 5 percent were *A. aerogenes* or *E. freundii* species (*Enterobacter-Citrobacter* by contemporary taxonomic standards). Similar results were presented by Taylor [*13*] who reported the very early findings of MacConkey in Britain and Clemesha in India—both of whom showed that *E. coli* comprised greater than 90 percent of the total number of organisms isolated from feces.

The predominance of *E. coli* in feces is not reflected in the distribution of coliforms found in sewage. Table 2 is illustrative of the numbers and distribution of coliforms found in raw, primary, and secondary effluents from a trickling filter sewage treatment plant which receives mainly domestic wastewater. It is significant that, in raw wastewater, *E. coli* makes up only about one third to one fourth of the coliforms detected, while the other two thirds to three fourths were identified presumptively as belonging to the genera *Klebsiella*, *Enterobacter*, and *Citrobacter*. Carr [*14*] has reported finding about 25 percent *E. coli*, 25 percent *Klebsiella*, and 50 percent *Enterobacter-Citrobacter*. There are three possible explanations for this change in the coliform distribution between the lower gastrointestinal tract and the treatment plant. They are (*a*) a 99 percent die-off of *E. coli* between the time of entrance into the sewerage system and the time of sampling at the influent to the sewage treatment plant with a concomitant survival of the non*E. coli* coliforms, (*b*) a multiplication of the *Enterobacter*, *Klebsiella*, and *Citrobacter* during transport to the plant, and (*c*) surface runoff of these latter types into combined sewerage systems to the point where they outnumber *E. coli* densities. If the *E. coli* portion of the coliform population was dying at a disproportionate rate relative to the other coliforms or if other coliforms grow, then it might be presumed that in a small wastewater system,

TABLE 1—*Percentage distribution of genera of coliforms in human feces.*

			Percent of Coliforms		
Study Year	Number of Samples	Number of Colonies Examined	*E. coli*	*Klebsiella* species	*Enterobacter/ Citrobacter*
1975	13	438	99.99	0.01	. . .
1976	15	285	89	5	6
Totals	28	723	96.8	1.5	1.7

TABLE 2—*Coliform distribution in sewage treatment plant effluents.*

Indicator Organism	Bacterial Densities at Points of Sampling					
	Raw Sewage		Primary Treated Effluent		Secondary Treated Effluent	
	Density[a]	TC,%[b]	Density	TC, %	Density	TC, %
E. coli	19.5	21.8	29.4	29.2	10	33.4
Klebsiella species	37.2	41.5	47.2	46.8	8.2	27.3
Citrobacter/ *Enterobacter* species	32.9	36.7	24.2	24.0	11.8	39.3

[a]All densities multiplied by 1×10^{-3}/ml.
[b]TC = total coliform.

such as a septic tank, the ratio of the various coliforms would reflect sewage treatment plant rather than fecal samples. This hypothesis was tested by assaying septic tank samples in the same manner as the sewage treatment plant samples. The results from the examination of two septic tank systems are shown in Table 3; it is readily apparent that, at least with these closed systems, the distribution of coliforms more closely resembles that of fecal material rather than raw sewage. The alternative hypothesis, that the non-*E. coli* biotypes are the result of runoff, is more difficult to examine empirically. However, the literature contains much data showing that members of the *Klebsiella-Citrobacter-Enterobacter* genera can be easily found in soils and vegetation [*15,16*]. Thus, this would also be a viable alternative.

Another source of the high percentage of coliforms, other than *E. coli*, found in sewage treatment plant effluent might be the fecal waste of animals which makes its way to the plant through combined sewer systems. If animal fecal wastes are the source of these bacteria, then the frequency of *E. coli* relative to other coliforms should be low. This premise was tested by examining fecal samples from 78 domestic animals; the results are given in Table 4. Overall, 94 percent of 1627 colonies examined were *E. coli*; 2 percent were *Klebsiella* species, and 4 percent were *Cibrobacter* / *Enterobacter* species. It is quite obvious that the distribution of coliforms in animal fecal wastes does not resemble that found in sewage treatment plants. Thus, one can conclude that this reservoir most likely is not the main source of coliforms other than *E. coli*. These data support the use of the proposed definition designating *E. coli* as the indicator of fecal pollution.

TABLE 3—*Distribution of coliforms in two septic tanks.*

Septic Tank	Number of Colonies Examined	Percent of Total Coliforms		
		E. coli	*Klebsiella* species	*Enterobacter*/ *Citrobacter*
1	80	79	11	10
2	123	90	0	10

TABLE 4—*Distribution of coliforms in animals.*

Animal	Number of Animals Examined	Number of Colonies Examined	Percent of Total Coliforms		
			E. coli	*Klebsiella* species	*Enterobacter*/*Citrobacter*
Chicken	11	201	90	1	9
Cows	15	264	99.9	. . .	0.1
Sheep	10	192	97	. . .	3
Goats	8	129	92	8	. . .
Pigs	15	399	83.5	6.8	9.7
Dogs	9	233	91	. . .	9
Cats	7	185	100	. . .	. . .
Horses	3	24	100	. . .	. . .
Totals	78	1627	94	2	4

Significance of Finding *E. coli* in Surface Waters

The significance of the presence of *E. coli* in aquatic environments is that the water in question has been contaminated with fecal material from warm-blooded animals. As pointed up previously, this genus is found in the feces of men and animals in higher densities and with greater frequency than any other coliform. However, the demonstration of *E. coli* as the principal coliform found in the feces of men and animals should not generate an overconfidence in the ability of this species to meet all of the prerequisites of a good indicator. Characteristics such as the survival of the indicator relative to that of pathogens and the relationship between the degree of pollution and the level of indicator also should receive consideration. An indirect measure of these characteristics can be found in epidemiological studies relating the level of an indicator bacterium to the incidence of illness in a water resource user population. This relationship should, in effect, not only define the indicator of choice, but it should also give an indication of the degree of association of the relationships mentioned previously.

Our laboratory has taken the epidemiological approach to the problem of selecting an indicator of fecal pollution [*17*]. One of the initial premises at the inception of the study was that it would be conducted without preconceived ideas about a "best" bacterial indicator. Proceeding on this premise, several bacteria associated with human fecal wastes were examined to determine if the levels at which they were found in marine bathing waters was related to the gastrointestinal illness level in the swimming population. The preliminary results [*18*] from the microbiological-epidemiological study indicate that two bacteria, *E. coli* and enterococci, showed the best correlation with illness rates. The finding that *E. coli* shows the best relationship to gastrointestinal illness, rather than members of the Klebsiella-Enterobacter-Citrobacter group, should not be too surprising in view of its position of dominance in the distribution of coliforms found in fecal wastes from humans and warm-blooded animals.

The significance of the presence of *E. coli* in surface waters is that fecal contamination due to humans or other warm-blooded animals has occurred and therefore a potential health hazard risk from microbial or viral enteric pathogens does exist.

Enumeration Methodology

Since it would be quite unreasonable to propose a definition for a bacteriological indicator for fecal pollution without a method for its detection and enumeration, a literature survey of the available membrane filter and most probable number (MPN) methods for *E. coli* was made. The results of the survey indicated that all of the methods had two common characteristics: the use of elevated temperature as a selective agent against other coliforms and some background organisms and a biochemical reaction to differentiate *E. coli* from those coliforms not inhibited by the elevated temperature.

An obvious question arises at this point. If *E. coli* is to be used as an indicator of fecal pollution, does aerogenesis or growth at the elevated temperature allow all of the cells in a given sample to be detected? It is, in fact, well known that aerogenesis at an elevated temperature is not an absolute characteristic of *E. coli*. This genus was examined for its ability to produce gas at 44.5°C by Mishra et al (1018 strains) [*19*], Pugsley et al (64 strains) [*20*], and Halls and Ayres (607 strains) [*21*]. They found, respectively, 15, 9, and 12 percent of the strains they examined to be anaerogenic at this elevated temperature. Thus, even though gas formation at an elevated temperature will not detect all *E. coli*, it detects enough to warrant using the methods to be proposed. In order to determine if any significant growth inconsistencies might arise due to using an elevated temperature enumeration technique for quantifying *E. coli*, the 28 fecal samples mentioned previously were examined using an elevated temperature (44.5°C) membrane filter procedure [*22*]. The results shown in Table 5 indicate that there were no significant differences between the densities of *E. coli* incubated at 35°C and those incubated at 44.5°C.

TABLE 5—*Effect of incubation temperature on* E. coli *recoveries from fecal samples.*

	1975 Study	1976 Study
Number of Samples	13	15
E. coli (35°C)[b]	73.3[a]	13.7[a]
Thermotolerant *E. coli* (44.5°C)[c]	68.2[a]	19.3[a]

[a]Geometric mean; counts are per gram × 10^6.
[b]Enumerated by mC procedure, Ref *12*.
[c]Enumerated by mTEC procedure, Ref *22*.

The fecal nature of those *E. coli* that are anaerogenic or will not grow at the elevated temperature remains unanswered; however, as pointed up by Pugsley, Evison, and James [*20*], there is no evidence that these strains are not of fecal origin. Approached from another point of view, growth or aerogenesis at an elevated incubation temperature are constant characteristics of *E. coli* and not necessarily an indication of their fecal origin.

Five MPN methods which claim to enumerate *E. coli* have been found in the literature. The method proposed by Andrews and Presnell [*23*] measures coliforms other than *E. coli* as shown by their results. This has been confirmed in a private communication with one of the authors as well as with others who have used this method. Two of the methods [*24,25*] use conventional presumptive and elevated temperature MPN steps with the addition of a third set of peptone tubes which are also incubated at an elevated temperature and subsequently tested for indole production. The other two methods, one by Mara [*26*] and the other by Pugsley et al [*20*], employ a single medium after the presumptive test for detection of gas at 44°C and a subsequent test for indole production. The method proposed by Pugsley et al is an interesting departure from the traditional means of differentiating coliforms in that mannitol rather than lactose is the carbon source.

These latter four methods, all of which were developed in England, appear to work well in that country as well as in Norway and Denmark. However, they have not been extensively evaluated in North America. This task, hopefully, will be accomplished in the very near future.

Membrane filter techniques for the enumeration of *E. coli* are not quite as numerous as MPN methods. The Teepol medium proposed by Halls and Ayres [*21*] for enumeration of *E. coli* in seawater utilizes elevated temperature for its selectivity. The specificity of their medium appears to be excellent; however, it is not known whether the method was tested in water containing high densities of *Klebsiella*.

Another method is that described by Delaney, McCarthy, and Grasso [*27*]. The medium contains only tryptone, bile, and agar. After incubation at 44.5°C for 20 to 24 h, the membrane is transferred to a pad saturated with a modified Erhlich's indole reagent. Indole positive colonies become red, and these are designated presumptively as *E. coli*. The detection of *E. coli* relative to accepted methods appeared to be very good. Good results were also obtained by Anderson and Baird-Parker [*28*] in their examination of foods with this method.

The third membrane filter method for *E. coli* was presented by Dufour, Strickland, and Cabelli at the 9th National Shellfish Sanitation Workshop in the Summer of 1975 [*22*]. Initial evaluation of this procedure indicated that 94 percent of the presumptive *E. coli* colonies were confirmed. However, the method has been tested only in the northeastern areas of the United States, and, thus, a judgment as to its true value will have to await further evaluation.

Our laboratory has begun an evaluation of these membrane filter methods which claim to enumerate *E. coli*, and the preliminary results of this comparison are shown in Table 6. It is immediately obvious that while the Teepol medium has the highest presumptive recovery rate, the verification rate for *E. coli* was very low. Many, but not all, of the false positive presumptive counts were verified as *Klebsiella* species. Other genera were probably able to grow because of the 4 h resuscitation period at 30°C prior to incubation at 44°C. The tryptone bile agar medium was much more specific than the Teepol medium, but the detection rate was very low. This was no doubt due to the nutritionally deficient character of this medium relative to the other two and also the lack of a resuscitation period. *Klebsiella* proved to be one of the principal causes of false positive counts with this medium also. Thirty percent of the false positives were indole producing *Klebsiella*. The membrane filter procedure for thermotolerant *E. coli* (mTEC) showed the best detection rate of the three media, probably because it was designed specifically to segregate *E. coli* from *Klebsiella* species.

The results indicate that at least two possible *E. coli* membrane filter

TABLE 6—*Relative recovery of* E. coli *by membrane filter methods.*

Sample Number	*E. coli* counts on: mTEC Presumptive	mTEC Verified[a]	TBA Presumptive	TBA Verified	Teepol Presumptive	Teepol Verified
1	45	43	29	23	34	34
2	20	17	24	18	26	9
3	11	9	12	5	42	12
4	2	2	4	3	44	4
5	73	73	22	16	47	46
6	21	21	9	7	20	14
7	26	26	6	6	32	10
Totals	198	191	106	78	245	129
E. coli, %	96		73		52	

[a]Verified as citrate negative, urease negative, oxidase negative, and indole positive.

enumeration techniques may be available. Further evaluation is obviously necessary since the number of samples described here was extremely small.

Concluding Remarks

There is no worldwide agreement as to what bacterium or group of bacteria should be used to indicate the presence of fecal contamination. Some countries, such as Denmark, Belgium, England, and France, use *E. coli* as the indicator of choice. Conversely, countries such as the United States, Canada, and others use fecal coliforms, that is, those coliforms that produce gas from lactose at 44.5°C. *E. coli* is preferred by the former countries because of its high specificity as an indicator of fecal pollution. The latter countries place greater emphasis on simple methodology (no requirement for ancillary tests) and legal acceptance and appear willing to sacrifice somewhat on specificity. This presentation has been an attempt to point up the virtues of using *E. coli* as an indicator of fecal contamination in surface waters in place of the heterogeneous fecal coliform group. The high specificity of *E. coli* as an index of the presence of fecal contamination and the preliminary epidemiological evidence that it may be the best bacterial indicator for assessing health hazard risk in surface waters polluted with the fecal wastes of humans and other warm-blooded animals strongly recommends its acceptance. The use of this bacterium as the indicator of choice for determining the presence of fecal contamination by those countries not using it will also allow for more meaningful comparisons of data collected by investigators all over the world.

References

[*1*] *Standard Methods for the Examination of Water and Wastewater*, 13th ed., American Public Health Association, Washington, D.C., 1971.

[*2*] Prescott, S. C., Winslow, E. E., and McCrady, M. H., *Water Bacteriology*, John Wiley & Sons, New York, 1946, p. 78, 170.

[*3*] Eijkman, C. *Central Blatt für Bakteriologie, Abth. I. Orig.* Vol. 37, 1904, p. 742.

[*4*] Hajna, A. A. and Perry, A. C., *Sewage Works Journal*, Vol. 10, 1938, p. 261.

[*5*] Hajna, A. A. and Perry, A. C., *American Journal of Public Health*, Vol. 33, 1943, p. 269.

[*6*] Geldreich, E. E., Clark, H. F., Kabler, P. W., Huff, C. B., and Bordner, R. H., *Applied Microbiology*, Vol. 6, 1958, p. 347.

[*7*] Proceedings Seminar on the Significance of Fecal Coliform in Industrial Wastes, R. H. Bordner and B. J. Carroll, Eds., 4–5 May 1972, Denver, Colo., EPA Technical Report 3, Environmental Protection Agency.

[*8*] Nunez, W. J. and Colmer, A. R., *Applied Microbiology*, Vol. 16, 1968, p. 1875.

[*9*] Dufour, A. P. and Cabelli, V. J., *Journal of the Water Pollution Control Federation*, Vol. 48, 1976, p. 872.

[*10*] Bergey's Manual of Determinative Bacteriology, 8th ed., R. E. Buchanan and N. E. Gibbons, Eds., Williams and Wilkins Co., Baltimore, Md., 1974.

[*11*] Davis, T. J. and Matsen, J. M., *Journal of Infectious Disease*, Vol. 130, 1974, p. 402.

[*12*] Dufour, A. P. and Cabelli, V. J., *Applied Microbiology*, Vol. 26, 1975, p. 826.

[*13*] *The Examination of Waters and Water Supplies*, 7th ed., Windle Taylor, Ed., J & A Churchill Ltd., London, England, 1958, p. 425.

[*14*] Carr, D. H., "A Study of Differentiated Coliforms in Treated and Untreated Wastewater," thesis, Tufts University, 1975.

[15] Duncan, D. W. and Razzell, W. D., *Applied Microbiology*, Vol. 24, 1972, p. 933.
[16] Geldreich, E. E., Huff, C. B., Bordner, R. H., Kabler, P. W., and Clark, H. F., *Journal of Applied Bacteriology*, Vol. 25, 1962, p. 8.
[17] Cabelli, V. J., Levin, M. A., Dufour, A. P. and McCabe, L. J., "The Development of Criteria for Recreational Waters," Paper No. 7, International Symposium on Discharge Sewage from Sea Outfalls, London, 1974.
[18] Cabelli, V. J., Dufour, A. P., Levin, M. A., and Haberman, P. W., The Impact of Pollution on Marine Bathing Beaches: An Epidemiological Study, In: Proc. Symposium on Middle Atlantic Continental Shelf and the New York Bight, Vol. 2, M. Grant Gross Ed., The American Soc. Limnology and Oceanography, Inc., New York City, 1976.
[19] Mishra, R. P., Joshi, S. R. and Panicker, P. R. V. C., *Water Research*, Vol. 2, 1968, p. 575.
[20] Pugsley, A. P., Evison, W. M., and James, A., *Water Research*, Vol. 7, 1973, p. 1431.
[21] Halls, S. and Ayres, P. A., *Journal of Applied Bacteriology*, Vol. 37, 1974, p. 105.
[22] Dufour, A. P., Strickland, E. R., and Cabelli, V. J., "A Membrane Filter Procedure for Enumerating Thermotolerant *E. coli*," Proceedings, 9th National Shellfish Sanitation Workshop, D. S. Wilt Ed., June 25, 1975, Charleston, S.C., U.S. Food and Drug Administration.
[23] Andrews, W. H. and Presnell, M. W., *Applied Microbiology*, Vol. 23, 1972, p. 521.
[24] Mackenzie, E. F. W., Taylor, E. W. and Gilbert, W. F., *Journal of General Microbiology*, Vol. 2, 1948, p. 197.
[25] *The Bacteriological Examination of Water Supplies*, Reports on Public Health and Medical Subjects No. 71. Her Majesty's Stationary Office, London, England, 1968.
[26] Mara, D. D., *Journal of Hygiene*, Vol. 71, 1973, p. 783.
[27] Delaney, J. W., McCarthy, J. A. and Grasso, R. J., *Water and Sewage Works*, Vol. 109, 1962, p. 289.
[28] Anderson, J. M. and Baird-Parker, A. C., *Journal of Applied Bacteriology*, Vol. 39, 1975, p. 111.

W. N. Mack[1]

Total Coliform Bacteria

REFERENCE: Mack, W. N., **"Total Coliform Bacteria,"** *Bacterial Indicators / Health Hazards Associated With Water, ASTM STP 635*, A. W. Hoadley and B. J. Dutka, Eds., American Society for Testing and Materials, 1977, pp. 59–64.

ABSTRACT: Although the total coliform bacterial count has its shortcomings, it is our best indicator of the health hazards associated with drinking water that we have today. Some of these defects can be corrected by additional bacterial tests to determine interference or identify noncoliform lactose fermenting organisms. Occasionally drinking water samples containing viruses have given negative tests for coliform bacteria when tested by the standard methods. Concentration methods used to recover viruses in water samples have shown that the coliform organisms were present but in an insufficient number to be indicated by the routine standard methods.

Suggestions are given for the search for a better method to detect bacterial pollution in drinking water.

KEY WORDS: bacteria, water, coliform bacteria

The coliform bacterial count is probably the most frequently used bacterial test performed today. For years, it has been used as the indicator organism for fecal pollution of water. This group of organisms continues to be used as an indicator of the reduction of pathogenic bacteria in wastewaters, but microbiologists have asked, "Isn't there a better, more reliable indicator of fecal pollution for our drinking water and what is the significance of coliform bacteria present in water?"

The development of the total coliform test as an indicator of water pollution began after the discovery by Eberth in 1880 that *Salmonella typhi* produced typhoid fever in man, and, in 1884, Gaffky cultivated the typhoid organism [*1*].[2] It was realized at the time that the typhoid bacillus could be transmitted by the water route, and attempts were made to isolate the organisms from water samples. While examining a stool sample, Escherich in 1885 found many organisms that appeared similar to the typhoid organism, but later these organisms were distinguished as different from the typhoid bacillus, one of which was named *Bacillus coli* by Escherich which later became *Escherichia coli*. The abundant numbers of these organisms in

[1]Department of Microbiology and Institute of Water Research, Michigan State University, East Lansing, Mich. 48824.

[2]The italic numbers in brackets refer to the list of references appended to this paper.

the stool soon made it clear that *E. coli* was a common inhabitant of the intestinal tract, and this explained its being found in large numbers in water receiving fecal pollution. The deduction was that if the coli organism was not present in a drinking water supply, there was no fecal pollution and the water supply was free of typhoid and cholera organisms. That this deduction is not always true is of little consequence considering that the coli count has been our indication of fecal pollution in drinking water supplies for three quarters of a century, and some of us survived.

It soon became apparent that other gram-negative, lactose fermenting organisms, in addition to *E. coli*, were present in both stool samples and fecal polluted water. *Klebsiella pneumoniae*, or Friedlander's bacillus, and *K. rhinoscleromatis* were recorded in 1882. Again, Escherich in 1885 described *Aerobacter aerogenes*, and Jordan in 1890 described *A. cloacae*. Thus, the three principal genera of *Eschrichia*, *Aerobacter*, and *Klebsiella* were soon being called the coliform group of organisms [*2*]. The term "coliform" has been used by the British microbiologists [*3–11*] at least since 1901, and Breed and Norton [*12*] in the United States suggested that coliform be used to describe the lactose fermenting bacteria found in polluted water and that the organisms be used as a measure of pollution. Other terms used to describe the group were coli-aerogenes [*13*], Escherichia-Aerobacter [14], and the colon group [*15*]. In 1937, Jordan suggested that the term "coliform be used to describe this group of organisms [*15*].

Atypical coliform bacteria were soon described by many bacteriologists, especially those working with water samples [*2*]. The atypical forms were of two types: the first were those forms giving peculiar reactions—for instance, chromogenic [*16*], capsules [*17*], sugar-tolerant [*18*], serological reactions with other organisms [*19*], etc.—and the second were those giving irregularities with the utilization of lactose. This problem was especially true with water analysis when lactose was not fermented within the 48-h period—or the slow fermenters. Some atypical strains fail to ferment lactose; others produced acid, but failed to produce gas. Strains were found to ferment lactose at room temperature but failed to do so at 37°C [*2*]. Malcolm in 1938 [*20*] found 3 percent of the 1636 cultures he examined to be atypical, and Kline [*21*] found 126 of the 325 strains (38 percent) that he studied were atypical organisms. Mackie [*8*] found that immune serum to certain *E. coli* types had little or no reaction to either strains of organisms that had the same cultural characteristics as those used for immunization. For a complete review of the cultural and biochemical variations in the coliform group of bacteria, the reader is referred to the excellent review in 1939 by Leland Parr [*2*]. In the eighth edition (1974) of *Bergey's Manual*, Buchanan and Gibbons [*22*] list *A. aerogenes* as *K. pneumoniae* as a consequence of the demonstration by Edwards [*23–24*] and Kauffmann [*25*] that strains of *A. aerogenes* that grew at 37°C were indistinguishable from *Klebsiella*. This change may be welcomed by the water microbiologist as it eliminates some of the soil coliform organisms.

The 14th edition (1975) of *Standard Methods of Water Analysis* [*44*] classifies the coliform group as *E. coli*, *Enterobacter* (*Aerobacter*) *aerogenes*, and *Citrobacter* (Escherichia) *freundii* species. All of the total coliform organisms will not ferment lactose 100 percent of the time. Yet another variable is the test methods. Because of the two test procedures for the coliform organisms, two definitions are required. The multiple tube dilution method using the most probable number (MPN) depends on the ability of an organism to ferment lactose producing lactase and permease. Organisms within the coliform group that do not ferment lactose have lost their ability to produce permease. When using the multiple tube dilution method, the definition of the total coliform organisms reads "All of the aerobic and facultative anaerobic gram-negative, non-spore forming, rod-shaped bacteria that ferment lactose with gas formation within 48 hours at 35°C."

The second test procedure for the total coliform group of organisms is the membrane filter technique (MF) and depends upon the breakdown of lactose, producing acetaldehyde which oxidizes sodium sulfite-acetaldehyde which in turn oxidizes the sodium sulfite-basic fuchsin complex in the medium and results in the green metallic sheen of the colonies. A definition of the total coliform organisms using the MF technique is that it "comprises all the aerobic and facultative anaerobic, gram-negative, nonspore-forming, rod-shaped bacteria that produce a dark colony with metalic sheen within 24 hours on an Endo-type medium containing lactose."

Frequently the lactose fermenting *Aeromonas* bacteria are found in water supplies, and they mimic the coliform group [*26–29*]. They are especially troublesome in the warmer months in surface water. The oxidase-positive Aeromonas group can be distinguished by the oxidase reaction as all coliforms are oxidase negative. For this reason, the D19.01.04 Task Group on Microorganisms of Health and Sanitary Significance of ASTM Committee D-19 on Water recommended that oxidase negative be incorporated into any definition of the coliform group including those already given previously regardless of the procedure used to detect the coliform. Dr. Shotts of the University of Georgia has already presented excellent information on this organism.

The total coliform organisms can be found everywhere in nature; this does not include the fecal coliform bacteria. It has been known for many years that the nonfecal coliform bacteria can grow in water containing very small amounts of organics. In 1937, Bigger [*30*] found that the coliform organisms multiplied in inoculated autoclaved water samples, and Hendricks and Morrison [*31*] found that several enteric bacteria, including *E. coli*, were capable of multiplication in cold mountain streams. Goodrich et al [*32*] also demonstrated that the coliform organisms were present in large numbers in mountain streams draining watersheds which were essentially untouched by man. The coli-aerogenes group have been found in most types of soil by Geldreich et al [*32*]. The numbers of organisms in undisturbed soil were small, but, in polluted soils, the number of organisms was as large as 49 000/g of soil. The coliform group also have been found in insects [*34–35*], on vegetation [*35*], fish [*36*], in arctic soils and water [*37*], and forest and farm produce [*38*]. The coliform organisms occasionally multiply on leather washers, wood, in swimming pool ropes, and jute packing as well as bacterial film development inside of pipes [*44*].

The total coliform are used to test for drinking water quality, and their presence in any drinking water should initiate an immediate search for a contaminating source. Their known presence should never be ignored. Their presence in wastewater or untreated surface waters is to be expected and does not have the significance of the presence of the fecal coliform.

Occasionally *Salmonella*, *Shigella*, enteric viruses, or pathogenic protozoa are found in drinking water in which few or no coliform indicator organisms are present by using the standard bacteriological methods. Dutka [*39*] has reviewed these outbreaks in which there were only a few total coliform organisms present in water samples that also contained pathogens. In one described by Muller [*40*], salmonellae were found in a coliform-free water supply. In Michigan, virus was isolated from 2.5 gal (9.48 litre) of unchlorinated drinking water by Mack et al [*41*]. The water sample came from a well that had previously been known to contain coliform organisms, but there were none found in the 2.5 gal, as tested by standard methods and before concentration. After concentration to recover virus, coliform bacteria were found in the concentrate.

In a prepublished report by Wellings et al [*42*] from Florida, virus was recovered after 100 gal (378 litre) of water from a chlorinated water supply was concentrated. Before concentration for virus recovery, no coliform organisms were found in the water supply; however, after concentration, the standard test showed that coliform bacteria were present.

Here are two outbreaks where following gastroenteritis, standard test procedure did not indicate that the drinking water was contaminated, but, following concentration of the sample to recover one infectious tissue culture dose of virus, the concentrate did contain coliform bacteria. Following outbreaks of illness in which the water supply is implicated, why do we test only 100 ml of water? In both of these outbreaks, the indicator organism was present but not found because of the small number of organisms in a large volume of water. The 14th edition of *Standard Methods of Water Analysis* [*44*] recommends sampling 400 litre (105.7 gal) of water to recover virus. How did we get started sampling only 55.5 ml with the multiple dilution test and 100 ml with the membrane filtration method? Geldreich [*43*] has shown that there will be interference in detecting low densities of total coliform organisms if the water sample contains 500 or more noncoliform organisms per millilitre.

Dutka and Tobin [*45*] demonstrated recently the major variations obtained in growing the coliform organism from various types of water samples and used four procedures for growing lactose fermenting oxidase negative populations. They found that each medium combination and procedure has a specific selective basis and that there is no universal coliform estimation procedure and that the local conditions will often dictate which method and medium will give the greatest estimate of number of organisms.

There is a general dissatisfaction with the total coliforms as an index organism as well as the various procedures to estimate its population in various water samples. No one has provided us with a better indicator organism than the total coliform bacteria for drinking water. The major criticisms against the coliforms are that they are not detected when they are in limited numbers in large volumes of water or in the presence of interfering non-coliform bacteria. To some extent, these shortcomings could and should be corrected as follows.

1. Concentrate the few coliforms present in large volumes of water before testing for their presence.

2. Determine the total bacterial count of the water sample to detect interference.

3. If we are to continue the use of lactose fermenting bacteria as indicators, we should develop means of detecting *E. coli* specifically, so that it can be rapidly separated from other lactose fermenting coliforms.

As for other indicators for the detection of bacterial pollution in water, it might be the time and place to begin an investigation of bacterial leakage material or enzymes as a better index of water purity and eliminate the uncertainty of growth of an organism in a restricted environment or medium.

References

[*1*] Wilson, G. S. and Miles, A. A., *Topley and Wilson's Principles of Bacteriology and Immunology*, 4th ed., Williams and Wilkins, Baltimore, 1957, p. 1720.

[*2*] Parr, Leland, *Bacteriological Reviews*, Vol. 3, 1939, pp. 1–48.

[*3*] Arkwright, J. A., *Journal of Hygiene*, Vol. 13, 1913, pp. 68–86.

[*4*] Buxton, P. A., *Journal of Hygiene*, Vol. 19, 1920–21, pp. 68–71.

[*5*] Horrocks, W. H., *An Introduction to the Bacteriological Examination of Water*, J. and A. Churchill, London, England, 1901.

[*6*] Hay, H. R., *Journal of Hygiene*, Vol. 32, 1932, pp. 240–257.

[*7*] Lewis, C. J., *Birmingham Medical Review*, Vol. 81, 1917, pp. 1–10.

[*8*] Mackie, T. J., *Transactions*, Royal Society South Africa, Vol. 9, 1921, pp. 315–366.

[*9*] Revis, C., "The Stability of the Physiological Properties of Coliform Organisms," *Zantr. Bakt. Parasitenk*, Vol. 2, 1910, pp. 161–178.

[10] Robinson, A. L., *Journal of the Royal Naval Medical Service*, Vol. 14, 1928, pp. 104–117.

[*11*] Stewart, M. J., *Journal of Hygiene*, Vol. 16, 1918, pp. 291–316.

[*12*] Breed, R. S. and Norton, J. F., *American Journal of Public Health*, Vol. 27, 1937, pp. 560–563.

[*13*] *Standard Methods of Water Analysis*, 8th Ed., American Public Health Association and American Water Works Association, 1936.

[*14*] *Standard Methods of Water Analysis*, 6th ed., American Public Health Association, 1934.

[15] Jordan, H. E., *Journal of the American Water Works Association*, Vol. 29, 1937, pp. 1999–2000.

[*16*] Oesterle, P., "*Bacterium coli flavum*," *Zentr. Bakt. Parasitenk*, Vol. 134, 1935, pp. 115–118.

[17] Parr, L. W., *Proceedings,* Society for Experimental Biology and Medicine, Vol. 31, 1933-34, pp. 226-227.
[18] James, L. H., *Journal of Bacteriology*, Vol. 19, 1930, pp. 145-148.
[19] Habs, H. and Arjona, E., "Ueber einem Stam von *Bacterium coli* mit Antigenbeziehungen zur Salmonellagruppe," *Zentr. Bakt. Parasitenk*, Vol. 133, 1934-35, 204-209.
[20] Malcolm. J. F., *Journal of Hygiene*, Vol. 38, 1938, pp. 395-423.
[21] Kline, E. K., "The Colon Group of Bacteria in Milk," 19th Ann. Rept. Intern., Dairy Milk Inspections, 1930.
[22] Buchanan, R. E. and Gibbons, N. E., *Bergey's Manual of Determinative Bacteriology*, 8th ed., Williams and Wilkins, Baltimore, 1957, p. 322.
[23] Edwards, P. R., *Journal of Bacteriology*, Vol. 15, 1928, pp. 245-266.
[24] Edwards, P. R., *Journal of Bacteriology*, Vol. 17, 1929, pp. 339-353.
[25] Kauffmann, F., *Acta Pathologica Et Microbiologica Scandinavica*, Vol. 26, 1949, pp. 381-406.
[26] Bonde, G. J., *Health Laboratory Science*, Vol. 3, 1966, pp. 124-128.
[27] Ptak, D. J., Ginsburg, W., and Willey, B. F., "Aeromonas, the Great Masquerader," *Proceedings*, American Water Works Association Water Quality Conference, 2-3 Dec. 1974.
[28] Bell, J. B. and Vanderpost, J. M., *Proceedings*, 16th Conference of Great Lakes Res., Vol. 15, 1973, p. 20.
[29] Lupo, L., Strickland, E., Dufour, A. and Cabelli, V., "The Effect of Oxidase Positive Bacteria on Total Coliform Density Estimates." personal communication.
[30] Bigger, J. W., *Journal of Pathology and Bacteriology*, Vol. 44, 1937, pp. 167-211.
[31] Hendricks, C. W. and Morrison, S. M., *Water Research*, Vol. 1, 1967, pp. 567-576.
[32] Goodrich, T. D., Stuart, D. G., Bissonnette, G. K., and Walter, W. G., *Montana Academy of Science*, Vol. 30, 1970, pp. 59-65.
[33] Geldreich, E. E., Huff, C. B., Bordner, R. H., Kabler, P. W., and Clark, H. F., *Journal of Applied Bacteriology*, Vol. 25, 1962, pp. 87-93.
[34] Steinhaus, E. A., *Journal of Bacteriology*, Vol. 42, 1941, pp. 757-789.
[35] Geldreich, E. E., Kenner, B. A., and Kabler, P. W., *Applied Microbiology* Vol. 12, 1964, pp. 63-69.
[36] Geldreich, E. E. and Clarke, N. A., *Applied Microbiology*, Vol. 14, 1966, pp. 429-437.
[37] Boyd, W. L. and Boyd, J. W., *Canadian Journal of Microbiology*, Vol. 8, 1962, pp. 189-192.
[38] Duncan, D. W. and Razzell, W. E., *Applied Microbiology*, Vol. 24, 1972, pp. 933-938.
[39] Dutka, B. J., *Journal of Environmental Health*, Vol. 36, 1973, pp. 39-46.
[40] Muller, VonGertrud, *Archiv für Hygiene und Bakteriologie*, Vol. 148, 1964, pp. 321-327.
[41] Mack, W. N., Lu, Yue-Shoung, and Coohon, D. B., *Health Services Reports*, Vol. 87, 1972, pp. 271-274.
[42] Wellings, F. M., Mountain, C. W., Lewis, A. L., Nitzkin, J. L., Saslaw, M. S., and Graves, R. A., "Isolation of Enterovirus from Chorinated Tap Water," *Journal of Infectious Disease*, in press.
[43] Geldreich, E. E., Is the Total Count Necessary? AWWA 1st Water Quality Technical Conference, 2 Dec. 1973, Cincinnati, Ohio, American Water Works Association.
[44] *Standard Methods of Water Analysis*, 14th ed., American Public Health Association, 1975.
[45] Dutka, B. J. and Tobin, S. E., *Canadian Journal of Microbiology*, Vol. 22, 1976, pp. 630-635.

V. J. Cabelli[1]

Clostridium perfringens as a Water Quality Indicator

REFERENCE: **Cabelli, V.J., "*Clostridium perfringens* as a Water Quality Indicator,"** *Bacterial Indicators / Health Hazards Associated With Water, ASTM STP 635*, A. W. Hoadley and B. J. Dutka, Eds., American Society for Testing and Materials, 1977, pp. 65–79.

ABSTRACT: *Clostridium perfringens* and the more general group to which it belongs, "sulfite-reducing, spore forming anaerobes," are considered with regard to their use as health effects, water quality indicators. The groups are examined with regard to their identification, their importance, the characteristics which impinge on their use as indicators, monitoring applications, available evaluation data, and enumeration methods. *C. perfringens* has specific and limited applications as a water quality indicator. It appears to be the fecal indicator of choice for measuring remote or intermittant pollution and in situations where resistance to disinfectants and environmental stress is at a premium.

KEY WORDS: bacteria, water pollution, coliform bacteria

Clostridium perfringens was suggested as an indicator of the pollution of water with fecal wastes in the late 1890s [*1*].[2] Nevertheless, since the 1930s, it has been used rarely as such in the United States. On the other hand, some European workers have used *C. perfringens*, or a group of clostridia which includes this species, as water quality indicators in certain specific cases [*2–6*], notably in the examination of potable waters and particularly those from ground sources [*6*]. In some parts of France, there are drinking water standards based on the density of "sulfite-reducing, spore-forming anaerobes" [*6*].

The lack of interest in the United States in *C. perfringens* as a water quality indicator probably stems from a number of factors. Included are the popularity of the coliform test, methodological problems in recovering and identifying anaerobes, and the uncertainty as to which of the many species of *Clostridium* should be enumerated. However, the major factor has been the

[1]Chief, Recreational Water Quality Criteria Program, Marine Field Station, Health Effects Research Laboratory-Cin., U. S. Environmental Protection Agency, West Kingston, R. I. 02892.

[2]The italic numbers in brackets refer to the list of references appended to this paper.

ubiquity of *C. perfringens* in nature, primarily because it is a spore former. The "American" position was aptly expressed by Levine [*7*], who noted that the use of anaerobic spore-forming bacteria as an index of the fecal pollution of water is undesirable because of their extreme resistance, abundance in decomposing organic matter and soil, and failure to occur in numbers that correlate with the results of a sanitary survey.

Levine's conclusions, modified to refer to recent fecal pollution and to conditions where there is significant soil runoff or resuspension of sedimented material back into the water column, appears to be consistent with the facts. However, there are situations where the detection of remote as well as recent pollution is desirable, soil runoff and resuspension of bottom sediments are not significant, and the survival properties of the water quality indicator are at a premium. This paper is intended as a critical but brief reexamination of *C. perfringens* as a water quality indicator under such circumstances—this rather than an extensive literature review since the latter is beyond the scope of this short status report. For the latter, the reader is referred to Bonde's book [*3*], since it provides an excellent review and analysis of the literature along with the findings from his own investigations.

Description of Indicator

Functional Definition

C. perfringens can be described functionally as a gram-positive, anaerobic, spore-forming, nonmotile, rod-shaped bacterium which ferments lactose, sucrose, and inositol with the production of gas, produces "stormy fermentation of milk," reduces sulfite to hydrogen sulfide (H_2S), reduces nitrate, hydrolyzes gelatin, and produces lecithinase and acid phosphatase.

Identification Problems

There are more than 60 species of *Clostridium* [*8,9*]. Of these, *C. perfringens* is the one most consistently associated with human fecal wastes [*10–12*], and the initial intent was to enumerate this organism specifically [*13*]. However, enumeration and identification methods which are rapid and facile enough for routine use were not available for identification to the species level. Aside from spore formation, anaerobiosis, and Gram's stain, the characteristics most widely used to characterize the groups of clostridia used as water quality indicators have been stormy fermentation of milk and sulfite reduction.

Stormy fermentaiton of a milk-water sample mixture, heated to 80°C for 10 min and incubated under anaerobic conditions at 37°C, was one of the earliest detection and identification methods used [*14*]. However, it can be used efficiently only with liquid cultures. For enumeration, this requires a most probable number (MPN) procedure with the attendant problems of precision and logistics. This reaction continues to be used as a confirmation method for sulfite reduction in the British MPN method [*15*]. It is significant that only three species of *Clostridium* in addition to *C. perfringens* produce stormy fermentation of milk [*9*]. Thus, the early methods were, in fact, rather specific for the enumeration of *C. perfringens*. Starting in 1925 with the work of Wilson and Blair [*16*], stormy fermentation of milk was largely replaced by sulfite reduction as the major identifying characteristic of the group to be enumerated, hence the designation of the group as sulfite-reducing, spore-forming anaerobes (SSA). According to *Bergey's Manual of Determinative Bacteriology* [*8*], of the 56 species of *Clostridium* which will grow on ordinary media, 22 to 30 will produce H_2S. However, this does not necessarily mean that all these strains are capable of producing H_2S from the reduction of sulfite. Bonde [*3*] reports that, with pasteurized and unpasteurized samples of feces, raw sewage, sludge and treated sewage assayed in his sulfite-alum medium incubated at 48°C, typical morphology and stormy fermentation of milk were observed in over 90 percent of the isolates. However, with pasteurized and unpasteurized samples of receiving water, the values were 37.5 and 75.6 percent, respectively. When a negative mannitol fermentation test also was required [*9*], lower confirmation frequencies were generally obtained (Table 1). Only in the pasteurized samples of feces and raw sewage was there a large percentage of the H_2S positive isolates which met all three of the requirements for *C. perfringens*. Based upon these data, it would seem that the change to the SSA group is questionable, especially since it was done largely as a matter of convenience and in the absence of data showing that the source of those sulfite-reducing species other than *C. perfringens* is consistently and specifically the feces of warm-blooded animals, particularly man.

The foregoing notwithstanding, the position taken by most European investigators [*6*] is that the group enumerated should be sulfite-reducing, spore-forming anaerobes and that further identification to *C. perfringens* is optional with the user. However, some investigators require further characterization so that the indicator enumerated is to all intents and purposes *C. perfringens*. In the British methods manual for "The Bacteriological Examination of Water Supplies" [*15*], the enumeration procedure requires stormy fermentation of milk in addition to sulfide reduction. Bonde [*3*]

TABLE 1—*Confirmation of typical* C. perfringens *isolates obtained by Bonde's sulfite-alum agar method, incubation temperature 48°C* [3].

	Pasteurized			Unpasteurized		
	No. of	Confirmation, %		No. of	Confirmation, %	
Sample Source	Colonies	*a*	*b*	Colonies	*a*	*b*
Feces	19	100.0	100.0	20	100.0	30.0
Raw sewage	90	99.1	88.9	131	98.0	24.4
Sludge	157	99.0	81.5	381	95.5	60.9
Treated sewage	119	92.9	73.1	257	96.1	39.7
Receiving water	40	37.5	30.0	266	75.6	54.1

[a]Confirmed by typical morphology and stormy fermentation of milk.
[b]Confirmed by typical morphology, stormy fermentation of milk, and no fermentation of mannitol.

contends that the SSA group is too broad and that the enumeration of clostridia as water quality indicators should be limited to *C. perfringens*. He utilizes incubation at 48° C in a "lean" medium and, as required, confirmation by the milk test to accomplish this end [*2*]. Food microbiologists in the United States and Canada [*17–21*] incorporate antibiotics such as cycloserine to inhibit some of the clostridia other than *C. perfringens* and then confirm the sulfite reducers by the absence of motility, the reduction of nitrate, and, in some cases, the fermentation of lactose.

Importance

C. perfringens is a major cause of wound infections and gas gangrene and has been implicated in a number of outbreaks of food intoxication. Shortly after it was first described by Welch and Nuttall in 1892 [*22*], Klein [*14*] and Klein and Houston [*13*] associated the organism with feces, recovered it from sewage, described the characteristic stormy fermentation of milk, and suggested its use as an indicator of the fecal pollution of water. Although *C. perfringens* is a pathogen, there are no data suggesting that waterborne transmission is significant in the epidemiology of diseases caused by this organism. Its importance to water microbiologists is as a potential fecal indicator and a potential conservative tracer in field studies. It is one of the organisms which causes false positive reactions in the presumptive coliform MPN test and which is largely eliminated in the confirmed test [*23*].

Characteristics of the Organism Impinging on Its Use as a Water Quality Indicator

C. perfringens strains can be differentiated by certain variable biochemical characteristics, their antigenic patterns, and the toxins they produce. However, there are no data to suggest that the *C. perfringens* populations in the environment differ significantly from those found in the colon of warm-blooded animals, although it has been reported [*10*] that *C. perfringens* strains from the intestinal tract of man are exclusively type A. The significant factors are (*a*) the consistency with which the organism can be recovered from normal human feces [*10*], an observation questioned by some workers [*11,12*]; (*b*) the moderate densities found in feces [*10,12*] and sewage relative to those for *Escherichia coli* (Table 2); (*c*) the production of spores and their apparent ubiquity in nature; and (*d*) the anaerobic requirement for growth.

Density in Sewage

When colony counting procedures are used, the lower densities of *C. perfringens* spores in sewage relative to those for *E. coli* require the examination of larger quantities of receiving water for obtaining estimates of comparable precision. This, coupled with the greater survival of the spore, should narrow the range of *C. perfringens* densities found in environmental samples and, hence, the pollution discriminating ability of the indicator system. This can be seen from Fig. 1. Over the 16 km "downstream" from

TABLE 2—*A comparison of* C. perfringens *and* E. coli *densities in raw, treated, and chlorinated sewage.* [a]

			Log$_{10}$ Density/100 ml for					
			E. coli			*C. perfringens*		
STP No.[b]	Treatment[c]	Residual Cl_2, ppm	Raw[e]	Treated[f]	Chlorinated[g]	Raw	Treated	Chlorinated
1	P	1.5	6.301	6.146	<0	4.114	4.903	4.000
2	P, TF, SF	1.0	6.431	5.568	<0	5.041	3.079	1.505
3	P, AS	0.25	7.255	5.982	0.301	5.362	4.447	4.079
4	. . .	2.5	7.380	7.380	0.477	5.230	5.30	4.771
5	P, AS	2.5	6.724	4.491	<0	4.778	3.778	3.000
6	P, AS	2.5	6.969	5.982	2.041	5.146	4.431	4.279
Mean	. . .	. . .	6.843	5.925	0.470	4.945	4.311	3.606

[a]Data provided by J. Bisson and J. Miescier.
[b]Sewage treatment plant number.
[c]P—primary; TF—trickling filter; AS—activated sludge; SF—sand filtration.
[e]Influent to plant.
[f]Following treatment as indicated but before chlorination.
[g]After chlorination.

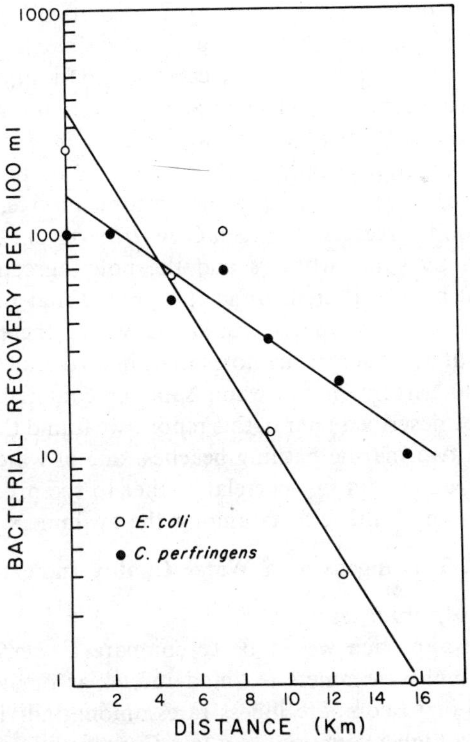

FIG. 1—*The recoveries of* C. perfringens *and* E. coli *at a series of stations "downstream" from a cluster of pollution sources discharging into an estuary; the distances given are those from the first sampling station.*

station 1 there was more than a 2 log reduction in the *E. coli* density but only a 1 log reduction in the *C. perfringens* density. However, the distribution of *C. perfringens* spores in the aquatic environment may be more analogous to that of the enteroviruses in terms of their survival therein.

Anaerobiosis

The anaerobic characteristic also is a mixed blessing. On one hand, it restricts the terrestrial or aquatic environments wherein the organism can multiply—a desirable feature of a fecal indicator—and, on the other hand, it complicates the enumeration and identification of the organism.

Spore-formation

Undoubtedly, the spore forming ability of *C. perfringens* has had the most far reaching effect on the appropriateness of the organism as a water quality indicator.

Although it is probable that there are anaerobic, ecological niches in terrestrial and aquatic environments wherein *C. perfringens* can multiply, there is no evidence that this occurs to any significant extent, certainly not in most waters. Furthermore, the widespread distribution of *C. perfringens* in soils and air [*24*] as well as estuarine and coastal waters and their underlying sediments [*24–28*], in most cases, can be attributed to the persistence of the spores in these environments over long periods of time. It follows, therefore, that spores of *C. perfringens* should be carried into the aquatic environment with the runoff from urban as well as rural areas even though the contamination of those areas with fecal wastes was remote in distance or time.

Except under conditions of heavy turbulence or upwelling, *C. perfringens* spores would be expected to settle to the bottom, survive, and accumulate. Therefore, the good agreement between *C. perfringens* densities and sanitary surveys reported by some workers and the poor agreement reported by others would not be unexpected. Bonde [*26*] noted that *C. perfringens* was not recovered in areas of the Sargasso sea where there was very rapid renewal of the water. However, he notes that, in a zone of upwelling off the coast of Peru, 40 percent of the water samples contained the organisms.

In a study to be described later in this report, we found that the *C. perfringens* densities at two marine bathing beaches, one of which had more surf activity than the other, did not correlate either to the proximity of known pollution sources or health effects among the swimmers.

Evaluation of C. Perfringens as a Water Quality Indicator

Epidemiological Approaches

The preferable approach would be to compare *C. perfringens* densities to illness rates among swimmers at "moderately" as opposed to "slightly" polluted beaches or to compare illness rates among individuals that drink water which has a high as opposed to a low *C. perfringens* density. The less direct approach would be to compare *C. perfringens* to enteropathogen (especially enteropathogenic viruses) densities in a variety of areas including some subject to remote or intermittent pollution from fecal sources. To the best of the author's knowledge, with one exception, data are not available using either of these two approaches. The one exception was an epidemiological study conducted at our laboratory in which *C. perfringens* densities and symptom rates among swimmers relative to nonswimmers were compared at a "barely acceptable" as opposed to a "relatively unpolluted" beach; the former was closer to known sources of municipal sewage wastes while the latter had more surf action. The rate of gastrointestinal symptoms was significantly higher for swimmers than for nonswimmers at the barely acceptable but not at the relatively unpolluted beach. The densities of a number of potential water quality indicators, including *E. coli* and enterococci, were higher at the barely acceptable than at the relatively unpolluted beach; however, the densities of *C. perfringens* at the two beaches were neither appreciably nor significantly different (Table 3). These findings would suggest that *C. perfringens* is not an appropriate water quality indicator for marine recreational waters.

TABLE 3—C. perfringens *densities at "barely acceptable" (BA) and "relatively unpolluted" (RU) beaches.*

Trial No.	Mean Density/100 ml at BA	Mean Density/100 ml at RU
1	66	351
2	24	31
3	5	4
4	10	2
5	10	11
6	47	4
Log Mean	18.2	12.6

Comparison to Other Indicators

A more expedient but considerably less meaningful approach towards evaluating *C. perfringens* as a water quality indicator has been to compare the densities of this organism to those of other indicators, notably total coliforms, fecal coliforms, and *E. coli* [*1,26*]. The fallacy of this approach is obvious. If agreement between the densities of the two indicators is not obtained, the question of which is the better one remains unresolved. On the other hand, if good agreement is obtained, a marked improvement in assay logistics would be the only justification for replacing a more widely used and accepted indicator system with one less so. Areas subject to remote or intermittent pollution should be included in such comparisons.

Studies conducted in the 1930s and 1940s comparing the densities of sulfite-reducing clostridia to coliform densities were reviewed by Prescott [*1*] and Bonde [*3*]. The degree of agreement appears to vary with the type of sample (potable waters, ground waters, surface waters, sediments), the specific locations from which the samples were collected, and, possibly, the clostridial biotypes being enumerated. The findings from a study conducted by Matheson [*29*] in waters near Hamilton, Ontario, illustrate additional problems involved in such comparisons. The average density of sulfite-reducing clostridia for each of seven general locations paralleled its coliform counterpart. However, this was not the case when the relationship was examined in individual samples. The density of sulfite-reducing clostridia was less than 10/100 ml in more than 40 percent of those water samples from a Lake Ontario bay which had coliform densities in excess of 100/100 ml of water. Conversely, in an area subject to minimal pollution, relatively high densities of sulfite reducers were obtained when the coliform levels were relatively low. The real question remained unanswered; that is, which indicator more closely reflected potential health hazards to users, especially those hazards associated with enteroviruses in polluted waters. From his comparisons of *C. perfringens* and *E. coli* densities in a number of marine water samples, Bonde [*30*] concluded that no distinct and constant relationship exists between the numbers of *E. coli* and *C. perfringens* (Fig. 2).

Application in Monitoring Water Quality

Although it generally limits the use of *C. perfringens* as a water quality indicator, the refractory nature of the spore may provide some specific applications for use of this indicator system. The absence of *C. perfringens* in some reasonable quantity of drinking water would indicate the absence of either recent or remote fecal contamination of the raw source or adequate disinfection. The presence of the organism is another matter; certainly it need not indicate a recent source of fecal pollution, nor fecal pollution at all if there are significant extrafecal sources of this organism in the environment. More information is needed concerning the frequency and importance of such sources. Furthermore, it is unlikely that significant numbers of fecal pathogens will be present when *C. perfringens* spores cannot be recovered from some reasonable quantity of drinking water. Thus, the *C. perfringens* system would seem to have some distinct advantages over the coliform indicators currently used for monitoring the quality of drinking waters in this country. Unlike the residual chlorine determination, a *C. perfringens* test could be used with waters disinfected by other means or when no disinfection at all is practiced; that is, some deep artesian wells and bottled waters. One of the more attractive features of *C. perfringens* as an indicator of drinking water quality is that, except for some very unusual circumstances, there would be little concern about the time required to return the water samples to the laboratory for assay. Bonde [*3*], after reviewing the results and conclusions of Wilson and Blair [*16*], Buttiaux [*31*], Willis [*32*], and Taylor [*5*], concluded that, "absence of *Cl. perfringens* in a 100 ml sample may, however, be considered a requirement which good-quality drinking

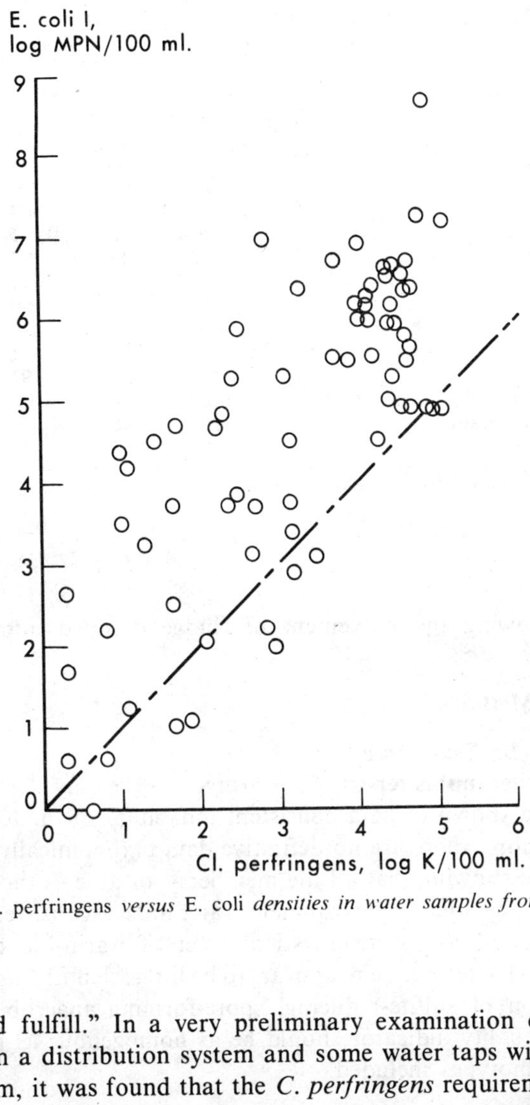

FIG. 2—C. perfringens *versus* E. coli *densities in water samples from Bonde* [30].

water should fulfill." In a very preliminary examination of some water samples from a distribution system and some water taps within a hospital on the system, it was found that the *C. perfringens* requirement as just set forth would have been met less frequently than that for the coliforms (Table 4). In only one instance were coliforms detected in the absence of *C. perfringens* densities in excess of 1/100 ml of water. More data are needed.

C. perfringens is not appropriate as a water quality indicator for recreational waters, and it is unlikely that it would be of much value for shellfish-growing areas since, even with low levels of intermittent pollution, the spores could be expected to settle, survive, and accumulate in the sediments where the shellfish reside.

Two additional applications warrant consideration. *C. perfringens* spores could serve as a conservative tracer in pathogen die-off studies and as a

TABLE 4—C. perfringens *densities in a water supply system.*

Description and Location	No. Samples	Residual chlorine, ppm	Indicator Density [a] Total Coliform	C. perfringens [b]	Standard Plate Count [c]
Raw	2	. . .	ND[d]	>20	370,570
Finished	2	0.4	<1	1	<1
Distribution system	6	0.3 to 0.1	<1	<3	<80
Distribution system	1	0.3	3	5	240
Distribution system	1	0.1	19	3	15
Distribution system	1	< 0.1	<1	7	480
Hospital, various locations	7	<0.1	<1	<3	<80
Hospital, defoamer	1	<0.1	2	<1	440
Hospital, defoamer	1	<0.1	<1	<1	700
Hospital, defoamer	1	<0.1	<1	92	30
Hospital, laundry tub	1	<0.1	4	3	80
Hospital, physical therapy	1	<0.1	46	1	2
Hospital, defoamer	1	<0.1	<1	7	>3000

[a]All *P. aeruginosa* densities <1/100 ml.
[b]Per 100 ml.
[c]Per ml.
[d]No data.

means of following the movement of sludge dumped into the aquatic environment.

Enumeration Methods

Population To Be Enumerated

As noted earlier in this report, *C. perfringens* is the only clostridial species which has been shown to be a consistent inhabitant of the fecal wastes of man. Furthermore, there are no definitive data taxonomically defining the SSA group and showing that all the members—or at least those frequently recovered from the aquatic environment—are, in fact, consistent and exclusive inhabitants of the gastrointestinal tract of warm-blooded animals, especially man. Therefore, there appears to be little scientific justification for the enumeration of sulfite-reducing, spore-forming anaerobes, especially since a water quality indicator should be as homogenous as possible with regard to the biotypes included.

Pretreatment of Sample

The *C. perfringens* work group of the International Standards Organization in a draft of a recent report [*6*] recommends heating the water samples at 80°C for 10 min so that vegetative cells are destoyed, presumably to eliminate much of the background microbial flora. However, Bonde [*3,27*] using his procedure, recommends the examination of unheated as well as heated water samples because the differential between the results from these two procedures provides some information on the immediacy of the pollution source. From his comparison of the recoveries from pasteurized versus unpasteurized samples of raw sewage, sludge, treated sewage, lagooned sewage, and receiving waters, Bonde [3] concluded that there were large numbers of *C. perfringens* vegetative cells in all these materials. The presence of viable vegetative cells during sewage treatment is understandable, in the receiving waters at some distance from the source less so. Bonde reports that vegetative cell densities which were significantly higher than the spore densities occurred more frequently in the samples from receiving waters and lagoon-treated sewage than in those from the other sources. An alternate explanation for his data with receiving waters might be the destruction of heat sensitive spores during his treatment process (80°C for 5 min). We found that 30 to 50 percent of the organisms in 4 to 6 month old buffered saline spore suspensions could not be recovered when the suspensions were heated at 70°C for 15 min. This variable, which is unacceptable when *C. perfringens* is to be used as a conservative tracer, might be eliminated by the use of a lower inactivation temperature, the substitution of an ethanol pretreatment [33], or the incorporation of selective antibiotics [16] in the medium as a substitute for heat treatment.

Methodology

Four enumeration methods are briefly described in the Appendix to this report. They are the MPN method set forth by the British in "The Bacteriological Examination of Water Supplies" [15], Bonde's [2] pour tube method, Hauschild and Hilsheimer's [16] pour plate procedure (it was developed for the examination of foods but should be applicable to the examination of water and wastewaters), and a membrane filter (MF) method recommended by the International Standard Organization [5] for the examination of drinking water.

The first three methods enumerate *C. perfringens*. The MF procedure measures the SSA group. Each of these methods has one or more problems associated with its use.

The MPN procedure requires subculturing positive tubes (those which have a black precipitate) to a milk medium if *C. perfringens* is to be enumerated. It suffers from the precision problems characteristic of MPN procedures and requires 48 h for completion. However, it is the method most amenable to the examination of sediments, faunal tissue, and turbid water samples.

The pour plate method requires the picking of suspect colonies into a modification of Angelloti's [17] nitrate motility medium; hence, results are not available for 48 h. In addition, this procedure has not been evaluated when applied to the examination of water and wastewater. Using a double strength medium one could examine up to 10 ml of water in a standard (100 by 15 mm) plate.

Bonde's notes that his pour tube method, as described, generally does not require (it is not especially amenable to) the picking of colonies for confirmation because of the selectivity of the 48°C incubation temperature. However, some preliminary findings from our own laboratory would suggest that the recovery of *C. perfringens* is markedly reduced at this high incubation temperature. The use of a pour tube markedly restricts the quantity of sample examined.

The MF method as described enumerates sulfite-reducing, spore-forming anaerobes. In our opinion, this is a major disadvantage. It is a "sandwich" procedure wherein the filter is placed on a base medium and then overlayed with two successive layers, the first of which contains glucose and the sulfite for the H_2S indicator system. As reported, this removes two of the advantages of MF over pour methods. They are the ability to examine colonial morphology and the ease of picking colonies for confirmation. As an MF method, it permits the examination of relatively large volumes of water.

Data on the performance characteristics are not available for some of these methods and are meager for some of the others. Hauschild and Hilsheimer [*16*] report quantitative recovery with their method in pure culture studies, and Bonde [*3*] notes confirmation frequencies of 37 percent to 100 percent (see Table 1) depending whether one is examining sewage or water samples.

In our laboratory we are completing the evaluation of a MF procedure we developed for the enumeration of *C. perfringens*. It is a 24-h method which utilizes antibiotics and an elevated incubation temperature (45°C) to reduce overgrowth by "background" flora and some other clostridia. It does not require picking of colonies for confirmation since this is accomplished by the differential features of the medium and, if required, *in situ* tests.

Summary and Conclusions

In summary, the enumeration of *C. perfringens* spores as a water quality indicator appears to have specific and limited applications, primarily as an adjunct to the commonly used coliform and fecal streptococcus indicators. As Bonde [*2,3*] points out, it appears to be a microbial indicator of choice when the requirement is for measuring remote and intermittent sources of pollution reaching an area (such as, tracing sewage sludge dumped into bodies of water or the infiltration of ground water supplies with fecal wastes) and in instances where other microbial indicators are rapidly destroyed, such as chlorinated water supplies. In addition, it has promise as a conservative tracer for use in field studies on the movement and decay of pathogenic microorganisms in sewage and sludge.

With reference only to those applications of the indicator system noted previously, there still remain several incompletely answered questions. To what extent, if any, do extrafecal sources contribute to the *C. perfringens* densities in the aquatic environment? How long will the vegetative cells of the organism remain viable therein? How do the survival kinetics of *C. perfringens* spores (in sewage) compare to those of the enteroviruses? In addition, more work is needed on enumeration methodology, especially the determination of the performance characteristic of those methods now available.

APPENDIX

Enumeration Methods

British MPN Method

This method is recommended by the British Department of Health and Social Security [15]. The water sample should be heated at 75°C for 10 min to destroy non-sporing organisms. One volume of 50 ml should be added to 50 ml of double-strength [*34*] differential reinforced clostridial medium (DRCM) in a 4 oz. (114 ml) screw-capped bottle. Five volumes of 10 ml should be added to separate 10-ml volumes of double-strength DRCM in 1 oz. (28 ml) screw-capped bottles. Separate volumes of 1 ml, and further 10-fold dilutions if necessary, should be added to 5, 25-ml volumes of single strength DRCM in 1-oz. (28-ml) bottles. All the bottles should be topped up if necessary with further single-strength DRCM to bring the level of liquid up to the neck of the bottle, leaving a small air space. The bottles should then be incubated at 37°C for 28 h. A positive reaction will be shown by blackening of the medium due to reduction of the sulphite and precipitation of ferrous sulphide.

A positive reaction may be produced by any clostridium. A loopful from each positive culture should be inoculated into a tube of litmus milk which has been freshly steamed and cooled. These tubes are then incubated at 37°C for 48 h. Those containing *C. perfringens* generally will produce a "stormy clot" in which the milk is acidified and coagulated; the clot is disrupted by gas and often blown to the top of the tube. Growth in the litmus milk is improved by adding to each tube, immediately before subculture, a short length of iron wire sterilized by heating to redness. A MPN can be read from the probability tables in the same way as for coliform organisms.

Bonde's Pour Tube Method

The medium consists of 1 percent meat extract, 1 percent peptone, and 0.75 percent agar; pH 7.2. It is distributed in 10-ml amounts to tubes which are autoclaved at 121°C for 15 min. Before use, the tubes are heated to melt the agar and drive off the air; 2 ml of a 1 percent solution of anhydrous sodium sulphite and two drops of a 5 percent iron alum solution are added, and 5 ml quantities of the sample or an appropriate dilution thereof are distributed to each of the tubes. After rapid cooling, the tubes are incubated at 48°C for 24 h. The black colonies are counted. A count of pasteurized samples could give a separate count for spores (heating to 80°C for 5 min). Bonde notes that the same medium and incubation temperature may be applied with MF and a double layer.

Pour Plate Procedures of Hauschild and Hilsheimer

This procedure [*17*] uses a modification of the tryptose-sulfite-cycloserine TSC medium of Harmon, et al (in turn a modification of the Shahidi-Ferguson medium) in which the egg is eliminated, and a pour plate is used instead of a spread plate with overlay. Confirmation is obtained by stabbing suspected isolates into nitrate motility (MN) agar supplemented with glycerol and galactose and in lactose-gelatin medium. The isolates must be nonmotile.

Membrane Filter Method [6]

Ten millilitres of nutrient agar (3 g meat extract, 10 g peptone, 5 g sodium chloride, and 15 g agar in 1 litre distilled water) autoclaved at 121°C for 20 min is poured into a 100 by 15 mm petri dish. Just prior to use, the plate is dried at 37°C for 30 min. Before filtration, 100 ml of the water sample is heated to 75°C for 20 min in an

Erlenmeyer flask. An appropriate quantity of the water sample then is passed through a MF (pore size 0.45 mm), and the filter is placed on the surface of the nutrient agar plate. After the plate containing the filter has been dried for 30 min at 37°C, the filter is carefully and slowly overlayed with 18 ml of a sulfite-glucose iron agar (1 litre of nutrient agar base is supplemented with 20 glucose, 1 ml of a 10 percent sodium sulfite solution, and 5 drops of an 8 percent iron sulfate solution; it is adjusted to a pH of 7.6 with sodium hydroxide and autoclaved at 121°C for 20 min). After the overlay has hardened, an additional 15 ml of nutrient agar is poured over the first overlay. When the second overlay has hardened, the plates are incubated anaerobically at 37°C for 24 to 44 h. All black colonies are counted.

References

[*1*] Prescott, S. C., Winslow, C. A., and MacCrady, M., *Water Bacteriology*, 6th ed., Wiley, New York, 1945, p. 215–223.

[*2*] Bonde, G. J., *Health Laboratory Science*, Vol, 3, 1966, p. 124.

[*3*] Bonde, G. J., *Bacterial Indicators of Water Pollution*, 2nd ed., Teknisk Forlao, Copenhagen, Denmark, 1963.

[*4*] Buttiaux, R. and Mossel, D. A. A., *Journal of Applied Bacteriology*, Vol. 24, 1961, p. 353.

[*5*] Taylor, E. W., *The Examination of Water and Water Supplies*, 7th ed., Churchill, London, England, 1958, p. 468.

[*6*] International Standards Organization, Draft Report of SC4/WG5 Meeting on Sulfite-Reducing Spore-Forming Anaerobes (clostridial), 16 Jan. 1975, Berlin.

[*7*] Levine, M., "Bacteria Fermenting Lactose and Their Significance in Water Analysis," Bulletin 62, Iowa State College of Agriculture and Mechanical Arts Official Publication 20, Vol. 31, 1921.

[*8*] Buchanan, R. E. and Gibbons, N. E., *Bergey's Manual of Determinative Bacteriology*, 8th ed., Williams and Wilkins, Baltimore, 1974, pp. 551–575.

[*9*] Holderman, L. V. and Moore, W. E. C., *Anaerobe Laboratory Manual*, 2nd ed., Virginia Polytechnic Institute Anaerobe Laboratory, Blacksburg, Virginia, 1973, p. 67–89.

[*10*] Rosebury, T., *Microorganisms Indigenous to Man*, McGraw-Hill, New York, 1962, p. 87–90, 332–335.

[*11*] Akama, K. and Otani, S., *Japan Journal of Medical Science*, Vol. 23, 1970, p. 161.

[*12*] Haenel, H., *American Journal of Clinical Nutrition*, Vol. 23, 1970, p. 1425.

[*13*] Klein, E. and Houston, A. C., *Twenty-eighth Annual Report of the Local Government Board Containing the Report of the Medical Officer*, Supplement, 1899.

[*14*] Klein, E., *Twenty-seventh Annual Report of the Local Government Board Containing the Report of the Medical Officer*. Supplement, 1898.

[*15*] "The Bacteriological Examination of Water Supplies," *Reports on Public Health and Medical Subjects*, 4th ed., Report No. 71, Her Majesty's Stationary Office, London, 1969.

[*16*] Wilson, W. J. and McV. Blair, E. M., *Journal of Hygiene*, Vol. 24, 1925, p. 111.

[*17*] Hauschild, A. H. W. and Hilsheemer, R., *Applied Microbiology*, Vol. 27, 1974, p. 78.

[*18*] Angelotti, R., Hall, H. E., Foter, M. J., and Lewis, K. H., *Applied Microbiology*, Vol. 10, 1962, p. 193.

[*19*] Marshall, R. S., Steenbergen, J. F., and McClung, L. S., *Applied Microbiology*, Vol. 13, 1965, p. 559.

[*20*] Shahidi, S. A. and Ferguson, A. R., *Applied Microbiology*, Vol. 21, 1971, p. 500.

[*21*] Harmon, S. M., Kautter, D. A., and Peeler, J. T., *Applied Microbiology*, Vol. 22, 1971, p. 688.

[*22*] Welch, W. H. and Nuttall, G. H. F., "A Gas-Producing Bacillus (Bacillus aerogenes capsulatus, N.S.) Capable of Rapid Developement in the Blood Vessels After Death," *Bulletin of the Johns Hopkins Hospital*, July, 1892.

[*23*] Porter, R., McCleskey, C. S., and Levine, M., *Journal of Bacteriology*, Vol. 33, 1937, p. 163.

[*24*] Sidorenko, G. I., *Journal of Hygiene, Epidemiology, Microbiology, and Immunology, Moscow*, Vol. 11, 1967, p. 171.

[*25*] Davies, J. A., *Journal of Applied Bacteriology*, Vol. 32, 1967, p. 164.

[*26*] Bonde, G. J., "Studies on the Dispersion and Disappearance Phenomena of Enteric Bacteria in the Marine Environment. *Rev. Int. Oceanogra. Med. Tome*, Vol. 9, 1968, p. 17.
[*27*] Smith, L. DS., *Canadian Journal of Microbiology*, Vol. 14, 1968, pp. 1301–1304.
[*28*] Matches, J. R. and Liston, J., *Canadian Journal of Microbiology*, Vol. 20, 1974, p. 1.
[*29*] Matheson, D. H., *Canadian Public Health Journal*, Vol. 28, 1937, p. 241.
[*30*] Bonde, G. J. in *International Symposium on Discharge on Sewage from Sea Outfalls*, Pergamon Press, London, 1975.
[*31*] Buttiaux, R., *L'analyse Bacteriologique des Eaux de Consommation*, 1st ed., Paris, France, 1951.
[*32*] Willis, A. T., *Journal of Applied Bacteriology*, Vol. 19, 1956, p. 105.
[*33*] Johnston, R., Harmon, S., and Lautter, D., *Journal of Bacteriology*, Vol. 88, 1964, p. 1522.
[*34*] Gibbs, B. M. and Freame, B., *Journal of Applied Bacteriology*, Vol. 28, 1965, p. 95.

A. W. Hoadley[1]

Potential Health Hazards Associated With *Pseudomonas aeruginosa* in Water

REFERENCE: Hoadley, A. W., **"Potential Health Hazards Associated With *Pseudomonas aeruginosa*,"** *Bacterial Indicators/Health Hazards Associated With Water, ASTM STP 635*, A. W. Hoadley and B. J. Dutka, Eds., American Society for Testing and Materials, 1977, pp. 80–114.

ABSTRACT: *Pseudomonas aeruginosa* is an opportunistic pathogen of man and animals which may be spread by water. The major source of *P. aeruginosa* in waters appears to be fecal wastes of man and animals associated with man, although growth may occur under certain conditions. The demonstration of the species in surface waters suggests the influence of man, and its numbers reflect the degree of pollution. However, they survive only for short periods, and there frequently appears to be little relation between populations of *P. aeruginosa* and those of other pathogens or fecal indicators.

The value of *P. aeruginosa* as an indicator of potential health hazards associated with water must be judged on the basis of its own role as a waterborne pathogen. In this paper, the role of *P. aeruginosa* as a waterborne pathogen, its sources, and its behavior in aquatic environments are reviewed. Its isolation from drinking waters, farm water supplies, swimming pool waters, whirlpool waters, and surface recreational waters should be regarded with concern. However, while authorities have recommended limitations on *P. aeruginosa* in waters used for various purposes, few epidemiologic studies have been undertaken upon which to base standards, and few standards have been established.

KEY WORDS: bacteria, water, coliform bacteria

Pseudomonas aeruginosa is considered generally to be a ubiquitous bacterial inhabitant of surface waters and soil. It is noted for its biochemical versatility and its resistance to antibacterial agents, and it may infect a variety of plants in addition to man and animals. It is also of some significance as a spoilage organism attacking many exotic materials and a slime former interfering with jet fuel injection systems and many industrial processes. Because water may play a major part in the dissemination of the species, our present understanding of its behavior in the aqueous environment, the role of water in its spread, and factors affecting its use in the assessment of water quality are considered in the following discussion.

[1]Professor, School of Civil Engineering, Georgia Institute of Technology, Atlanta, Ga. 30332.

P. aeruginosa and the Fluorescent Pseudomonas Group

P. aeruginosa is the type species of the genus *Pseudomonas*, which consists of gram-negative, rod-shaped bacteria, motile by means of polar flagella and exhibiting respiratory, but never fermentative, metabolism. Representatives of the genus are strict aerobes save for those species which utilize nitrate as a terminal electron acceptor. Typically, *Pseudomonas* species require no growth factors and can multiply in mineral media containing single organic compounds which serve as sole source of carbon and energy. Furthermore, many members of the genus can utilize an exceptionally wide range of organic substrates [*1–6*].[2]

Among the aerobic pseudomonads are species which produce diffusible fluorescent pigments. The fluorescent pseudomonads and four related nonfluorescent species belong to a single DNA-DNA and rRNA-DNA homology complex which is distinct from other *Pseudomonas* species [*7,8*]. The fluorescent complex includes saprophytic species which occur in soil and aquatic environments and exhibit a positive arginine dihydrolase reaction and phytopathogenic species which fail to exhibit a positive arginine dihydrolase reaction [*2,5,6*].

Three saprophytic fluorescent species and four closely related nonfluorescent species are of interest in the present discussion. *P. putida* and *P. fluorescens* form a distinct cluster, members of which possess polar tufts of flagella and are unable to grow at 41°C. The two species differ from one another in their respective inability and ability to hydrolyze gelatin. In contrast to *P. putida* and *P. fluorescens*, *P. aeruginosa* and the related nonfluorescent species, *P. stutzeri*, *P. mendocina*, *P. alcaligenes*, and *P. pseudoalcaligenes*, possess single polar flagella and are capable of growing at 41°C. The latter group of species (*P. aeruginosa* and related nonfluorescent species) constitute a second cluster within the fluorescent group [*8*]. *P. aeruginosa* and *P. mendocina* are very "tight" species, exhibiting a very high degree of internal homogeneity. With the exception of *P. pseudoalcaligenes*, the species in this cluster exhibit a high degree of homogeneity [*7*]. *P. pseudoalcaligenes* constitutes a heterogeneous group with respect to the arginine dihydrolase reaction and the accumulation of poly-β-hydroxybutyrate.

The type species of the genus, *P. aeruginosa*, is of particular interest for reasons described in the following section. Most *P. aeruginosa* strains produce a greenish-blue color in growth media as a result of the production of fluorescent pigment and a blue phenazine pigment, pyocyanin. Some strains produce a brownish red diffusible pigment, pyorubin, which turns dark in time. Characteristically, *P. aeruginosa* strains are oxidase positive, denitrify nitrate, hydrolyze gelatin and casein (but not starch), hemolyze blood, oxidize gluconate, and utilize acetamide. The production of nitritase, gelatinase, and caseinase, the oxidation of gluconate to 2-ketogluconate, and the utilization of acetamide, consistent characteristics among pyocyanogenic

[2]The italic numbers in brackets refer to the list of references appended to this paper.

strains, are said to be variable among apyocyanogenic strains [*9,10*]. Furthermore, susceptibility to tetracycline, streptomycin, and kanamycin are said to be variable among apyocyanogenic strains [*11,12*].

It has been suggested [*13*], however, that variability among apyocyanogenic strains may be caused by the inclusion of unidentified fluorescent *Pseudomonas* (UFP) strains and that apyocyanogenic *P. aeruginosa* strains, like pyocyanogenic strains, exhibit little variability. Characteristic reactions of *P. aeruginosa* and major UFP groups are included in Table 1. While most *P. aeruginosa* strains may be typed serologically using *P. aeruginosa* antisera and according to pyocins produced, UFP strains cannot as a rule be typed by these techniques. Furthermore, analysis of cellular fatty acids suggests a relationship to the related nonfluorescent species—*P. alcaligenes*, *P. pseudoalcaligenes*, *P. stutzeri*, and *P. mendocina*—and competition experiments suggest a substantial degree of homology with *P. mendocina*.

Significance of *P. aeruginosa* and the Fluorescent Pseudomonas Group

While the primary interest here lies with *P. aeruginosa*, other saprophytic species of the fluorescent group (including related nonfluorescent species) may be of economic and public health importance. *P. fluorescens* and *P. putida*, while they ordinarily do not grow at 37°C when inoculated into laboratory media preincubated at that temperature, can, occasionally, cause

TABLE 1—*Selected differential characteristics of* P. aeruginosa *and major UFP*[a] *groups* [13].

	P. aeruginosa				UFP Strains			
	Pyocyanogenic (14 Strains)		Apyocyanogenic (40 Strains)		Group I (67 Strains)		Group III[b] (46 Strains)	
Characteristic	Sign[c]	% Positive	Sign	% Positive	Sign	% Positive	Sign	% Positive
Gelatin Hydrolysis	+	100	+	88	+	100	+	100
Casein Hydrolysis	+	93	+	90	+	87	+	96
Denitrification	+	100	+	100	–	0	–	2
Gluconate Oxidation	+	100	+	95	–	0	–	2
Utilization of								
Mannitol	+	100	+	93	–	0	–	0
Gluconate	+	100	+	93	–	2	–	0
Acetamide	+	100	+	100	V	2	–	0
Resistance to								
Carbenicillin	V	36	V	41	R	89	R	96
Streptomycin	V	79	V	73	S	2	S	2
Tetracycline	R	100	R	100	V	75	V	82
Kanamycin	R	93	R	88	S	2	S	0

[a]UFP = unidentified fluorescent *Pseudomonas*.
[b]Differs from UFP I strains by possession of a polar tuft of flagella.
[c]Sign: + positive (>85% positive).
– negative (<15% positive).
R resistant (>85% resistant).
S sensitive (<15% resistant).
V variable (15 to 85% positive or resistant).

infections in man [*14,15,16*]. Such infections are very rare, however, and these species comprised less than 1 percent of all pseudomonads isolated in one university hospital [*16*]. *P. fluorescens* also can cause spoilage of foods during storage. Of interest in the present discussion is the production of proteolytic defects and rancidity in butter by *P. fluorescens* [*17*], the ability to produce pigmented slime and objectionable odors in cottage cheese [*18*], and grow in and cause deterioration of milk quality [*19*].

The nonfluorescent species *P. stutzeri* and *P. mendocina* normally appear to be of minor importance as pathogens, although *P. stutzeri* has been isolated from infections of man, usually in mixed culture [*14*]. *P. alcaligenes* and *P. pseudoalcaligenes*, while they may be isolated frequently from human specimens, appear not to be of general importance as the etiologic agents of infections in man [*14*].

We are interested here primarily in *P. aeruginosa*, a species of considerable versatility and a pathogen of major significance. *P. aeruginosa* is capable of growth on many petroleum products [*20*], utilizing volatile hydrocarbons of low-molecular weight preferentially as sole carbon and energy source [*21*]. It has been isolated from oils and environments containing oils, including oil well brines [*22,23*], cutting oils, in which it may cause foul odors [*24–27*], jet fuels, in which it may grow and produce slimes [*21,28–30*], and soils in oil producing areas [*31*] or soils receiving oil separator sludges [*32*]. *P. aeruginosa* possesses esterases enabling it to hydrolyze nonionic surfactants used as emulsifying agents leading to separation in many products, particularly cosmetics [*33*], pharmaceutical products [*34*], including ophthalmic solutions and ointments [*35–40*], steroid creams [*41*], and hand lotions and creams [*42*]. This property also has conferred upon *P. aeruginosa* the ability to cause deterioration of some paints [*43*] and plasticizers [*44–46*]. Its ability to produce slimes enables it to interfere with the manufacture of cosmetics, paint, varnish, chemicals, photographic materials, paint brushes [*47*], and paper [*48*].

As a pathogen, *P. aeruginosa* can cause infections in a variety of plants, insects, and warm-blooded animals. In 1941, Elrod and Brown [*49*] reported that *P. polycolor* which caused bacterial leaf spot disease in tobacco, displayed a high degree of virulence when injected intraperitoneally into small laboratory animals and considered this species identical to *P. aeruginosa*. More recent studies have supported this conclusion [*6,50,51*], although the two species appear to differ from one another in several respects [*6*]. The association of *P. aeruginosa* with plants, both potted plants and plants under cultivation in the field, has been demonstrated [*52,53*], and the species is capable of infecting many economic plants, including tobacco, onions, cucumbers, potatoes, lettuce, cabbage, and sugarcane [*50,54,55*].

P. aeruginosa is well established as a pathogen of grasshoppers and other insects [*56–63*], and its use in the biological control of grasshoppers has been considered [*64–66*].

The record of *P. aeruginosa* as a pathogen of domestic and other warm-blooded animals is of more immediate interest. The organism is reported to produce a variety of infections including pneumonia and gastroenteritis in laboratory animals [*67–70*], zoo animals [*71*], fur-bearing animals [*67, 72–78*], poultry [*79–85*], dogs [*86–89*], sheep [*90, 91*], swine [*91–93*], horses [*94,95*], and the bovine animal [*96–124*]. *P. aeruginosa* may cause mastitis in cows, and its frequency has increased since the early 1950s [*91,109*]. *P. aeruginosa* can be recovered often from infected udders [*122,124*], and it is associated commonly with mastitis which can cause extensive damage to dairy herds. Milk from infected cows may contain *P. aeruginosa* [*91,125*]. Schalm and Lasmanis [*126*] reported isolation of the species from 0.07 percent of milk samples from a dairy herd used in mastitis control experiments. The frequency of occurrence undoubtedly is higher among most herds. Probably the most frequently cited instance of contamination by *P. aeruginosa* and its consequences was described by Hunter and Ensign [*127*] who reported more than 400 cases of diarrhea in the general population and 24 cases among 278 infants born at one hospital. Aside from the obvious public health importance of *P. aeruginosa* in milk, hemorrhagic gastroenteritis in calves has been associated with the feeding of milk contaminated with *P. aeruginosa* [*91*].

In man, *P. aeruginosa* may cause a wide variety of infections [*128*], the importance of which has increased steadily in recent years [*129,131*], probably in large measure as a consequence of their resistance to a wide range of antibiotics. The organism is a particular problem in hospitals where it causes approximately 9 percent of nosocomial infections [*132*]. *P. aeruginosa* infections are most frequent and dangerous in nurseries [*133*] and among patients with cancer [*134*], burns [*135,136*], and tracheostomies [*137*]. The highest rates of infection in burns units are at the sites of burns which become infected at a rate of about 25 percent [*135*]. Among patients with acute nonlymphocytic leukemia at one cancer research center, 50 percent eventually became colonized, and, once colonized, septicemia occurred in about two thirds [*134*].

Outside the hospital, *P. aeruginosa* also may cause infections which are of special interest. *P. aeruginosa* is noted for its association with outer ear infections. It is not generally considered a normal inhabitant of the healthy ear [*138–142*], where its frequency of isolation is said to lie generally between 0.5 and 1.5 percent [*139,140,142*]. More recent studies have indicated, however, that the frequency may increase to between 10 and 20 percent among swimmers and among the general population in hot, humid summer weather [*143,144*]. In otitis externa, on the other hand, *P. aeruginosa* is more prominent [*138–140,144–149*]. Singer and Hardy and their collaborators [*139,140*] isolated the species from 65.5 to 80 percent of acute outer ear infections over a three-year period. Frequency of isolation appears to be related to the severity of infection. Senturia [*149*] reported its isolation from 100 percent of severely infected, 63.6 percent of moderately infected, and 38.1 percent of mildly infected outer ears. This observation was confirmed by Hoadley and Knight [*144*]. Complications which may follow external otitis have been reviewed by Forkner [*128*].

Since 1970, reports have appeared in the literature describing three epidemic outbreaks of previously undescribed pruritic, pustular skin rash caused by *P. aeruginosa*, serotype 11 [*150–154*]. While there exists a limited literature concerning *Pseudomonas* skin rashes, the subject is of interest here in view of its association with the water environment (see the following discussion of whirlpool baths).

Methods for the Enumeration of *P. aeruginosa* in Water

Most Probable Number Procedures

A variety of methods has been employed to enumerate or confirm the presence of *P. aeruginosa* in waters and wastes. Most probable numbers (MPNs) have been employed most often. Ringen and Drake [*155*] employed the technique first, using a modification of Burton's medium [*156*] to which was added pyocyanin to select for *P. aeruginosa*. Following 96 h incubation at 37°C, cells were transferred from the enrichment medium to agar slants of the basal medium without pyocyanin for confirmation of pyocyanin production. Reitler and Seligmann [*157*] inoculated samples of drinking water into bile salt lactose peptone water which they incubated for 48 h at 37°C. After incubation, cells were streaked onto "plain agar" plates which were observed for pyocyanin production after incubation at 37°C for 24 h. Schubert and Blum [*158*] recently have employed nutrient broth containing malachite green dye (1:100 000) incubated at 37°C to enrich for *P. aeruginosa*. Clark and Vlassoff [*159*] utilized enrichments in MacConkey broth modified by the addition of 5 g Tryptone per litre to demonstrate the presence or absence of *P. aeruginosa* and other indicator bacteria in drinking water samples simultaneously. Enrichments in the modified broth were incubated for up to five days at 35°C, after which cells were transferred to the liquid medium 10 of Drake [*161*] which in turn was incubated at 35°C for two to four days. Cells from tubes exhibiting fluorescence were streaked on plates of MacConkey agar which was incubated for 24 h at 35°C and examined for typical colonies of *P. aeruginosa*.

The most frequently employed medium for the determination of dilution counts in the United States is asparagine broth. Unfortunately, several formulations are in use (Table 2). While no published comparative studies of enrichment broths include Drake's medium 10, recent investigations of Highsmith and Abshire [*164*] demonstrated that the modified asparagine medium described by Favero et al [*162*] was far superior to that described in the 13th edition of *Standard Methods* [*163*]. Inoculated tubes of Drake's medium 10 or modified asparagine broth have been incubated at 39 or 37°C for up to four days. Cultures are examined daily for fluorescence under ultraviolet light. Cells from tubes exhibiting fluorescence are tested for their ability to utilize acetamide as the sole source of carbon and nitrogen [*162*].

TABLE 2—*Composition of asparagine enrichment broths for the enumeration of* P. aeruginosa.

Component	Medium 10 of Drake [*161*]	Asparagine broth of Favero [*162*]	Asparagine broth of Standard Methods [*163*]
Asparagine	2 g	3 g	2 g
Proline	1 g	. . .	. . .
K_2HPO_4	1 g	1 g	1 g
KH^2PO_4	. . .	. . .	10 g
$MgSO_47H_2O$	. . .	. . .	0.5 g
$MgSO_4$	0.5 g	0.5 g	. . .
K_2SO_4	10 g	. . .	. . .
Glycerol	. . .	. . .	8 ml
Ethanol	2% w/v	. . .	. . .
Distilled water	1 litre	1 litre	1 litre

UFP strains capable of growth at 41°C may grow and produce fluorescence in Drake's medium 10 and frequently constitute in excess of 90 percent of the fluorescent population recovered from surface waters in that medium [*165*]. Dutka and Kwan [*166*] were able to confirm the identity of isolates from only 72 to 80.4 percent of fluorescent tubes of Drake's medium 10 inoculated with raw and treated sewage and from 64 to 79.2 percent of tubes inoculated with surface waters when tubes were incubated at 38°C. Higher confirmation rates were obtained when tubes were incubated at 42°C. Occasionally UFP strains, either alone or in mixed culture, will exhibit positive acetamide reactions. It has been suggested [*167*], therefore, that pyocyanin production be confirmed on slants of King's A medium [*168*].

Robinton [*169*] has compared recoveries of *P. aeruginosa* from river waters in Drake's medium 10 followed by confirmation in acetamide broth and testing of isolates picked from MacConkey agar plates and in Trypticase Soy broth followed by confirmation of isolates picked from MacConkey agar plates. *P. aeruginosa* was isolated most frequently from MacConkey agar plates that were streaked from enrichments in Trypticase Soy broth incubated at 42°C. Mossel and Indacochea [*170*] employed enrichment in a rich peptone medium containing crystal violet kanamycin, and tylosin, followed by streaking on a modified cetrimide medium (GMAC agar) of Brown and Lowbury [*171*] which contained acetamide and phenol red indicator and in which mannitol replaced half the glycerol. The medium was incubated at 42°C, and colonies surrounded by red zones were counted. Recoveries of *P. aeruginosa* from sewage and lake water were slightly higher when this procedure was employed than when Drake's medium was employed.

Recently, several workers have reported isolation of *P. aeruginosa* from *Salmonella* enrichment broths. Gundstrup [*172*] observed that salmonellae and *P. aeruginosa* grew simultaneously in tetrathionate enrichment broth. Grunnet, Gundstrup, and Bonde [*173*] reported optimal recovery of *P. aeruginosa* from sewage in tetrathionate broth incubated for two to four days at 42°C and streaked on cetrimide agar which in turn was incubated at 42°C.

Earlier, Nemedi and Lanyi [*174*] employed brilliant green-selenite broth incubated at 37°C for 48 h from which cells were streaked on brilliant green agar plates which in turn were incubated at 42°C for 24 h to enumerate *P. aeruginosa* in raw and finished waters, swimming waters, and sewage. Similarly, Kenner [*175,176*] reported successful recovery of *P. aeruginosa* sewage, sewage sludge, and potable water in dulcitol selenite broth incubated at 40°C for 24 to 48 h followed by streaking on xylose lysine desoxycholate (XLD) agar which was incubated at 37°C for 24 h.

Direct Plating Techniques

Selenka [*177*] employed a medium containing 0.2 percent triphenyltetrazolium chloride incubated at 37°C for direct plating of sewage and polluted waters. Red colonies exhibiting positive cytochrome oxidase reactions were examined using selected tests to confirm their identity as *P. aeruginosa*. Selenka reported that two thirds of the cytochrome oxidase positive colonies appearing on plates inoculated with river water were *P. aeruginosa*. Hoadley and McCoy [*178*], however, reported formation of colonies by only about 25 percent of the cells of a *P. aeruginosa* test strain inoculated on Selenka's medium. The use of selective plating media, such as Selenka's medium, the solid medium of Drake [*161*] which contains $10^{-3}M$ cadmium chloride, or the cetrimide agar of Brown and Lowbury [*171*], for the enumeration of *P. aeruginosa* in surface waters may lead to erroneous counts as some cells that are injured during suspension may not form colonies [*179*]. However, Mossel and Indacochea [*170*] reported counts of *P. aeruginosa* in sewage and lake water which were plated directly onto GMAC agar that were nearly as high as dilution counts obtained with Drake's medium 10. Furthermore, Carson et al [*180*] reported that while recovery of subcultured cells was dependent upon the diluent employed and recoveries in asparagine and acetamide broths without inhibitors were poor, good recoveries of naturally occurring cells may reflect their greater resistance to injury. This observation was reported also in an earlier publication [*181*].

Membrane Filter Techniques

Several media have been developed for use with membrane filters. Drake [*161*] employed a modification of King's A medium containing 0.05 percent hexadecyltrimethyl ammonium bromide to inhibit growth of unwanted species. Brodsky and Nixon [*182*] employed nonfluorescent black membrane filters to enumerate *P. aeruginosa* in swimming pool waters. Filters were incubated at 42°C for 24 h on MacConkey agar plates. Following incubation, plates were allowed to stand at room temperature to permit development of fluorescence. Fluorescent colonies were counted under ultraviolet light. This technique, modified by preincubation for 4 h at 30°C followed by incubation for 20 h at 42°C, is employed by the Thames Water Authority [*183*]. Confirmation of the identity of colonies is carried out by testing for the oxidase reaction and growth characteristics on milk agar at 42°C. However, overgrowth by other gram-negative bacteria occurs on occasion, interfering with the application of the technique to the enumeration of *P. aeruginosa* in water supplies [*183*]. Similarly, Dutka and Kwan [*166*] found that when coliform counts exceeded counts of *P. aeruginosa* by a factor of 100, background growth interfered with development of *P. aeruginosa* colonies. The technique was considered of little value for the enumeration of *P. aeruginosa* in surface waters and effluents. Lantos et al [*184*] incubated filters on ENDO agar for 24 and 48 h at 37°C. Colonies were streaked on nutrient agar and deoxycholate citrate agar plates. Following incubation, typical colonies were subjected to selected diagnostic tests.

The membrane filter procedure employed most often in the United States is that of Levin and Cabelli [*185*]. These authors proposed a solid medium (mPA, Table 3) containing sulfapyridine, kanamycin, nalidixic acid, and

TABLE 3—*Composition of mPA medium and mPA medium B.*

Component	mPA Medium	mPA Medium B
L-lysine hydrochloride	0.5 g	0.5 g
NaCl	0.5 g	0.5 g
$MgSO_4$	. . .	0.15 g
Yeast extract	0.2 g	0.2 g
Xylose	0.25 g	0.25 g
Sucrose	0.125 g	0.125 g
Lactose	0.125 g	0.125 g
Sodium thiosulfate	0.68 g	0.5 g
Ferric ammonium citrate	0.08 g	0.08 g
Phenol red	0.008 g	0.008 g
Agar	1.5 g	1.5 g
Distilled water	100 ml	100 ml
Following autoclaving at 121°C for 15 min and cooling to 55 to 60°C, adjust pH to 7.2 ± 0.1 and add the following.		
Sulfapyridine	17.6 mg	17.6 mg
Kanamycin	0.85 mg	0.85 mg
Nalidixic acid	3.7 mg	3.7 mg
Actidione	15.0 mg	15.0 mg

actidione to reduce background growth in addition to indicator systems by which to distinguish colonies of lactose, sucrose, xylose fermentors, and hydrogen sulfide (H_2S) producers from *P. aeruginosa*. Membrane filters were placed on the agar medium and incubated at 41.5°C for 48 h. Ninety-two percent of viable *P. aeruginosa* cells seeded in estuary and fresh waters and "stressed" by storage were recovered on the mPA medium; background counts were reduced by at least three orders of magnitude. At least 90 percent of typical colonies (0.8 to 2.2 mm in diameter, flat, with light outer rims and brownish to greenish-black centers) were confirmed as *P. aeruginosa*, and no more than 10 percent of colonies not designated as *P. aeruginosa* could be confirmed as the species. It should be noted that UFP strains which produce fluorescence in tubes of Drake's medium probably seldom form colonies on mPA medium since they are susceptible to kanamycin.

Dutka and Kwan [*166*] confirmed the specificity of mPA medium and were able to verify from 92 to 99 percent of typical colonies. Between 2.7 and 10 percent of atypical colonies were identified as *P. aeruginosa*. By the addition of 1.5 g / litre magnesium sulfate ($MgSO_4$) and the reduction in the concentration of sodium thiosulfate to 5 g / litre (mPA medium B, Table 3) and extending the incubation period to four days, these workers were able to improve recoveries of *P. aeruginosa* significantly. On the basis of comparative studies of the membrane filter procedure employing both the mPA medium and mPA medium B and the MPN procedure employing Drake's medium 10, Dutka and Kwan concluded that both membrane filter techniques provided better recovery. On the other hand, Carson et al [*180*] compared the mPA procedure with the MPN procedure employing asparagine broth according to Favero [*162*]. On the basis of their studies, these authors concluded that the two systems were comparable when assaying samples from hospitals, but, in river waters and sewage, recoveries were significantly higher by the asparagine broth MPN procedure.

The Role of Water in the Spread of *P. aeruginosa*

Drinking Waters

In view of its role as a pathogen, the appearance of *P. aeruginosa* in distribution systems must be regarded with concern. It must be of particular concern in water supplies to hospitals where patients in nurseries and burns units and immunologically suppressed patients or patients receiving inhalation therapy constitute a highly vulnerable population.

A number of workers investigating outbreaks of *Pseudomonas* infections in hospitals have demonstrated the organism in humidifier water and supposedly sterile distilled water [*186*]. While contamination of moist environments in hospitals probably can be traced ordinarily to fecal carriers or infected patients, *P. aeruginosa* may enter the hospital directly or indirectly by way of water supplies. Contamination of the hospital environment results in the development of reservoirs of *P. aeruginosa* which appear not to occur in domestic settings [*187*]. Sinks, baths, washbasin traps, and sponges frequently harbor *P. aeruginosa* in hospitals, although some investigators have not found sinks to be the source of strains causing outbreaks [*188,189*], and Barrie [*190*] exposed blood agar plates to the endotracheal tube of a resuscitation device—the water column of which contained approximately 2.5×10^6 *P. aeruginosa* / ml. *P. aeruginosa* failed to contaminate the plates. On the other hand, the contamination of the hospital environment with this organism does represent a serious hazard to patients. Wilson et al [*191*] reported the isolation of *P. aeruginosa* from faucet aerators of a nursery and surgical scrub sinks. Both hot and cold water run through one of the aerators yielded *P. aeruginosa*. These investigators considered water to be the probable source of the organism causing infections to infants. Similarly, Cross et al [*192*] and Kohn [*193,194*] considered faucet aerators and sinks, respectively, to be the source of *P. aeruginosa* causing infections in patients. Other investigators have associated outbreaks in hospitals with contaminated evacuators, suction equipment, resuscitators, oxygen bubblers, and humidifier reservoirs of incubators [*188,190,195–198*]. Kresky [*199*] reported contamination, mainly by *P. aeruginosa* and *Klebsiella pneumoniae*, of sterile water in squeeze bottles, 100 percent; hexachloraphene dispensers, 60 percent; aerators in sinks, 80 percent; Isolette humidifier systems, 70 percent; and Vapo jet water bottles in a nursery, 50 percent. The species was isolated also from 90 percent of samples from aerators in scrub sinks in the delivery room. In addition, it was isolated from the nasopharynx or cord, or both, in 10 percent of 350 infants examined over a period of years in the same hospital. The hazard to patients is clear. The availability of solvents in the environment which could support growth of *P. aeruginosa* on aerators may explain their isolation in the hospital but not at home.

While Edmondson et al [*200*] recommended the use of chlorinated tap water in nebulizer jars because it retarded multiplication of *P. aeruginosa*, Grieble et al [*201*] recommended that chlorinated tap water not be used in humidifier reservoirs because it supported luxuriant growth of *Pseudomonas*. The ability of *P. aeruginosa* to grow in dialysis fluid in artificial kidney machines and distilled water is well established [*162,202,203*] and presents obvious hazards to patients.

P. aeruginosa may be transmitted to patients in hospitals with water or food, which in turn may be contaminated by water, enriched soils, or human contact. Kominos et al [*204*] and Shooter et al [*205*] demonstrated that vegetables may be the source of *P. aeruginosa* strains causing infections in hospitalized patients, and Shooter et al [*205*] demonstrated similar types in the stool of a patient and peppermint water he had been drinking. Hunter and Ensign [*127*] attributed an epidemic of diarrhea in a newborn nursery to contamination of the milk supply. At one milk plant supplying the hospital, *P. aeruginosa* entered the pasteurized milk from a leaking water pipe wrapped with a rag which dripped into the cooling vats. In more recent episodes, water supplies have been shown to constitute reservoirs of *P. aeruginosa* and sources of contamination in hospitals. *P. aeruginosa* was isolated from the taps of eight washing sinks in an operating theater of one hospital, the cold water supply of which consisted of open roof tanks which were fouled by birds and contained two dead pigeons [*206*]. Weber et al [*207*] reported an epidemic of *P. aeruginosa* infections in a newborn nursery which was caused by a well water supply contaminated by seepage of sewage and infiltration of contaminated stream water.

Gastrointestinal infections with *P. aeruginosa* have been documented, particularly among infants. There is some evidence that such infections can be caused by consumption of contaminated foods or water. Lartigau, in 1898 [*208*], first described two outbreaks of gastroenteritis involving 15 individuals of whom four died. *P. aeruginosa* was isolated from fecal material and from each of five well waters used by the individuals involved. Taylor [*209*] described outbreaks of diarrhea among inhabitants of several cottages and noted the "possibly significant finding" of *P. aeruginosa* in the shallow well water supplies and, in another case, its isolation from water from bore holes in chalk under suspicion of causing paratyphoid fever. Barnes [*210*] has reported several outbreaks of gastroenteritis among naval personnel during which *P. aeruginosa* was isolated from food as well as clinical material. Schiavone and Passerini [*211*] expressed the opinion that *P. aeruginosa* may play an important role in waterborne epidemic outbreaks of gastroenteritis.

As a rule, *P. aeruginosa* can be isolated only occasionally from drinking waters, especially treated drinking waters, unless gross contamination is present (Table 4). Contamination of the Szeged waters [*214*] resulting in the isolation of *P. aeruginosa* from 34.5 percent of samples occurred as a result of infiltration of sewage into underground storage basins. Hungarian supplies were designated as satisfactory or unsatisfactory in Table 4 on the basis of coliform counts. *P. aeruginosa* was isolated relatively frequently also by Reitler and Seligmann [*157*] from northern Israel water supplies. There was, however, no relation between populations of *P. aeruginosa* and those of *Escherichia coli*. Shubert and Blum [*212*] demonstrated its presence in a relatively large fraction of water samples from distribution systems. Again, there was no correlation with the isolation of coliforms. Kenner and Clark

TABLE 4—*Occurrence of* P. aeruginosa *in drinking water samples.*

Source of Samples	No. of Samples	No. Positive	Percent Positive	Reference
Water supplies in northern Israel	1 000	241	24.1	[*157*]
Water supplies in Germany				
Public water supplies				
ground waters	194	4	2.06	[*212*]
distribution systems	216	16	13.5	[*212*]
Individual water supplies	27	1	3.7	[*212*]
Water supplies in Hungary				
Satisfactory				
well supplies	2 774	31	1.1	[*213*]
spring supplies	123	4	3.3	[*213*]
finished waters	2 785	10	0.4	[*213*]
Unsatisfactory				
well supplies	2 519	54	2.1	[*213*]
spring supplies	47	3	6.4	[*213*]
finished waters	971	14	1.4	[*213*]
Budapest municipal supply				
membrane filter	34 420	88	0.2	[*213*]
most probable number	353	12	3.4	[*213*]
Szeged uncontaminated finished water	374	6	1.6	[*214*]
contaminated finished water	133	46	34.5	[*214*]
Finished waters in southern Ontario	14 486	57	0.4	[*160*]

[*176*] reported the isolation of *P. aeruginosa* from 17 of 20 samples from water supplies in the United States, including wells, cisterns, and small municipal supplies. Fecal coliforms were not detected in most supplies investigated, although fecal streptococci were detected in some.

Hoadley and Cheng [*179*] reported that *P. aeruginosa* cells suspended in tap water in Atlanta, Georgia, underwent injury which prevented recovery on the highly selective solid medium of Drake [*161*] and caused a rapid decline in the viable count determined on Trypticase Soy agar (Fig. 1). The tap waters employed in these studies were free of residual chlorine when taken from the tap. Thus, tap waters may be toxic to subcultured cells of *P. aeruginosa*, which tends to support the majority of published findings. On the other hand, in view of the demonstration by Favero et al [*162*] of growth after 48 h in distilled water of *P. aeruginosa* grown initially in a rich medium (Fig. 2), the possibility of multiplication following an initial period of decline in some nontoxic finished waters cannot be excluded. Indeed, naturally occurring cells have been shown by Favero and his colleagues [*180,181,202, 203*] to exhibit greater resistance than subcultured cells. The occurrence of *P. aeruginosa* in bottled drinking waters, first reported in 1928 [*215*] but reported since by investigators in Germany and Brazil [*212,216*], is evidence that the organism can grow in drinking water in containers, although Geldreich et al [*216a*] demonstrated the species in less than 1 percent of bottled water samples in the United States.

Greer, Tenney, and Nyan [*215*] considered that *P. aeruginosa* in drinking waters could not be ignored, and its isolation should condemn a water for drinking purposes. More recent authors have supported the earlier suggestion that the presence of *P. aeruginosa* in a drinking water should not be ignored. Buttiaux [*217*] stated that while the search for *P. aeruginosa* in

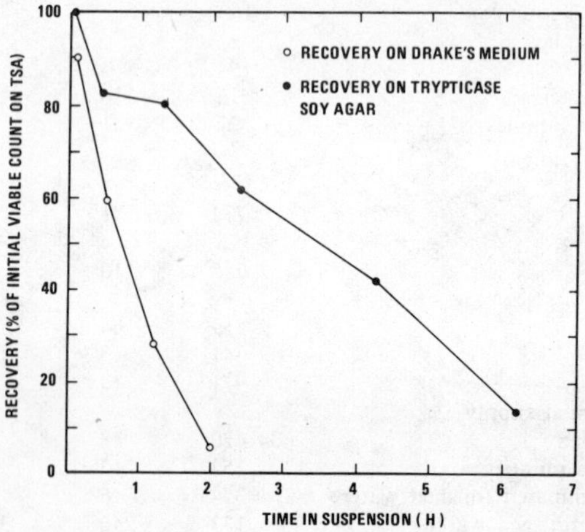

FIG. 1—*Recovery of* P. aeruginosa *ATCC 10145 from tap water* [179].

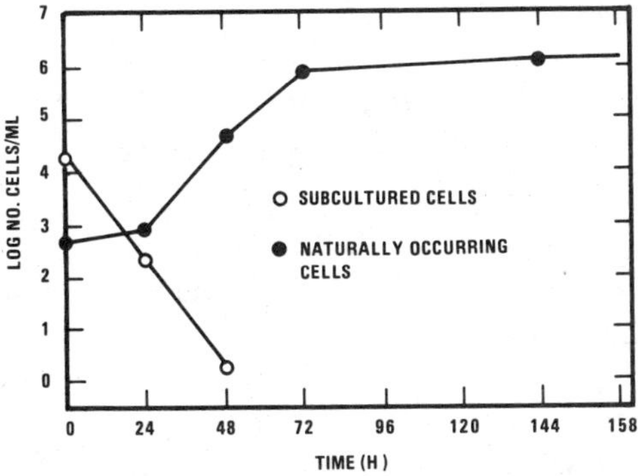

FIG. 2—*Behavior of subcultured and naturally occurring* P. aeruginosa *in buffered distilled water at 25° C* [162].

drinking waters often would not be necessary, the presence of this potential pathogen indicated serious contamination. He felt that this occurred only in the presence of pollution and that it was always accompanied by numerous enteric bacteria. Taylor [*209*] made a similar statement but supported the view of Reitler and Seligmann [*157*] that because of its association with human fecal matter and its role as a pathogen, tests for its presence should be included in the routine examination of the water where a preliminary survey shows that the organism may occur.

Lanyi [*213*] in 1966 stated that the Hungarian National Standard for the Bacteriological Examination of Water prescribes that the presence of *P. aeruginosa* indicates pollution and, therefore, any drinking water that contains this organism is of unsatisfactory quality. Regulation No. MSZ 22901–71 of the Drinking Water Standards [*218*] states that "drinking water should be qualified unsatisfactory if *P. aeruginosa* is present in samples of 260 ml cholorinated piped water; 100 ml non-chlorinated piped water, driven well or mineral water; 50 ml of dug well water." Most recently, Schubert and Blum [*212*] concluded that the correlation of *P. aeruginosa* to *E. coli* and coliforms was so weak that a separate test for *P. aeruginosa* in the sanitary evaluation of water samples appeared necessary.

Water Supplies for Food Processing

In the processing of milk and dairy products, *P. fluorescens* present in the water supply may contaminate the product, causing proteolytic defects and rancidity. Pseudomonads present initially in milk stored at low temperatures reduce the keeping quality [*219*]. Witter [*17*] directed attention to water used in the washing of butter and cottage cheese as a source of serious contamination, and Harmon [*221*] considered the water supply of greatest significance as a source of contamination with spoilage organisms at cheese plants. Olson et al [*220*] pointed up the importance of chlorination of water supplies, especially for the manufacture of butter and cottage cheese. While water unfit for drinking is certainly unfit for use in food processing, water suitable for drinking purposes may not be satisfactory for use in the manufacture of butter [*222,223*], and total counts and counts of proteolytic and lipolytic bacteria [*222*] must be employed to judge quality.

Water Supply for Animals

Water has been implicated as the source of mastitis caused by *P. aeruginosa* in cows. Pickens et al [*96*] described a gravity water supply from a tank in a barn housing an infected dairy herd which was contaminated with apparent *P. aeruginosa*. Later Cherrington and Gildow [*97*] attributed a recurrent outbreak of mastitis to *P. aeruginosa* in the water supply, since improvements resulted in elimination of the problem. More recently, other authors have attributed outbreaks of mastitis in dairy herds to high concentrations of *P. aeruginosa* in water supplies used for washing udders [*107,117, 120,123*], and Hoadley and McCoy [*224*] cited greater than normal losses of calves at a dairy farm at which some cows apparently had acquired a *Pseudomonas* mastitis following wading in marsh and lake waters during an algal bloom. Mushin and Ziv [*225*] observed a high incidence (80.1 percent) of pyocin type 1 *P. aeruginosa* associated with bovine mastitis of farms in Israel which reflected a predominance of the type in the environment, including water samples.

Other reported waterborne outbreaks of *P. aeruginosa* infections in animals include pneumonia in calves [*118*] and infections of mink [*76,77*], chinchillas [*67*], and rabbits [*78*].

Swimming Pools and Whirlpool Baths

The significance of *P. aeruginosa* in swimming pools and whirlpool baths has received increasing attention in recent years. Interest in the organism is related to its association with outer ear infections among swimmers and with skin rashes among users of whirlpool baths.

The association of *P. aeruginosa* with outer ear infections is well established (see previous discussion). The frequency of isolation increases with severity of infection [*144,149*]. The frequency of isolation from infected outer ears also increases among swimmers [*144*]. Furthermore, the incidence of external otitis among swimmers appear to be from two and one half to five times that among nonswimmers [*144*].

Senturia [*226*], discussing the contribution of swimming and diving to the establishment of diffuse external otitis, proposed that during the pre-inflammatory stage many of the protective secretions from the surface of the auditory canal were removed by a douche-like action of the water. It is possible also that in hot, humid weather changes may occur in the amounts and constituents of the cutaneous cover. In the acute inflammatory stage, more lipids are removed and water is absorbed. Consequent swelling chokes the apopilosebaceous and apocrine gland orifices, causing a lack of antibacterial substances on the skin surface. Following loss of air and oily covering, follicular pores become permeable, allowing water and bacteria to enter the ducts of apocrine and sebaceous glands and eventually to enter the gland cells. Scratching to relieve itching can result in contamination introduced on the fingers. Bacteria present in the water also may be introduced into the auditory canal.

Recently, Wright and Alexander [*143*] reported the absence of gram-negative bacteria in the ear canals of nonswimmers but were able to isolate *P. aeruginosa* and other gram-negative bacteria from 6 to 20 percent of divers' ears between the third and thirteenth days of a 15-day period of observation during which the gram-positive flora gave way to a predominantly gram-negative mixed flora after diving began. By the tenth day, the gram-positive flora had reestablished itself. The observations of Wright and Alexander lend support to the notion that tissue maceration and absorption of water may be of importance in predisposing the ear canal to infection. Further support for this theory was the apparent relation between the degree of water exposure in the ear canal and the incidence of external otitis. *P. aeruginosa* was the organism most frequently associated with the disease among the divers, and symptoms appeared to be related closely to the ratio of gram-positive to gram-negative bacteria. The results of Wright and Alexander might also explain a tendency for outer ear infections to occur with greatest frequency at the beginning of the swimming season.

Several reports suggest the direct role of swimming waters in the transmission of *P. aeruginosa* causing outer ear infections. Cothran and Hatlen [*227*] reported isolation of *P. aeruginosa* from swimmng pool water and from the infected outer ears of swimmers at the pool. Favero et al [*228*] later reported that by phage typing, J. C. Hoff had established the identity of strains isolated from the pool water and from infected outer ears. Seyfried [*229*] reported the isolation of *P. aeruginosa* serotype 11 from a private pool water and the infected outer ear of a swimmer who had used the pool. Similarly, Hoadley, Ajello, and Masterson [*230*] reported the isolation of *P. aeruginosa* serotype 3, pyocin type 621424 from a pool water on two occasions, from the infected outer ears of two swimmers, and from both the infected outer ear and rectum of a third swimmer at the pool. It is of interest to note that the latter investigators also isolated UFP strains from pool waters, the healthy outer ear of a swimmer, and the infected outer ear of a nonswimmer.

While it is not known to have caused skin infections among users of swimming pools, *P. aeruginosa* has been the cause of a pruritic skin rash among users of whirlpool baths at motels [*150–154*]. *Pseudomonas* skin rashes were reported first by Hoiyo-Tomoka et al [*231*] who demonstrated the development of high populations on normal skin in a superhydrated environment. The severity of rashes was related to the density of the *Pseudomonas* populations, and damage to the skin was severe both clinically and histologically, although the bacteria did not invade living tissue.

Outbreaks of skin rash have been reported among users of whirlpool baths at three motels since 1972 [*150–154*]. The affected persons were mostly children and attack rates varied from 53 to 85 percent. The onset of symptoms occurred from one to three days following the use of whirlpools, and the rash generally subsided within one week. *P. aeruginosa* serotype 11 was isolated from the skin of two bathers in one outbreak, and each of two and four bathers cultured in the second and third outbreaks. Serotype 4 was isolated from the affected skin of the second bather in the first outbreak. Isolates of serotype 11 were recovered from pool waters at the sites of the second and third outbreaks [*154*].

Because *P. aeruginosa* is associated commonly with outer ear infections and because swimming is known to increase both the risk of acquiring outer ear infections and the risk of *P. aeruginosa* involvement [*144*], many investigations of swimming pools have been undertaken in recent years which establish this species as a frequent inhabitant of pool waters (Table 5), often exhibiting resistance to disinfection [*161,174,182,212,227-230,232-234a*].

The relationships between numbers of *P. aeruginosa* or its occurrence in swimming pool waters and coliforms is not always clear. For example, the data of Black et al [*233*] might at first suggest a positive correlation, but, upon closer examination, such a conclusion appears unjustified. The data of Nemedi and Lanyi [*174*] do suggest, however, that when *P. aeruginosa* is present in swimming pool waters coliforms will be present as well, although *P. aeruginosa* often outnumber coliforms. On the other hand, results of Shubert and Blum [*212*] and Botzenhart, Thofern, and Hunefeld [*234*] failed to demonstrate a positive correlation. While *P. aeruginosa* may enter swimming pools in association with coliforms from skin surfaces contaminated with fecal material (or with fecal material when small children are present), more likely sources may be unrelated to sources of coliforms. Carriage into

TABLE 5—*Demonstration of* P. aeruginosa *in swimming pool waters.*

Location of Pools	No. of Samples	No. Positive	Percent Positive	Reference
Hungary				
Membrane filter technique	235	8	3.4	[*174*]
Selenite enrichment technique	161	136	84.4	[*174*]
Canada				
Membrane filter-MacConkey agar technique	130	75	6.64	[*182*]
Drake's membrane filter technique	710	57	8.03	[*182*]
Germany				
Nutrient broth-malachite green enrichment	144	15	10.42	[*212*]
Coliform enrichments	207	14	6.76	[*234*]
Coliform enrichments	359	29	8.08	[*234a*]
Florida				
Asparagine broth-acetamide enrichments	192	31	16.16	[*233*]
Drake's medium	17	5	29.4	[*230*]

pools from the environment is possible. Balacescu and Grün [*235*] isolated *P. aeruginosa* from eight of 218 smears taken from benches at swimming pools, and Kush and Hoadley [*236*] suggested that *P. aeruginosa* carried into whirlpool baths from environmental surfaces by bathers represented a source of new strains establishing themselves in the water (see discussion of whirlpool baths). Nemedi and Lanyi [*174*] showed that the frequency of *P. aeruginosa* in swimming pool waters was greater than that in inflowing water, which might suggest growth in the pool waters and growth in slime on pool surfaces also would seem possible. Of greatest potential significance is urine, however. Hoadley et al [*230*], investigating pools at an institution for the mentally retarded, suggested that such gross contamination of pool waters would explain the predominance of single UFP types or *P. aeruginosa* serotypes and pyocin types which they observed in individual samples. Predominant serotypes isolated from swimming pools by Nemedi and Lanyi [*174*] were: type 3(Habs type 2), 15.8 percent; type 7(Habs type 11), 13.7 percent; type 6(Habs type 4), 10.9 percent; and type 4 (Habs type 1), 10.2 percent.

Maintenance of chlorine residuals is important in the control of *P. aeruginosa* populations in swimming pools. Favero et al [*228*] observed *P. aeruginosa* ordinarily was not demonstrated in pool waters unless chlorine residuals were low. The bacteria were rarely isolated when free chlorine residuals exceeded 0.5 mg / litre. Similarly, Black et al [*233*] isolated *P. aeruginosa* from 43.8 percent of pools exhibiting no residual chlorine, 27.3 percent of pools in which the total chlorine residual was less than 0.3 mg/litre, 10.4 percent of pools in which the total residual was greater than 0.3 mg / litre but whose free residual was less than 0.3 mg / litre, but from only 1.8 percent of pools in which the free residual chlorine concentration exceeded 0.3 mg / litre.

Favero et al [*228*] reported a high incidence of *P. aeruginosa* in private pools in which sodium dichloroisocyanurate was used, and Black et al [*233*] isolated *P. aeruginosa* from 21.4 percent of pool waters containing cyanuric acid stabilized chlorine. In their earliest studies, Keirn and Putnam [*232*] reported isolation of *P. aeruginosa* from one pool only after using quaternary ammonium compounds. Earlier, Skadhauge and Fogh [*237,238*] and Quisno and Foter [*239*] reported that *P. aeruginosa* was affected little by low concentration of cetylpyridinium chloride which was not practical for use in swimming pools. Fitzgerald and DerVartanian [*240*], in more extensive studies demonstrated that at lower concentrations of chlorine, cyanuric acid caused lengthening of the time required to kill *P. aeruginosa*. Furthermore, the presence of ammonia reduced the effectiveness of chlorine as a bactericidal agent. It has been shown that if ammonium acetate is present in commercial benzalkonium chloride, growth of *P. aeruginosa* may occur in solutions [*241*]. There is evidence that *P. aeruginosa* grown in the presence of benzalkonium chloride are less virulent than parent strains grown in the absence of the mixed quaternary ammonium compound [*242*].

The susceptibility of *P. aeruginosa* to chlorine relative to the susceptibility of other indicator bacteria is of interest. Butterfield et al [*242a,242b*], investigating the effects of chlorine and chloramines on *P. aeruginosa* and several enteric bacteria, showed that *E. coli* and *P. aeruginosa* were killed by chlorine at the same rate at pH 7, but, at higher pH values, *P. aeruginosa* exhibited greater resistance than *E. coli* or *S. typhi*. However, Favero et al [*228*] reported that *P. aeruginosa* was very much more susceptible to free chlorine than *E. coli* or *S. aureus*. Similarly, Botzenart, Thofern, and Hünefeld [*234*] demonstrated that *P. aeruginosa* was killed more rapidly by chlorine than was *E. coli*, although the rates were closer than observed by Favero. Fitzgerald and DerVartanian [*240*] reported that *P. aeruginosa* appeared to be more sensitive than *S. faecalis* to chlorine in the presence of ammonia. It should be stressed, however, that naturally occurring cells of *P. aeruginosa* are more resistant to disinfecting agents than are the subcultured cells employed in most laboratory studies of disinfection [*181,203*] (Fig. 3).

In view of the importance of *P. aeruginosa* as a pathogen, its occurrence in swimming pool waters not properly disinfected, the frequent lack of correlation between its presence and the presence of other indicator species, and its relation to ear infections in swimmers, various investigators have recommended that it be viewed with concern in pools [*174,212,228,243*]. Favero et al [*228*] recommended that "the isolation of *P. aeruginosa* in 100 ml or less of swimming pool water would warrant closure of the pool until an adequate chlorine residual could be maintained." Fish [*243*] recommended that "once a swimming pool is recognized as a source of human infection, it should be

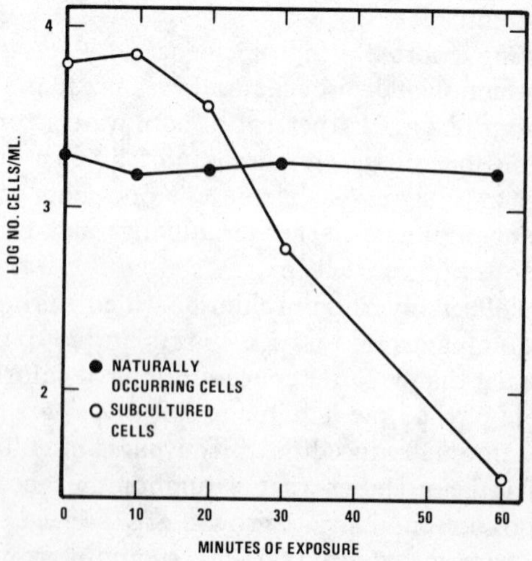

FIG. 3—*Survival of subcultured and naturally occurring* P. aeruginosa *in distilled water from mist therapy units at pH 6.5 and 25°C when exposed to 67 mg/litre ClO_2* [203].

closed and all programs terminated." The only official standard appears to be the proposed Hungarian regulation that a maximum of 10 *P. aeruginosa* / 100 ml be allowed in recirculating pools, and 100 *P. aeruginosa* / 100 ml be allowed in fill and draw pools [*218*].

Because *P. aeruginosa* has been associated with skin rashes among users of whirlpool baths, studies of its occurrence in whirlpool baths have been undertaken [*236*]. In a survey of eight whirlpool baths at health clubs and an apartment complex in Atlanta, Georgia, temperatures in the range of 36 to 42°C were encountered commonly, pH values varied between 6.5 and 7.8, and total chlorine residuals varied from 0.0 to 3.0 mg / litre. Concentrations of total organic carbon, organic nitrogen, and ammonia nitrogen, while they varied from pool to pool, remained relatively constant at individual pools, reaching 211, 91, and 17 mg / litre, respectively, at one pool. Populations of *P. aeruginosa* varied from less than two to more than 2400 / 100 ml. There was no relation demonstrated between populations of *P. aeruginosa* and those of presumptive coliforms. The relation between populations of *P. aeruginosa* and total chlorine residuals was only poorly defined. Whereas presumptive coliforms were demonstrated in only one of 13 samples in which the residual was 3 mg / litre or less and six of eleven samples in which the residual was 2 mg / litre or less, *P. aeruginosa* was isolated from five of the 13 samples containing 3 mg / litre of residual chlorine and ten of the eleven samples containing 2 mg / litre or less. The predominant serotypes of *P. aeruginosa* were 11, 6, and 4—serotype 11, which was isolated from 29.2 percent of the samples, being the one most frequently demonstrated.

Sampling at a single whirlpool bath throughout a single day of use revealed little variation in populations of *P. aeruginosa*. However, a continuous shift in the predominance of serotypes, and the emergence of serotype 11 as a predominant serotype only following its isolation from floor surfaces surrounding the bath, suggested that the organism may be carried into baths on the feet of bathers. Chlorination with approximately 50 mg / litre on one evening had no effect on populations of *P. aeruginosa* in the bath the following morning. However, populations increased from 31 *P. aeruginosa* / 100 ml at the start of the day to a high of 920 *P. aeruginosa* / 100 ml by 6:00 p.m. Serotypes 3 and 10, initially not demonstrated in pool waters, emerged following their isolation from floor surfaces. Again, among typable strains of *P. aeruginosa*, serotype 11 was demonstrated most frequently. The predominance of serotype 11 is of particular interest in view of its apparent association with skin rashes among users of whirlpool baths [*154*].

On the basis of their investigations of whirlpool baths, Kush and Hoadley [*236*] stressed the apparent need for regulations for the operation of baths and the maintenance of water quality. They concluded that existing regulations promulgated for swimming pools relating to coliform counts are of little value for the protection of bathers at whirlpool baths. The investigation of total plate counts was not undertaken. Furthermore, there arose some question concerning the determination of free residual chlorine concentrations as ordinarily conducted. Regulations relating to the content of organic matter, nitrogen concentrations, temperature, periodic draining of baths, disinfection of floor surfaces, and bather cleanliness were recommended, as well as the periodic examination of waters and floor surfaces for *P. aeruginosa.* Of particular interest are Japanese standards for heated baths which place limitations on organic matter (permanganate value less than 25 mg / litre and require that the temperature exceed 42°C and that the water be changed daily at least [*244*].

Shellfish Waters

Little interest has been shown in the contamination of shellfish by *P. aeruginosa.* Probably the first report of the species in shellfish was made by Houston [*245*], who in 1904 reported its isolation from 0.1 ml of stomach juice from 10 Penryn oysters. More recently, its isolation from mussels obtained from inland waters of India was reported [*246*]. Most recently, Denis [*247*] examined oysters and mussels from the Marennes-Oléron region on the coast of France. Bacteriophages of *P. aeruginosa* were demonstrated in 24.3 percent of 446 samples of oysters (12 oysters per sample) and 21 percent of 80 samples of mussels (20 mussels per sample) during 1973. Contamination by bacteriophages was infrequent between January and May but occurred in 21.4 to 69 percent of samples of oysters between July and December. Samples were examined for *P. aeruginosa* between July and December. *P. aeruginosa* was isolated from 48 percent of 89 samples of oysters and 74 percent of 35 samples of mussels. The demonstration of specific phages was unrelated to the isolation of *P. aeruginosa*, however. Serotype 3 (Pasteur Institute) was the predominant type, as it was in most hospitals of the region.

While Denis suggested that shellfish could constitute a source of *P. aeruginosa* which might colonize the intestinal tract of man, he did not explore possible threats to human health associated with the consumption of contaminated shellfish.

Surface and Recreational Waters

P. aeruginosa may reach drinking water supplies, water supplies for animals, industrial processes with raw surface water supplies, and shellfish waters with streams discharging into bays and estuaries. The significance of *P. aeruginosa* in these waters has been considered previously. The significance of *P. aeruginosa* in surface and marine waters used for recreational or irrigation purposes remains to be explored.

The role of swimming and *P. aeruginosa* in outer ear infections was discussed previously (see Swimming Pools and Whirlpool Baths). The role of surface swimming waters as a source of *P. aeruginosa* in outer ear infections has not been established, although the studies initiated by Cabelli et al [*248*] may provide insight into the importance of this species in marine recreational waters. *P. aeruginosa* infections of wounds sustained in contaminated aquatic environments have been described. Recently, Taplin [*249*] reported infections among survivors of a plane crash in the Everglades and in a 17-year-old girl severely injured in a boating accident.

The role of vegetables in transmission of *P. aeruginosa* to hospitalized patients reviewed previously suggests that the presence of the species in irrigation waters or wastes applied to agricultural land may be of concern. The roles of water and wastes have not been discussed in the literature, although Green et al [*53*] suggested that contamination occurred during harvest, handling, processing, and transit rather than during growth in fields. However, vegetables may come in contact with soil and insects, as well as with humans during harvesting, and contamination of soils by water, wastes, and sewage sludges (see following discussion of soil) may contribute to the potential spread by vegetables.

While *P. aeruginosa* often is considered a common inhabitant of soil and water, its presence in surface waters not influenced by man has been questioned [*250–252*]. Buttiaux [*217*] considered the species to be an inhabitant of extremely polluted waters only (where it was accompanied by numerous enterobacteria) and was never isolated by itself. Bonde [*252*] demonstrated *P. aeruginosa* when populations of thermotolerant coliforms exceeded 1000/100 ml, and Drake [*161*] considered that populations of 1 to 10 *P. aeruginosa*/100 ml may be expected in streams with low but definite levels of contamination.

Selenka and Ruschke [*253*] demonstrated 5 *P. aeruginosa*/litre in the surface water of the Bodensee and 2.7/litre at a depth of 60 m. Recently, Dutka [*254*] has demonstrated low populations in both inshore and offshore waters of lakes Erie, Ontario, Huron, and Georgian Bay. Greater populations in stream waters contaminated with sewage have been reported. Selenka [*177*] reported populations of from 200 to 700 *P. aeruginosa*/100 ml in surface waters, and Hoadley and McCoy [*255*], using a modification of Selenka's procedure, demonstrated 550 *P. aeruginosa*/100 ml in a stream water 0.32 km downstream from a sewage treatment plant outfall in Wisconsin but considered the actual population to be about four times that determined. The latter investigators failed to isolate the species 5.8 km downstream from the point of discharge.

On the basis of their later studies in Wisconsin streams and lakes, Hoadley, McCoy, and Rohlich [*251*] suggested that *P. aeruginosa* probably does not occur in waters not recently affected by human activity or that of domestic animals. Low-population levels of less than 100 organisms/100 ml were demonstrated in waters adjacent to human activity. Populations in excess of 100 organisms/100 ml occurred in waters receiving surface drainage from urban areas or recently contaminated by sewage. Populations of from 1000 to 10 000/100 ml were observed in small streams below sewage outfalls; however, populations were reduced rapidly in stream waters under field conditions, being diminished ordinarily by over 90 percent in 3 h. The rate of disappearance was related to temperature (Fig. 4). Botzenhart, Wolf,

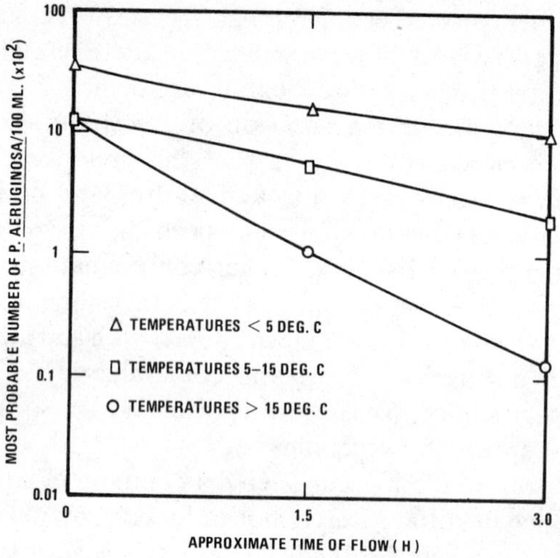

FIG. 4—*Median most probable numbers of* P. aeruginosa *in a stream receiving unchlorinated primary sewage* [25].

and Thofern [*252*] also observed the elimination of *P. aeruginosa* from surface waters but at very much lower rates. The latter investigators considered that, while *P. aeruginosa* occurs in open waters, it is only present following the discharge of sewage. Furthermore, Cabelli et al [*266*], on the basis of investigations undertaken at two freshwater-estuarine systems in Rhode Island and beaches in Lake Michigan, concluded that ratios of *P. aeruginosa* to fecal coliforms increase with distance from sources of pollution, suggesting greater survival of *P. aeruginosa* and the usefulness of the ratio, in conjunction with population densities, to measure the proximity of pollution sources. These investigators have suggested that when ratios of numbers of *P. aeruginosa* to fecal coliforms exceed 0.2 at high-population densities of each group, a nonfecal source is indicated. The finding of high-fecal coliform counts in excess of 1000 bacteria / 100 ml and counts of *P. aeruginosa* below 1 organism / 100 ml would suggest fecal pollution of animal origin. Populations of *P. aeruginosa* in excess of populations of fecal coliforms have been reported by Highsmith and Abshire [*164*] in both raw sewage and stream waters, however.

In view of the demonstration of 2.5×10^7 *P. aeruginosa* / 100 ml in urban stream water by Highsmith and Abshire [*164*] using the asparagine broth of Favero, in contrast to 2.7×10^2 / 100 ml when asparagine broth according to *Standard Methods* was employed, counts may be higher than previously reported. Similarly, Dutka and Kwan [*166*] reported MPNs of 0.2 and 0.37 *P. aeruginosa* / 100 ml in a lake and canal water employing Drake's medium 10 and 19 and 15 / 100 ml, respectively, when the mPA medium B membrane filtration technique was employed.

That domestic sewage was the major source of *P. aeruginosa* reaching surface waters had been demonstrated earlier by Hoadley, McCoy, and Rohlich [*167*], who also had demonstrated growth in raw sewage when oxygen was present. Lanyi et al [*213*] and Nemedi and Lanyi [*174*] demonstrated a striking similarity of *P. aeruginosa* serotypes in water samples from numerous sources, sewage, and human feces. Similar observations have been made by Hoadley and Ajello (Table 6) [*265*]. *Pseudomonas aeruginosa* appears to be associated primarily with man, and, although reported frequencies of intestinal carriage vary in different countries, the rate in the United States (Table 7) appears to be slightly in excess of 10 percent among healthy adults but increases in certain hospitalized patients [*127,258–260*]. Populations of *P. aeruginosa* in raw sewage in excess of 10^5 organisms/100 ml are common, and populations in excess of 10^6/100 ml have been demonstrated (Table 8). Secondary treatment of sewage appears to reduce numbers of *P. aeruginosa* by approximately 99 percent, although growth may occur in trickling filter plants treating protein rich slaughter house wastes [*167*]. Storm drainage from municipal areas, barnyard runoff, and industrial cooling waters may act as lesser sources of *P. aeruginosa* entering surface waters.

Domestic animals, especially young ruminants, may serve as minor sources of *P. aeruginosa* reaching surface waters. The organism appears not to be a normal inhabitant of the intestinal tracts of domestic animals, however, and is rapidly eliminated upon weaning of ruminants [*116,224*]. A reservoir of the organism, such as a human carrier or cows with mastitis, appears to be necessary, and carriers in dairy herds appear to occur commonly only on certain farms [*224*]. The organisms may be isolated from drainage at some farms [*91,167*].

Wastes from hospitals may be of further interest because hospital strains

TABLE 6—*Occurrence of* P. aeruginosa *immunotypes among isolates from surface waters and wastes.*

Immunotype			Percent of Isolates			
Fisher Immunotype	Habs Serotype	Lanyi Serotype	Lake Water [*262*]	Polluted Stream [*262*]	Surface Waters [*174*]	Sewage [*174*]
1	6	4a-c	37.5	28.1	21.9	17.0
2	11	7	5.9	13.1	13.1	8.2
3	2	3	2.8	4.2	9.8	7.5
4	1	6	8.7	9.3	10.7	12.3
5	10	2	1.7	0.7	4.2	9.5
6	7,8	5	2.1	1.4	10.7	12.3
7	5	3a,b	6.2	10.3	13.1	8.2
3,7	. . .	. . .	5.6	10.3	. . .	. . .
Other strong cross reactions	. . .	. . .	4.9	1.4	3.9	2.0
Rough	. . .	. . .	1.4	3.7	. . .	. . .
Untypable	. . .	. . .	23.2	17.3	. . .	. . .
Other serotypes	. . .	. . .	. . .	. . .	21.1	23.8

TABLE 7—*Frequency of isolation of* P. aeruginosa *from feces and rectal swabs of healthy adults in the United States.*

No. Examined	No. Positive	Percent Positive	Reference
103	16	15.6	[*127*]
100	11	11	[*155*]
273	32	11.7	[*256*]
235	28	11.9	[*257*]
52	6	11.5	[*167*]
104[a]	13	12.5	[*258*]

[a] Reconstructive patients in good health in a burns institute for children.

may bear resistance markers to certain antibiotics. In recent years, *P. aeruginosa* strains resistant to carbenicillin and gentamicin have appeared in hospitals [*263,264*]. While strains of *P. aeruginosa* isolated from hospital wastes appear not to be resistant to gentamicin, carbenicillin resistant strains have been demonstrated in both hospital wastes and stream waters receiving hospital wastes (Table 9) [*265*].

Organically enriched surface waters are capable of supporting growth of *P. aeruginosa* in the laboratory, and Hoadley, McCoy, and Rohlich [*251*]

TABLE 8—*Sources of* P. aeruginosa *in wastes and runoff reaching surface waters.*

Source	Median Count/100 ml	Reference
Sewage		
Raw domestic, U.S.A.	2.25×10^5	[*167*]
Treated domestic, U.S.A.	3.5×10^3	[*167*]
Raw domestic, Germany (includes hospital, meat packing wastes)	3.3×10^6	[*253*]
Raw domestic, Germany	1.3×10^5	[*252*]
Treated domestic, Germany	2.1×10^3	[*252*]
Raw domestic, Hungary	2.5×10^5	[*174*]
Raw domestic A, South Africa	8×10^5	[*261*]
Raw domestic B, South Africa	4×10^5	[*261*]
Raw hospital A, South Africa	7.1×10^6	[*261*]
Raw hospital B, South Africa	6.6×10^6	[*261*]
Raw hospital, U.S.A.	2.9×10^4	[*262*]
Treated hospital, U.S.A.	3.3×10^3	[*262*]
Raw sewage, Canada		
Drake's medium 10	9.2×10^4	[*166*]
mPA medium B	2.2×10^5	[*166*]
Treated sewage, Canada		
Drake's medium 10	5.4×10^2	[*166*]
mPA medium B	1.0×10^4	[*166*]
Raw sewage, U.S.A.		
Asparagine broth (Standard Methods)	4.2×10^4	[*164*]
Asparagine broth (Favero)	4.3×10^7	[*164*]
Meat packing wastes, U.S.A.	2.9×10^5	[*167*]
Storm drainage, U.S.A.	2.9×10^2	[*167*]
Barnyard runoff, U.S.A.	7.8×10^0	[*167*]
Cooling water, Germany	6.0×10^1	[*252*]

TABLE 9—*Isolation of* P. aeruginosa *strains resistant to carbenicillin from hospital wastes and a receiving stream* [264].

Source of Isolates	No. Tested	No. Resistant	Percent Resistant
Stream above outfall	35	0	0
Hospital wastes	109	20	18.4
Stream below outfall	186	24	12.9

suggested that, on the basis of this observation and the occasional recovery of the species from unpolluted but organically enriched surface waters at times of algal blooms when water temperatures exceeded 30°C, growth might occur when temperatures were high. Although such temperatures are rare in the northern United States, they are not uncommon in the southern part of the country, and a direct relationship between water temperatures and populations of *P. aeruginosa* has been demonstrated by Hoadley and Ajello [*165*]. On the basis of increases observed through an impoundment in Germany, Botzenhart, Wolf, and Thofern [*252*] also have suggested that in nutrient rich surface waters, growth may take place, and Cabelli et al [*266*] suggested that growth occurred in Lake Michigan waters in association with residual fecal pollution. The association of *P. aeruginosa* with algal blooms has been reported also by Foster [*267*].

The isolation of *P. aeruginosa* from unpolluted surface waters at low temperatures only following rains was observed by Hoadley and Ajello [*165*], suggesting the possibility that soil may constitute a reservoir of the bacteria reaching surface waters. The role of soil as a source of *P. aeruginosa* is not clear. However, a number of attempts to isolate the organism from soils have yielded few positive results [*155,167*]. *P. aeruginosa* has been isolated from soils contaminated with oil [*31,32*], agricultural soils in California [*53*], and soils in pots planted with ornamental plants in hospitals [*52*]. Of special interest was the isolation of carbenicillin resistant strains from agricultural soil in California [*53*], since such strains are associated with infections in hospitals, being selected during antibiotic therapy, and have been isolated from vegetables served to patients in hospitals. All of these are enriched soils, however, and are in contact with man.

Conclusions

P. aeruginosa, while it is often considered to be a ubiquitous bacterial inhabitant of surface waters and soil, appears to enter the environment mainly with human fecal wastes or with the fecal wastes of animals associated with man. The isolation of this organism from surface waters suggests the influence of man, and its numbers reflect the degree of pollution. On the other hand, growth in organically enriched surface waters appears to occur as temperatures approach and exceed 30°C, and it may be isolated from enriched soils, particularly cultivated soils.

P. aeruginosa is an opportunistic pathogen of man and animals which may be spread by water. The value of *P. aeruginosa* as an indicator of potential health hazards associated with water must be judged solely on the basis of the potential hazard associated with its own presence. There is no reason to suggest that *P. aeruginosa* might be a good indicator of the possible presence of other pathogens, and conventional criteria for the evaluation of indicator bacteria probably cannot be satisfied. While human wastes and runoff from urban areas are probably the major sources of *P. aeruginosa* reaching most surface waters, their relatively short-survival time would make them poor indicators of fecal pollution. On the other hand, conventional indicators of fecal contamination are of little general value to indicate the possible presence of *P. aeruginosa.* The isolation of *P. aeruginosa* from drinking waters, farm water supplies, swimming pool waters, whirlpool waters, and surface recreational waters should be regarded with concern, and precautions should be taken to prevent exposure of susceptible populations or to eliminate the organism. However, while various investigators have recommended examination of drinking waters for *P. aeruginosa* where it has been shown to occur, or closure of swimming pools from which it is isolated until proper chlorine residuals can be maintained, no epidemiologic studies have been conducted upon which to base the establishment of standards.

References

[*1*] Colwell, R. R., *Journal of General Microbiology*, Vol. 37, No. 2, Jan. 1965, pp. 181–194.
[*2*] Doudoroff, M. and Palleroni, N. J. in *Bergey's Manual of Determinative Bacteriology*, 8th ed., R. E. Buchanan and N. E. Gibbons, Ed., Williams & Wilkins Co., Baltimore, Md., 1974, p. 217.
[*3*] Jessen, O., Pseudomonas aeruginosa *and Other Green Fluorescent Pseudomonads: A Taxonomic Study*, Munksgaard, Copenhagen, 1965.
[*4*] Palleroni, N. J. in *Genetics and Biochemistry of* Pseudomonas, P. H. Clarke and M. H. Richmond, Ed., Wiley, New York, 1975, p. 1.
[*5*] Stanier, R. Y., Palleroni, N. J. and Doudoroff, M., *Journal of General Microbiology*, Vol. 43, No. 2, May 1966, pp. 159–271.
[*6*] Sands, D. C., Schroth, M. N. and Hildebrand, D. C., *Journal of Bacteriology*, Vol. 101, No. 1, Jan. 1970, pp. 9–23.
[*7*] Palleroni, N. J., Ballard, R. W., Ralston, E. and Doudoroff, M., *Journal of Bacteriology*, Vol. 110, No. 1, Apr. 1972, pp. 1–11.
[*8*] Palleroni, N. J., Kunisawa, R., Contopoulou, R. and Doudoroff, M., *International Journal of Systematic Bacteriology*, Vol. 23, No. 4, Oct. 1973, pp. 333–339.
[*9*] Gilardi, G., *Applied Microbiology*, Vol. 21, No. 3, Mar. 1971, pp. 414–419.
[*10*] Hugh, R. and Gilardi, G., in *Manual of Clinical Microbiology*, E. H. Lennette, E. H. Spaulding and J. P. Truant, Ed., American Society for Microbiology, 1974, Chapter 23, p. 250.
[*11*] Gilardi, G. *Applied Microbiology*, Vol. 22, No. 5, Nov. 1971, pp. 821–823.
[*12*] von Graevenitz, A. and Redys, J. J., *Health Laboratory Science*, Vol. 5, No. 2, April 1968, pp. 107–112.
[*13*] Ajello, G. and Hoadley, A. W., *Journal of Clinical Microbiology*, Vol. 4, Nov. 1976, pp. 443–449.
[*14*] von Graevenitz, A., *Progress in Clinical Pathology*, Vol. 5, 1973, pp. 185–218.
[*15*] von Graevenitz, A. and Weinstein, J., *Yale Journal of Biology and Medicine*, Vol. 44, No. 3, Dec. 1971, pp. 265–273.
[*16*] Blazevic, D. J., Koepcke, M. H. and Matsen, J. M., *Applied Microbiology*, Vol. 25, No. 1, Jan. 1973, pp. 107–110.
[*17*] Witter, L. D., *Journal of Dairy Science*, Vol. 44, No. 6, June 1961, pp. 983–1015.
[*18*] Elliker, P. R., *Food Engineering*, Vol. 26, No. 11, Nov. 1954, pp. 79–88.

[19] Overcast, W. W. and Adams, G. A., *Journal of Milk and Food Technology*, Vol. 29, No. 1, Jan. 1966, pp. 14–18.
[20] Bushnell, L. D. and Haas, H. F., *Journal of Bacteriology*, Vol. 41, No. 5, May 1941, pp. 653–673.
[21] Leathen, W. W. and Kinsel, N. A., *Developments in Industrial Microbiology*, Vol. 4, 1963, pp. 9–16.
[22] Iizuka, H. and Komagata, K., *Journal of General and Applied Microbiology*, Vol. 10, 1964, pp. 223–231.
[23] Vorobyeva, G. I., *Doklady Academ Navk SSSR*, Vol. 112, No. 4, 1957, pp. 763–765.
[24] Lee, M. and Chandler, A. C., *Journal of Bacteriology*, Vol. 41, No. 3, March 1941, pp. 373–386.
[25] Duffett, N. D., Gold, S. H. and Weirich, C. L., *Journal of Bacteriology*, Vol. 45, No. 1, Jan. 1943, pp. 37–38 (Abstract).
[26] Samuel-Maharajah, R., Pivnick, H., Engelhard, W. E. and Templeton, S., *Applied Microbiology*, Vol. 4, No. 6, Nov. 1956, pp. 293–299.
[27] Feisal, E. V. and Bennett, E. O., *Journal of Applied Bacteriology*, Vol. 24, No. 2, June 1961, pp. 125–130.
[28] Prince, A. E., *Developments in Industrial Microbiology*, Vol. 2, 1961, pp. 197–203.
[29] Kereluk, K. and Baxter, R. M., *Developments in Industrial Microbiology*, Vol. 4, 1963, pp. 235–244.
[30] Edmonds, P., *Applied Microbiology*, Vol. 13, No. 5, Sept. 1965, pp. 823–824.
[31] Kvasnikov, E. I. and Tin'yanova, N. Z., *Mikrobiologiya*, Vol. 43, No. 4, July-Aug. 1974, pp. 710–714.
[32] Jensen, V., *Oikos*, Vol. 26, No. 2, 1975, pp. 152–158.
[33] Manowitz, M., *Developments in Industrial Microbiology*, Vol. 2, 1961, pp. 65–71.
[34] Barr, M. and Tice, L. F., *Journal of the American Pharmaceutical Association*, Vol. 46, 1957, pp. 442–445.
[35] McCulloch, J. C., *Archives of Ophthalmology*, Vol. 29, No. 6, June 1943, pp. 924–935.
[36] Theodore, F. H., *American Journal of Ophthalmology*, Vol. 34, No. 12, Dec. 1951, pp. 1764.
[37] Theodore, F. H. and Minsky, H., *Journal of the American Medical Association*, Vol. 147, No. 14, Dec. 1 1951, pp. 1381.
[38] Bignell, J. L., *British Journal of Ophthalmology*, Vol. 35, No. 7, July 1951, pp. 419–423.
[39] Theodore, F. H. and Feinstein, R. R., *American Journal of Ophthalmology*, Vol. 35, No. 5, May 1952, pp. 656–659.
[40] Vaughn, D. G., Jr., *American Journal of Ophthalmology*, Vol. 39, No. 1, Jan. 1955, pp. 55–61.
[41] Noble, W. C. and Savin, J. A., *Lancet*, Vol. 1, Feb. 12, 1966, pp. 347–349.
[42] Ayliffe, S. H., Lowbury, E. J. L., Hamilton, J. G., Small, J. M., Asheshov, E. A. and Parker, M. T., *Lancet*, Vol. 2, Aug. 21, 1965, pp. 365–369.
[43] Ross, R. T., *Developments in Industrial Microbiology*, Vol. 6, 1964, pp. 149–163.
[44] Stahl, W. H. and Pessen, H., *Applied Microbiology*, Vol. 1, No. 1, Jan. 1953, pp. 30–35.
[45] Bejuki, W. M., *Developments in Industrial Microbiology*, Vol. 2, 1961, pp. 263–270.
[46] Burgess, R. and Darby, A. E., *British Plastics*, Vol. 38, No. 3, March 1965, pp. 165–169.
[47] Poynter, S. F. B. and Mead, G. C., *Journal of Applied Bacteriology*, Vol. 27, No. 1, March 1964, pp. 182–195.
[48] Bouveng, H. O. Brenner, I., and Lindberg, B., *Acta Chemica Scandinavica*, Vol. 19, No. 4, 1965, pp. 1003–1004.
[49] Elrod, R. P. and Braun, A. C., *Science*, Vol. 94, No. 2448, Nov. 28 1941, pp. 520–521.
[50] Elrod, R. P. and Braun, A. C., *Journal of Bacteriology*, Vol. 44, No. 6, Dec. 1942, pp. 633–645.
[51] Hoff, J. C. and Drake, C. H., *Journal of Bacteriology*, Vol. 80, No. 3, Sept. 1960, pp. 420–4.
[52] Cho, J. J., Green, S. K., Schroth, M. N. and Kominos, S. D., *Phytopathology*, Vol. 63, No. 10, Oct. 1973, p. 1215 (Abstract).
[53] Green, S. K., Schroth, M. N., Cho, J. J., Kominos, S. D. and Vitanza-Jack, V. B., *Applied Microbiology*, Vol. 28, No. 6, Dec. 1974, pp. 987–991.
[54] Nemec, B., *Rozpravy Ceske Akademi Ved a Umeni*, Vol. 2, 1929, pp. 1–7.
[55] Desai, S. V., *Indian Journal of Agricultural Science*, Vol. 5, 1935, pp. 387–392.
[56] Duncan, J. T., *Parasitology*, Vol. 18, No. 2, June 1926, pp. 238–252.

[57] Bucher, G. E., *Journal of Insect Pathology*, Vol. 2, 1960, pp. 172–195.
[58] Bucher, G. E. in *Insect Pathology*, Vol. 2, E. A. Steinhaus, Ed., Academic Press, New York, 1963, pp. 117–147.
[59] Lysenko, O., *Journal of Insect Pathology*, Vol. 5, 1963, pp. 78–82.
[60] Lysenko, O., *Journal of Insect Pathology*, Vol. 5, 1963, pp. 83–88.
[61] Lysenko, O., *Journal of Insect Pathology*, Vol. 5, 1963, pp. 89–93.
[62] Lysenko, O., *Journal of Insect Pathology*, Vol. 5, 1963, pp. 94–97.
[63] Ashrafi, S. H., Zuberi, R. I. and Hafia, S., *Journal of Invertebrate Pathology*, Vol. 7, June 1965, pp. 189–191.
[64] Bucher, G. E. and Stephens, J. M., *Canadian Journal of Microbiology*, Vol. 3, No. 4, June 1957, pp. 611–625.
[65] Stephens, J. M., *Canadian Journal of Microbiology*, Vol. 3, No. 7, Dec. 1957, pp. 995–1000.
[66] Stephens, J. M., *Canadian Journal of Microbiology*, Vol. 5, No. 1, Feb. 1959, pp. 73–77.
[67] Larrivee, G. P. and Elvehjem, C. A., *Journal of the American Veterinary Medical Association*, Vol. 124, No. 927, June 1954, pp. 447–455.
[68] Hightower, D., Uhrig, H. T., and Davis, J. I., *Laboratory Animal Care*, Vol. 16, April 1966, pp. 85–92.
[69] Flynn, R. J., Ainsworth, E. J. and Greco, I., U. S. Atomic Energy Commission Report ANL-6368, 1960, pp. 35–37.
[70] Flynn, R. J. and Greco, I., U. S. Atomic Energy Commission Report ANL-6368, 1960, pp. 40–42.
[71] Thomson, J. K., Priestly, F. W. and Polding, J. B., *Veterinary Record*, Vol. 61, 1949, pp. 341–
[72] Keagy, H. F. and Keagy, E. H., *Journal of the American Veterinary Medical Association*, Vol. 118, No. 886, Jan. 1951, pp. 35–37.
[73] Knox, B., *Nordisk Veterinaermedicin*, Vol. 5, 1953, pp. 731–760.
[74] Newberne, P. M., *North American Veterinarian*, Vol. 34, 1953, pp. 187–188, 191.
[75] Farrell, R. K., Leader, R. W. and Gorham, J. R., *Cornell Veterinarian*, Vol. 48, No. 4, Oct. 1958, pp. 378–384.
[76] Beckenhauer, W. H. and Miner, C. A., *Veterinary Record*, Vol. 55, 1960, pp. 55–56.
[77] Trautwein, G., Helmboldt, C. F. and Nielsen, S. W., *Journal of the American Veterinary Medical Association*, Vol. 140, March 1962, pp. 701–704.
[78] McDonald, R. A. and Pinheiro, A. F., *Journal of the American Veterinary Medical Association*, Vol. 151, No. 7, Oct. 1967, pp. 863–864.
[79] Essex, H. E., McKenney, F. D. and Mann, F. C., *Journal of the American Veterinary Medical Association*, Vol. 77, No. 2, Aug. 1930, pp. 174–184.
[80] Stafseth, H. J., *Poultry Science*, Vol. 18, 1939, pp. 412.
[81] Chute, H. L., *Canadian Journal of Comparative Medicine and Veterinary Science*, Vol. 13, No. 5, May 1949, pp. 112–115.
[82] Dardiri, A. H., Zaki, O. and Reid, W. M., *Poultry Science*, Vol. 34, 1955, pp. 327–330.
[83] Bean, K. C. and MacLaury, D. W., *Poultry Science*, Vol. 38, 1959, pp. 693–698.
[84] Niilo, L., *Canadian Journal of Comparative Medicine and Veterinary Science*, Vol. 23, No. 10, Oct. 1959, pp. 329–330, 335–337.
[85] Saxena, S. P. and Sawhney, A. N., *Indian Journal of Veterinary and Animal Husbandry Research*, Vol. 5, 1960, pp. 59–60.
[86] Farrag, H. and Mahmoud, A. H., *Journal of the American Veterinary Medical Association*, Vol. 122, No. 910, Jan. 1953, pp. 35–36.
[87] Catcott, E. J. and Griesemer, R. A., *American Journal of Veterinary Research*, Vol. 15, No. 55, April 1954, pp. 261–265.
[88] Jones, W. G., *Journal of the American Veterinary Medical Association*, Vol. 127, No. 944, Nov. 1955, pp. 442–444.
[89] Ticer, J. W., *Journal of the American Veterinary Medical Association*, Vol. 146, No. 7, April 1 1965, pp. 720–722.
[90] Hewitt, A. C. T., *Journal of the Department of Agriculture, Victoria*, Vol. 48, 1950, pp. 13–14.
[91] Nilsson, P. O. and Thorne, H., *Nordisk Veterinaermedicin*, Vol. 14, 1962, pp. 538–546.
[92] Birch, R. R. and Benner, J. W., *Cornell Veterinarian*, Vol. 10, 1920, pp. 176–189.
[93] Morrill, C. C., *North American Veterinarian*, Vol. 24, 1943, pp. 606–607.

[*94*] Doll, E. R., Bruner, D. W. and Kinkaid, A. S., *Journal of the American Veterinary Medical Association*, Vol. 114, No. 866, May 1949, pp. 292.
[*95*] Bain, R. V. S. and Osborne, V. E., *Veterinary Journal*, Vol. 39, 1963, pp. 230–232.
[*96*] Pickens, E. M., Welsh, M. F. and Poelma, L. J., *Cornell Veterinarian*, Vol. 16, 1926, pp. 186–202.
[*97*] Cherrington, V. A. and Gildow, E. M., *Journal of the American Veterinary Medical Association*, Vol. 79, No. 6, Dec. 7 1931, pp. 803–808.
[*98*] Lovell, R. and Highes, D. L., *Journal of Comparative Pathology and Therapeutics*, Vol. 48, 1935, pp. 267–284.
[*99*] Cone, F. J., *Journal of Agricultural Research*, Vol. 58, No. 2, Jan. 1939, pp. 141–147.
[*100*] Gunsalus, I. C., Salisbury, G. W. and Willett, E. L., *Journal of Dairy Science*, Vol. 24, 1941, pp. 911–919.
[*101*] Gunsalus, I. C., Campbell, J. J. R., Beak, G. H. and Salisbury, G. W., *Journal of Dairy Science*, Vol. 27, 1944, pp. 357–364.
[*102*] Plastridge, W. N. and Williams, L. F., *Cornell Veterinarian*, Vol. 38, No. 2, April 1948, pp. 165–181.
[*103*] Rollinson, D. H. L., *Veterinary Journal*, Vol. 104, 1948, pp. 108–111.
[*104*] Schalm, O. W., *Cornell Veterinarian*, Vol. 38, No. 2, April 1948, pp. 186–189.
[*105*] Prince, P. W., Almquist, J. O. and Reid, J. J., *Journal of Dairy Science*, Vol. 32, 1949, pp. 849–855.
[*106*] Schalm, O. W., Bankowski, R. A., Ormsbee, R. W. and Browne, T. W., *American Journal of Veterinary Research*, Vol. 10, No. 34, Jan. 1949, pp. 56–62.
[*107*] Tucker, E. W., *Cornell Veterinarian*, Vol. 40, No. 1, Jan. 1950, pp. 95–96.
[*108*] Tanner, A., *Nordisk Veterinaermedicin*, Vol. 4, 1952, pp. 655.
[*109*] Schalm, O. W., *Journal of the American Veterinary Medical Association*, Vol. 122, No. 915, June 1953, pp. 462–467.
[*110*] Tucker, E. W., *Cornell Veterinarian*, Vol. 44, No. 1, Jan. 1954, pp. 110–124.
[*111*] Barnes, L. E., *American Journal of Veterinary Research*, Vol. 16, No. 60, July 1955, pp. 386–390.
[*112*] Winter, H. and O'Connor, R. F., *Australian Veterinary Journal*, Vol. 33, 1957, pp. 83–87.
[*113*] Mora, A. and Cavrini, C., *Nuova Veterinaria*, Vol. 15, 1959, pp. 112.
[*114*] Mura, D. and Pisanu, S., *Veterinaria Italiana*, Vol. 12, 1961, pp. 844.
[*115*] Heidrich, H. J. and Renk, W., *Krankheiten der Milchdruse bei Haustieren*, Parey, Berlin, 1963.
[*116*] Matthews, P. R. J. and Fitzsimmons, W. M., *Research in Veterinary Science*, Vol. 5, No. 2, 1964, pp. 171–174.
[*117*] Schalm, O. W., Lasmanis, J. and Carroll, E. J., *American Journal of Veterinary Research*, Vol. 28, No. 124, May 1967, pp. 697–707.
[*118*] Prasad, B. M., Srivastava, C. P., Narayan, K. G. and Prasad, A. K., *Indian Journal of Animal Health*, Vol. 7, No. 1, 1968, pp. 51–54.
[*119*] Forray, A. and Százados, I., *Magyar Allatorvosok Lapja*, 1968, 153.
[*120*] Curtis, P. E., *Veterinary Record*, Vol. 84, No. 19, May 1969, pp. 476–477.
[*121*] Fuchs, W. H., Hornich, M. and Walter, F., *Monatshefte für Veterinarmedizin*, 1969, 317.
[*122*] Howell, D., *Veterinary Record*, Vol. 90, No. 23, June 1972, pp. 654–657.
[*123*] Malmo, J., Robinson, B. and Morris, R. S., *American Veterinary Journal*, Vol. 48, 1972, pp. 137–139.
[*124*] Százados, I. and Kádas, I., *Acta Veterinaria Academiae Scientiarum Hungaricae*, Vol. 22, No. 3, 1972, pp. 241–249.
[*125*] Packer, R. A., *North American Veterinarian*, Vol. 28, No. 9, 1947, pp. 591–592.
[*126*] Schalm, O. W. and Lasmanis, J., *American Journal of Veterinary Research*, Vol. 18, No. 69, 1957, pp. 778–784.
[*127*] Hunter, C. A. and Ensign, P. R., *American Journal of Public Health*, Vol. 37, No. 9, Sept. 1947, pp. 1166–1169.
[*128*] Forkner, C. E., *Pseudomonas aeruginosa Infections*, Grune and Stratton, New York, 1960.
[*129*] Finland, M., Jones, W. F., Jr. and Barnes, M. W., *Journal of the American Medical Association*, Vol. 170, No. 18, Aug. 29, 1959, pp. 2188–2197.

[*130*] Asay, L. D. and Koch, R., *New England Journal of Medicine*, Vol. 262, No. 21, May 26, 1960, pp. 1062–1066.

[*131*] Alexander, J. W. in *Proceedings of the International Conference on Nosocomial Infections*, P. S. Brachman and T. C. Eichoff, Ed., American Hospital Association, Chicago, 1971, pp. 103–111.

[*132*] Data for period Jan. 1970-Dec. 1973 provided by the Hospital Infections Branch, Bacterial Diseases Division, Bureau of Epidemiology, Center for Disease Control, Atlanta, Georgia.

[*133*] Morehead, C. D. and Houck, P. W., *American Journal of Diseases of Children*, Vol. 124, No. 10, Oct. 1972, pp. 564–570.

[*134*] Schimpff, S. C., Greene, W. H., Young, V. M. and Wiernik, P. H., *European Journal of Cancer*, Vol. 9, No. 6, June 1973, pp. 449–455.

[*135*] Bennett, J. V., *Journal of Infectious Diseases*, Vol. 130, Supplement, No. 11, Nov. 1974, pp. S4–S7.

[*136*] Pruitt, B. A., Jr., *Journal of Infectious Diseases*, Vol. 130, Supplement, No. 11, Nov. 1974, pp. S8–S13.

[*137*] Lowbury, E. J. L., Thom, B. J., Lilly, H. A., *Journal of Medical Microbiology*, Vol. 3, Feb. 1970, pp. 39–56.

[*138*] Friedman, H. S. and Hinkel, C. L., *Archives of Otolaryngology*, Vol. 33, No. 5, May 1941, pp. 749–757.

[*139*] Singer, D. E., Freeman, E., Hoffert, W. R., Keys, R. J., Mitchell, R. B. and Hardy, A. V., *Annals of Otolaryngology, Rhinology, and Laryngology*, Vol. 61, No. 2, June 1952, pp. 317–330.

[*140*] Hardy, A. V., Mitchell, R. B., Schreiber, M., Hoffert, W. R., Yawn, E. and Young, F., *Laryngoscope*, Vol. 64, No. 12, Dec. 1954, pp. 1020–1024.

[*141*] Perry, E. T. and Nichols, A. C., *Journal of Investigative Dermatology*, Vol. 27, No. 3, Sept. 1956, pp. 165–170.

[*142*] Senturia, B. H. and Liebmann, F. M., *Journal of Investigative Dermatology*, Vol. 27, No. 5, Nov. 1956, pp. 291–317.

[*143*] Wright, D. N. and Alexander, J. M., *Archives of Otolaryngology*, Vol. 99, No. 1, Jan. 1974, pp. 15–18.

[*144*] Hoadley, A. W. and Knight, D. E., *Archives of Environmental Health*, Vol. 30, No. 9, Sept. 1975, pp. 445–448.

[*145*] Daggett, W. I., *Journal of Larynogology and Otology*, Vol. 57, No. 10, Oct. 1942, pp. 427–446.

[*146*] Birrell, J. F., *British Medical Journal*, Vol. 2, 21 July 1945, pp. 80–82.

[*147*] Gill, W. D. and Gill, E. K., *Southern Medical Journal*, Vol. 43, No. 5, May 1950, pp. 428–431.

[*148*] Alonso, M., *Laryngoscope*, Vol. 61, No. 11, Nov. 1951, pp. 1114–1122.

[*149*] Senturia, B. H., *Laryngoscope*, Vol. 55, No. 6, June 1945, pp. 277–293.

[*150*] McCausland, R. S. and Cox, P. J., *Journal of Environmental Health*, Vol. 37, No. 5, March/April 1975, pp. 455–459.

[*151*] Center for Disease Control, *Morbidity and Mortality Weekly Report*, Vol. 24, No. 19, May 10, 1975, pp. 166, 171.

[*152*] Center for Disease Control, *Morbidity and Mortality Weekly Report*, Vol. 24, No. 41, Oct. 11, 1975, pp. 349–350.

[*153*] Washburn, J., Jacobson, J. A. and Marston, E., *Journal of the American Medical Association*, Vol. 235, No. 20, 17 May 1976, pp. 2205–2207.

[*154*] Jacobson, J. A., Hoadley, A. W. and Farmer, J. J., III, *American Journal of Public Health*, Vol. 66, No. 11, Nov. 1976, pp. 1092–1093.

[*155*] Ringen, L. M. and Drake, C. H., *Journal of Bacteriology*, Vol. 64, No. 6, Dec. 1952, pp. 841–845.

[*156*] Burton, M. O., Campbell, J. and Eagles, B. A., *Canadian Journal of Research*, Vol. 26, No. 1, Feb. 1948, pp. 15–22.

[*157*] Reitler, R. and Seligmann, R., *Journal of Applied Bacteriology*, Vol. 20, No. 2, 1957, pp. 145–150.

[*158*] Schubert, R. H. W. and Blum, U., *Zentralblatt für Hygiene, Parasitenkunde, und Bakteriologic*, I Abteilung Originale, Reihe B, Vol. 158, 1974, pp. 583–587.

[*159*] Clark, J. A.' *Canadian Journal of Microbiology*, Vol. 15, No. 7, July 1969, pp. 771–780.

[160] Clark, J. A. and Vlassoff, L. T., *Health Laboratory Science*, Vol. 10, No. 3, July 1973, pp. 163–172.
[161] Drake, C. H., *Health Laboratory Science*, Vol. 3, No. 1, Jan. 1966, pp. 10–19.
[161] Favero, M. S., Carson, L. A., Bond, W. W. and Petersen, N. J., *Science*, Vol. 173, No. 3999, Aug. 1971, pp. 836–838.
[163] *Standard Methods for the Examination of Water and Wastewater*, 13th ed., American Public Health Association, New York, 1971.
[164] Highsmith, A. K. and Abshire, R. L., *Applied Microbiology*, Vol. 30, No. 4, Oct. 1975, pp. 596–601.
[165] Hoadley, A. W. and Ajello, G. W., *Canadian Journal of Microbiology*, Vol. 18, No. 11, 1972, pp. 1769–1773.
[166] Dutka, B. J. and Kwan, K. K., "Confirmation of the Single-Step Membrane Filtration Procedure for Estimating *Pseudomonas aeruginosa* Densities in Water," unpublished report.
[167] Hoadley, A. W., McCoy, E. and Rohlich, G. A., *Archiv für Hygiene und Bakteriologie*, Vol. 152, 1968, pp. 328–338.
[168] King, E. O., Ward, M. K. and Raney, D. E., *Journal of Laboratory and Clinical Medicine*, Vol. 44, No. 2, Aug. 1954, pp. 301–307.
[169] Robinton, E. D. and Burk, C. J., "The Mill River and its Floodplain in Northampton and Williamsburg, Massachusetts: A Study of the Vascular Plant Flora, Vegetation, and the Presence of the Bacterial Family Pseudomonadaceae in Relation to Patterns of Land Use," Completion No. 72-4, Water Resources Research Center, Univ. of Massachusetts, 1972.
[170] Mossel, D. A. A. and Indacochea, L., *Journal of Medical Microbiology*, Vol. 4, No. 3, Aug. 1971, pp. 380–382.
[171] Brown, V. I. and Lowbury, E. J. L., *Journal of Clinical Pathology*, Vol. 18, No. 6, Nov. 1965, pp. 752–756.
[172] Gundstrup, A. S. P., *Health Laboratory Science*, Vol. 11, No. 1, Jan. 1974, pp. 25–27.
[173] Grunnet, K., Gundstrup, A. S. P. and Bonde, G. J., *Nordisk Veterinaermedizin*, Vol. 26, 1974, pp. 239–242.
[174] Nemedi, L. and Lanyi, B., *Acta Microbiologica Academiae Scientia Hungaricae*, Vol. 18, 1971, pp. 319–326.
[175] Kenner, B. A., Dotson, G. K. and Smith, J. E., "Simultaneous Quantitation of Salmonella Species and *Pseudomonas aeruginosa*," U. S. Environmental Protection Agency, Cincinnati, Ohio, Sept. 1971.
[176] Kenner, B. A. and Clark, H. P., *Journal of the Water Pollution Control Federation*, Vol. 46, No. 9, Sept. 1974, pp. 2163–2171.
[177] Selenka, F., *Archiv für Hygiene und Bakteriologie*, Vol. 144, 1960, pp. 627–634.
[178] Hoadley, A. W. and McCoy, E., *Health Laboratory Science*, Vol. 3, No. 1, Jan. 1966, pp. 20–32.
[179] Hoadley, A. W. and Cheng, C. M., *Journal of Applied Bacteriology*, Vol. 37, No. 1, March 1974, pp. 45–57.
[180] Carson, L. A., Petersen, N. J., Favero, M. S., Doto, I. L., Collins, D. E. and Levin, M. A., *Applied Microbiology*, Vol. 30, No. 6, Dec. 1975, pp. 935–942.
[181] Carson, L. A., Favero, M. S., Bond, W. W. and Petersen, N. J., *Applied Microbiology*, Vol. 23, No. 5, May 1972, pp. 863–869.
[182] Brodsky, M. H. and Nixon, M. C., *Applied Microbiology*, Vol. 27, No. 5, May 1974, pp. 938–943.
[183] Burman, N. J., personal communication.
[184] Lantos, J., Kiss, M., Lanyi, B. and Volgyesi, J., *Acta Microbiologica Acadamiae Scientia Hungaricae*, Vol. 16, 1969, pp. 333–336.
[185] Levin, M. A. and Cabelli, V. J., *Applied Microbiology*, Vol. 24, No. 6, Dec. 1972, pp. 864–870.
[186] Sever, J. L., *Pediatrics*, Vol. 24, No. 1, July 1959, pp. 50–53.
[187] Whitby, J. L. and Rampling, A., *Lancet*, Vol. 1, No. 1, Jan. 1972, pp. 15–17.
[188] Rubbo, S. D., Gardner, J. F. and Franklin, J. C., *Journal of Hygiene*, Vol. 64, No. 1, March 1966, pp. 121–128.
[189] Lowbury, E. J. L., Thom, B. T., Lilly, H. A., Babb, J. R. and Whittall, K., *Journal of Medical Microbiology*, Vol. 3, No. 1, Feb. 1970, pp. 39–56.

[*190*] Barrie, D., *Archives of Diseases in Childhood*, Vol. 40, No. 213, Oct. 1965, pp. 555–558.

[*191*] Wilson, M. G., Nelson, R. C., Phillips, L. H. and Boak, R. A., *Journal of the American Medical Association*, Vol. 175, No. 13, April 1 1961, pp. 1146–1148.

[*192*] Cross, D. F., Benchimol, A., and Dimond, E. G., *New England Journal of Medicine*, Vol. 274, No. 25, June 23, 1966, pp. 1430–1431.

[*193*] Kohn, J. in *Research in Burns*, A. B. Wallace and A. W. Wilkinson, Ed., Livingstone, Edinburgh, 1966, pp. 486.

[*194*] Edmonds, P., Suskind, R. R., MacMillan, B. G. and Holder, I. A., *Applied Microbiology*, Vol. 24, No. 2, Aug. 1972, pp. 219–225.

[*195*] Moore, B. and Forman, A., *Lancet*, Vol. 2, Oct. 29, 1966, pp. 929–930.

[*196*] Fierer, J., Taylor, P. M. and Gezon, H. M., *New England Journal of Medicine*, Vol. 276, No. 18, May 4, 1967, pp. 991–996.

[*197*] Falcao, D. P., Mendonca, C. P., Scrassolo, A., deAlmeida, B. B., Hart, L., Farmer, L. H., and Farmer J. J., III, *Lancet* 2, No. 1, July 1, 1972, pp. 38–40.

[*198*] Bobo, R. A., Newton, E. J., Jones, L. F., Farmer, L. H., and Farmer, J. J. III, *Applied Microbiology*, Vol. 25, No. 3, Mar. 1973, pp. 414–420.

[*199*] Kresky, B., *American Journal of Diseases in Children*, Vol. 107, No. 4, April 1964, pp. 363–369.

[*200*] Edmondson, E. B., Rainarz, J. A., Pierce, A. K. and Sanford, J. P., *American Journal of Diseases of Children*, Vol. 111, No. 4, April 1966, pp. 357–360.

[*201*] Grieble, H. G., Colton, F. R., Bird, T. J., Toigo, A, and Griffith, L. G., *New England Journal of Medicine*, Vol. 282, No. 10, March 5, 1970, pp. 531–535.

[*202*] Favero, M. S., Carson, L. A., Bond, W. W. and Petersen, N. J., *Applied Microbiology*, Vol. 28, No. 5, Nov. 1974, pp. 822–830.

[*203*] Favero, M. S., Petersen, N. J., Carson, L. A., Bond, W. W. and Hindman, S. H., *Health Laboratory Science*, Vol. 12, No. 4, Oct. 1975, pp. 321–334.

[*204*] Kominos, S. D., Copeland, C. E., Grosiak, B. and Postic, B., *Applied Microbiology*, Vol. 24, No. 4, Oct. 1972, pp. 567–570.

[*205*] Shooter, R. A., Cooke, E. M., Gaya, H., Kumar, P., Patel, N., Parker, M. T., Thom, B. T. and France, D. R., *Lancet*, Vol. 2, June 21, 1969, pp. 1227–1229.

[*206*] Thomas, M. E. M., Piper, E., and Maurer, I. M., *Journal of Hygiene*, Vol. 70, No. 1, March 1972, pp. 63–73.

[*207*] Weber, G. Werner, H. P. and Matschnigg, H., *Zentralblatt für Bakteriologie Parasitenkunde Infektionskrankheiten und Hygiene*, 1 Abt. Orig. B, Vol. 216, No. 2, Feb. 1971, pp. 210–214.

[*208*] Lartigau, A. J., *Journal of Experimental Medicine*, Vol. 3, No. 6, Nov. 1898, pp. 595–609.

[*209*] Taylor, E. W., *The Examination of Waters and Water Supplies* (Thresh, Beale, and Suckling), 6th ed., Little Brown and Co., Boston, Massachusetts, 1958.

[*210*] Barnes, L. A., U. S. Naval Medical Bulletin, Vol. 43, 1944, pp. 707–716.

[*211*] Schiavone, E. L. and Passerini, L. M. D., *Sem Medicine*, Vol. 111, No. 23, 1958, pp. 1151–1157.

[*212*] Shubert, R. H. W. and Blum, U., *GWF-Wasser/Abwasser*, Vol. 115, No. 5, 1974, pp. 224–226.

[*213*] Lanyi, B., Gregacs, M. and Adam, M. M., *Acta Microbiologica Academiae Scientia Hungaricae*, Vol. 13, 1966, pp. 319–326.

[*214*] Lanyi, B., *Acta Microbiologica Academiae Scientia Hungaricae*, Vol. 13, 1966, pp. 295–318.

[*215*] Greer, F. E., Tenny, F. O. and Nyan, F. V., *Journal of Infectious Diseases*, Vol. 42, No. 6, June 1928, pp. 537–544.

[*216*] Martin, M. T., personal communication.

[*216a*] Geldreich, E. E., Nash, H. D., Reasoner, D. J. and Taylor, R. H., *Journal of the American Water Works Association*, Vol. 67, No. 3, March 1975, pp. 117–124.

[*217*] Buttiaux, R., *L'Analyse Bacteriologique des Eaux de Consommation*, Editions Medicales Flammarion, Paris, 1951.

[*218*] Lanyi, B., personal communication.

[*219*] Gyllenberg, H., Eklund, E., Antila, M. and Vartiovaara, U., *Acta Agriculturae Scandinavica*, Vol. 10, 1960, pp. 50–64.

[*220*] Olson, J. C., Jr., Parker, R. B. and Mueller, W. S., *Journal of Milk and Food Technology*, Vol. 18, No. 8, Aug. 1955, pp. 200–203.

[*221*] Harmon, L. G., *Journal of Milk and Food Technology*, Vol. 26, No. 3, March 1963, pp. 86–89.
[*222*] Corley, R. T., Long, H. F. and Hammer, B. W., *Canadian Dairy and Ice Cream Journal*, Vol. 23, 1944, pp. 31–33, 68.
[*223*] Harmon, L. G., *Journal of Milk and Food Technology*, Vol. 20, No. 7, July 1957, pp. 196–199.
[*224*] Hoadley, A. W. and McCoy, E., *Cornell Veterinarian*, Vol. 58, No. 3, July 1968, pp. 354–363.
[*225*] Mushin, R. and Ziv, G., *Journal of Hygiene*, Vol. 71, No. 1, March 1973, pp. 113–122.
[*226*] Senturia, B. H., *Diseases of the External Ear*, C. C. Thomas, Springfield, 1957.
[*227*] Cothran, W. W. and Hatlen, J. B., *Student Medicine*, Vol. 10, 1962, pp. 493–502.
[*228*] Favero, M. S., Drake, C. H. and Randall, G. B., *Public Health Reports*, Vol. 79, No. 1, Jan. 1964, pp. 61–70.
[*229*] Seyfried, P. L., *Canadian Journal of Public Health*, Vol. 65, No. 1, Jan-Feb 1974, p. 55.
[*230*] Hoadley, A. W., Ajello, G. and Masterson, N., *Applied Microbiology*, Vol. 29, No. 4, April 1975, pp. 527–531.
[*231*] Hojyo-Tomoka, M. T., Marples, R. R. and Kligman, A. M., *Archives of Dermatology*, Vol. 107, No. 5, May 1973, pp. 723–727.
[*232*] Keirn, M. A. and Putnam, H. D., *Health Laboratory Science*, Vol. 5, No. 3, July 1968, pp. 180–193.
[*233*] Black, A. P., Keirn, M. A., Smith, J. J., Dykes, G. M., Jr. and Harlan, W. E., *American Journal of Public Health*, Vol. 60, No. 4, April 1970, pp. 740–750.
[*234*] Botzenhart, K., Thofern, E. and Hünefeld, U., *Medizinische Monatsschrift*, Vol. 26, No. 8, Aug. 1972, pp. 364–368.
[*234a*] Botzenhart, K., Thofern, E. and Kulpmann, W. R., *Offentliche Gesundheitswesen*, Vol. 36, 1974, pp. 326–331.
[*235*] Balacescu, C. and Grun, L., *Zentralblatt für Babteriologie, Parasitenkunde, Infektionskrankheiten und Hygiene*, I Abt. Orig. B, Vol. 160, No. 3, May 1975, pp. 292–296.
[*236*] Kush, B. J. and Hoadley, A. W., unpublished report.
[*237*] Skadhauge, K. and Fogh, J., *Nordisk Hygienick Tidskrift*, Vol. 33, 1951–1952, pp. 255–286.
[*238*] Skadhauge, K. and Fogh, J., *Acta Pathologica et Microbiologica Scandinavica*, Vol. 32, 1953, pp. 290–299.
[*239*] Quisno, R. and Foter, M. J., *Journal of Bacteriology*, Vol. 52, No. 1, July 1946, pp. 111–117.
[*240*] Fitzgerald, G. P. and DerVartanian, M. E., *Applied Microbiology*, Vol. 17, No. 3, Mar. 1969, pp. 415=421.
[*241*] Adair, F. W., Geftic, S. G. and Gelzer, J., *Applied Microbiology*, Vol. 18, No. 3, Sept. 1969, pp. 299–302.
[*242*] Adair, F. W., Liauw, H. L., Geftic, S. G. and Gelzer, J., *Journal of Clinical Microbiology*, Vol. 1, No. 2, Feb. 1975, pp. 175–179.
[*242a*] Butterfield, C. T., Wattie, E. W., Megregian, S. and Chambers, C. W., *U.S. Public Health Reports*, Vol. 58, No. 51, Dec. 17 1943, pp. 1837–1866.
[*242b*] Butterfield, C. T. and Wattie, E., *U.S. Public Health Reports*, Vol. 61. No. 6, Feb. 8 1946, pp. 157–192.
[*243*] Fish, N. A., *Canadian Journal of Public Health*, Vol. 60, No. 7, July 1969, pp. 279–281.
[*244*] Iwata, K., personal communication.
[*245*] Houston, A. C., *Royal Commission on Sewage Disposal*, 4th Report, Vol. 3, 1904, pp. 176–177.
[*246*] Sreenivasan, A. and Venkataraman, R., *Journal of General Microbiology*, Vol. 15, No. 2, Oct. 1956, pp. 241–247.
[*247*] Denis, F. A., *Canadian Journal of Microbiology*, Vol. 21, No. 7, July 1975, pp. 1055–1057.
[*248*] Cabelli, V. J., Levin, M. A., Dufour, A. P. and McCabe, L. J., Paper No. 7 presented at "International Symposium on Discharge of Sewage from Sea Outfalls," London, August 1974.
[*249*] Taplin, D., unpublished report.
[*250*] Bonde, G., *Bacterial Indicators of Water Pollution*, Teknisk Forlag, Copenhagen, 1963.
[*251*] Hoadley, A. W., McCoy, E. and Rohlich, G. A., *Archiv für Hygiene und Bakteriologie*, Vol. 152, 1968, pp. 339–344.

[252] Botzenhart, K., Wolf, R. and Thofern, E., *Zentralblatt für Bakteriologie, Parasitenkunde, InfektionsKraukeiten und Hygiene*, I Abt Orig. B, Vol. 161, No. 1, Sept. 1975, pp. 72–83.
[253] Selenka, F. and Ruschke, R., *Archiv für Hygiene und Bakteriologie*, Vol. 149, No. 3/4, 1965, pp. 273–287.
[254] Dutka, B. J., personal communication.
[255] Hoadley, A. W. and McCoy, E., *Health Laboratory Science*, Vol. 3, No. 1, Jan. 1966, pp. 20–32.
[256] Mills, G. Y. and Kagan, B. M., *Annals of Internal Medicine*, Vol. 40, No. 1, Jan. 1954, pp. 26–32.
[257] Sutter, V. L., Hurst, V. and Lane, C. W., *Health Laboratory Science*, Vol. 4, No. 4, Oct. 1967, pp. 245–249.
[258] Smith, R. F., Dayton, S. L., Chipps, D. D. and Blasi, D., *Health Laboratory Science*, Vol. 10, No. 3, July 1973, pp. 173–179.
[259] Lowbury, E. J. L. and Fox, J., *Journal of Hygiene*, Vol. 52, No. 3, Sept. 1954, pp. 403–416.
[260] Stoodley, B. J. and Thom, B. T., *Journal of Medical Microbiology*, Vol. 3, No. 3, Aug. 1970, pp. 367–375.
[261] Grabow, W. O. K. and Nupen, E. M., *Water Research*, Vol. 6, No. 12, Dec. 1972, pp. 1557–1563.
[262] Mooney, D. B. and Hoadley, A. W., unpublished data.
[263] Lowbury, E. J. L., Kidson, A., Lilly, H. A., Ayliffe, G. A. J. and Jones, R. J., *Lancet*, Vol. 2, No. 7618, Aug. 30 1969, pp. 448–452.
[264] Greene, W. H., Moody, M., Schrimpff, S., Young, V. M. and Wiernik, P. H., *Annals of Internal Medicine*, Vol. 79, No. 5, Nov. 1973, pp. 684–689.
[265] Hoadley, A. W. and Ajello, G., unpublished results.
[266] Cabelli, V. J., Kennedy, H. and Levin, M. A., *Journal of the Water Pollution Control Federation*, Vol. 48, No. 2, Feb. 1976, pp. 367–376.
[267] Foster, D. H., personal communication.

R. R. Colwell[1] *and J. Kaper*[1]

Vibrio Species as Bacterial Indicators of Potential Health Hazards Associated With Water

REFERENCE: Colwell, R. R. and Kaper, J., ***Vibrio* Species as Bacterial Indicators of Potential Health Hazards Associated With Water,"** *Bacterial Indicators/Health Hazards Associated With Water, ASTM STP 635*, A. W. Hoadley and B. J. Dutka, Eds., American Society for Testing and Materials, 1977, pp. 115–125.

ABSTRACT: *Vibrio* species have been suggested as indicators of water quality. Methods for the isolation, identification, and enumeration of *V. cholerae* and *V. parahaemolyticus* (Biotype 1, parahaemolyticus, and Biotype 2, alginolyticus) have been developed and successfully applied in analyses of water. Recent isolations of *V. cholerae* from a harbor in the United States suggest that greater attention should be paid to the *Vibrio* species as waterborne pathogens. The identification and classification of *V. cholerae* and the two biotypes of *V. parahaemolyticus* have been greatly advanced. Other related *Vibrio* species remain to be described, and practical methods for their enumeration are not yet available. Because *Vibrio* species occur throughout the world in bays, estuaries, and marine waters, it is imperative that the ecology of this bacterial group be elucidated.

KEY WORDS: water, bacteria, coliform bacteria, bacterial indicators, health hazards

Vibrios commonly are found in estuarine, coastal, and deep ocean water. However, many freshwater *Vibrio* species also have been identified. The salt requiring and freshwater vibrios include species pathogenic for man, for example, *V. parahaemolyticus* and *V. cholerae*, respectively. Cholera has been recognized as a disease of man since the turn of the century, with the infectious agent entering water directly from the infected host or indirectly via wastewater from areas in which are found clinical cases of cholera, persons in the incubation stage, or healthy carriers. Transmission of *V. parahaemolyticus*, on the other hand, generally occurs via contaminated seafoods, with the reservoir of *V. parahaemolyticus* strongly suspected to be estuarine crustaceans [*1*].[2] Massive epidemics of cholera have been docu-

[1]Professor of Microbiology and faculty research assistant, Department of Microbiology, University of Maryland, College Park, Md. 20742.

[2]The italic numbers in brackets refer to the list of references appended to this paper.

mented, including the recent pandemic of the El Tor biotype of *V. cholera*, and a number of serious but more restricted outbreaks of *V. parahaemolyticus* have occurred since the organism was first isolated about 25 years ago. Thus, pathogenic species of vibrios for many years have been known to be potential health hazards, particularly in endemic areas.

Until recently, *Vibrio* species have not been routinely identified in clinical laboratories. However, evidence of *Vibrio* species associated with food poisoning outbreaks, wound infections, and otitis media have alerted clinicians to vibrios as potential causative agents of disease [*2*]. Thus, the potential health hazard of *Vibrio* species transmitted via the water route has been documented. What is now needed is extensive and reliable knowledge of the identification, classification, pathogenicity, and ecology of these microbial species.

Taxonomy

The type species for the genus *Vibrio* is *V. cholerae* (Pacini, 1854) which is described as follows: short, curved or straight rods, single or united into spirals, that grow well and rapidly on the surfaces of standard culture media, and are asporogenous and gram-negative, produce L-lysine and L-ornithine decarboxylases, but not L-arginine dihydrolase or hydrogen sulfide (Kligler iron agar). Guanine-plus-cytosine (G+C) overall base composition of the deoxyribonucleic acid (DNA) is approximately 48 plus or minus 1 percent. *V. cholerae* includes strains that may or may not elicit the cholera-red (nitroso-indole) reaction, may or may not be hemolytic, may or may not be agglutinated by Gardner and Venkatraman O group I antiserum, and may or may not be lysed by Mukerjee *V. cholerae* bacteriophages I, II, III, IV, and V [*3, 4*].

V. cholerae strains possess a common H antigen and can be serologically grouped into 39 serotypes, according to their O antigens [*5*]. The non-agglutinating cholera vibrios (NAG) of Heiberg groups I and II also have been shown to cause cholera-like disease, with the result that the World Health Organization has recognized the NAG vibrios as choleriform disease causing agents.

Vibrio species are facultatively anaerobic, with both a respiratory (oxygen utilizing) and a fermentative metabolism. Vibrios can be readily isolated from saltwater and freshwater samples but are often confused with species of other genera, such as *Aeromonas*, *Spirillum*, etc. Therefore, it is necessary to distinguish among the several species of *Vibrio* and related genera. From Tables 1 and 2, it can be seen that several physiological and biochemical taxonomic tests should be made if proper identification and classification of the vibrios is to be achieved.

In addition to the five established species of *Vibrio*, that is, *V. cholerae*, *V. parahaemolyticus, V. anguillarum, V. marinus* (also known as *V. fischeri*), and *V. costicola*, for which characteristics are listed in Table 1,

TABLE 1—*Features useful in differentiating and characterizing species of the genus* Vibrio.[a]

Characteristic	*V. cholerae*	*V. parahaemolyticus*	*V. anguillarum*	*V. marinus*	*V. costicolus*
Rod shape	+	+	+	+	+
Motility	+	+	+	+	+
Single polar flagellum	+	+	+	v	+
Lophotrichous flagella	−	−	−	v	−
Gram reaction	−	−	−	−	−
Diffusible pigment	−	−	−	−	−
Luminescence	−	−	−	−	−
Pathogenicity for man or animals	+	+	+	−	−
DNA base composition (% G+C)	46 to 49	44 to 46	44 to 45	40 to 44	50
Indole reaction	+	+	+	−	−
Methyl red reaction	+	+	+	+	−
Voges-Proskauer reaction	+	v	+	−	+
Citrate utilization	+	+	−	−	−
Citrulline utilization	−	−	+	−	nt
Sensitivity					
0/129	+	+	+	+	v
novobiocin	+	+	+	+	+
penicillin, 10 units	+	−	−	−	nt
polymyxin, 300 units	v	−	v	+	nt
streptomycin, 10 μg	+	−	−	+	nt
Growth at					
0% NaCl	+	−	+	+	−
1% NaCl	+	+	+	+	−
7% NaCl	v	+	v	+	+
10% NaCl	−	+	−	−	+
5°C	−	−	+	+	+
20°C	+	+	+	+	+
37°C	+	+	v	−	v
42°C	+	+	−	−	−
Acid production from					
arabinose	−	+	−	−	−
inositol	−	−	−	v	−
mannitol	+	+	+	+	v
mannose	+	+	+	+	+
salicin	−	−	−	v	−
sucrose	+	v	+	v	+
Gelatin liquefaction	+	+	+	+	v
Hydrolysis					
casein	+	+	+	+	−
starch	+	+	+	v	−
Tween 80	+	+	+	+	+
H_2S production (on lead acetate agar)	−	−	−	−	−
Lecithinase (egg yolk)	+	+	nt	nt	v
Arginine dihydrolase	−	−	+	−	+
Lysine decarboxylase	+	+	−	+	−
Ornithine decarboxylase	+	+	−	−	−
Hemolysis	+	+	v	−	−

NOTE— + = positive or present;
− = negative or absent;
v = reaction varied among the strains tested;
nt = not tested.

[a]Colwell, R. R., *Handbook of Microbiology*, CRC Press Inc., Cleveland, Ohio, 1976, p. 99.

TABLE 2—*Differentiation of related genera frequently isolated from the same source in nature.*[a]

Characteristic	*Vibrio*	*Aeromonas*	*Plesiomonas*	*Photobacterium*	*Lucibacterium*	*Pseudomonas*	*Spirillum*	*Campylobacter*
Morphology	straight or curved rod	straight rod	straight rod	straight rod	straight or curved rod	straight rod	helical	spirally curved rod
Diffusible pigment	none	none[b]	none	none	none	none or green fluorescent	none or green fluorescent	none
Motility	+	+	+	+	+	+	+	+
Flagella	polar	polar	lophotrichous	polar	peritrichous (usually)	polar	lophotrichous	polar
Carbohydrate metabolism	fermentative	fermentative	fermentative	fermentative	fermentative	respiratory or not metabolized	respiratory	not metabolized
Gas production from carbohydrates	–	v	–	+	–	–	–	–
Luminescence	v	–	–	+	+	–	–	–
Oxidase	+	+	+	v	+	+	+	+
0/129 sensitivity	+	–	+	+	–	–	–	–
"Round bodies" or "cysts" produced	+	–	–	–	–	–	+	+

NOTE—+ = positive or present; – = negative or absent; v = variable.

[a]From Colwell, R. R., *Handbook of Microbiology*, CRC Press, Cleveland, Ohio. 1976, p. 102.

[b]Species of *Aeromonas* may produce a brown pigment.

other species epithets can be found in the literature. *V. alginolyticus* has been given separate species status by many workers, but it is designated as Biotype 2 of *V. parahaemolyticus* in the Eighth Edition of *Bergey's Manual of Determinative Bacteriology* [*3*]. The genus *Beneckea* was reported by Baumann et al [*6*] to contain vibrio-like organisms with peritrichous flagella. However, the genus and species listed by Baumann et al [*6*] are regarded to be of uncertain taxonomic position [*3*] and are not recognized by investigators most familiar with the genus *Vibrio*.[3]

Isolation

It is important to note that NAG vibrios, as well as agglutinable *V. cholerae*, may be discharged in the stool of diarrheic patients, and these vibrios can survive in water for weeks, if the water is of appropriate salt concentration and alkaline pH. Such conditions permit survival of vibrios in the environment for relatively long periods of time. Survival and distribution of *Vibrio* species has been studied in our laboratory, where data for several sites in the Chesapeake Bay (a major estuary of the East Coast of the United States) have been gathered for more than a decade. Large volumes of water from several sites, including the Jones Falls area in the northwest branch of Baltimore Harbor have been examined for the presence of *Vibrio* species including *V. cholerae*. In the studies accomplished to date, samples were collected by submersible pump, with the intake hose of the pump situated at a depth of approximately 3 m below the surface. The water samples were held, after collection, in sterile Nalgene carboys and were processed either immediately aboard ship or back at the laboratory within 1 h of sample collection. Celite, a filtering adjuvant prepared in stock solution and autoclaved at 15 lb pressure for 15 min, was added to the water samples to a final concentration of 1 g/litre. After thorough agitation, 1 litre volumes of water were filtered through a 0.45 μm filter, after which the filters and the accumulated Celite were placed into flasks containing 50 ml of enrichment broth.

Enrichment media used for isolation of *V. cholerae* from estuarine water samples were (*a*) alkaline peptone broth (10 g/1 peptone,[4] 10 g/1 sodium chloride (NaCl), 8.4 pH), and (*b*) taurocholate tellurite peptone water (10 g/1 Tryptone digest,[4] 10 g/1 NaCl, 5 g/1 sodium taurocholate, 1 g/1 sodium carbonate (Na_2CO_3), and 0.001 percent potassium tellurite, 9.2 pH). After inoculation, the enrichment flasks were incubated, using a step-wise procedure which proved useful in isolation of *Salmonella* species from estuarine samples [*7*] and involved primary incubation at ambient temperature (about 25°C) for 4 to 6 h, followed by 18 h incubation at 37°C, yielding a combined incubation period of approximately 24 h. This step-wise procedure was found to ease the transition of stressed bacteria from the natural environment to the artificial environment of growth media prepared in the laboratory.

[3]Sakazaki, R., personal communication.
[4]Difco Laboratories, Detroit, Mich.

Following incubation for 24 h in enrichment broth, the cultures were streaked onto TCBS agar.[4] Colonies demonstrating a yellow color typical of *V. cholerae*, or blue-green color typical of *Vibrio parahaemolyticus*, were screened by testing for presence of oxidase. Those colonies that were oxidase positive were further characterized by a series of biochemical tests or by the Analytab Products, Inc. (API) 20 system of Plainview, N. Y., and were identified to the species level using the API 20 enteric computer profile. Cultures identified as *V. cholerae* were sent to the Center for Disease Control (CDC), Atlanta, Ga. and the Vibrio Reference Laboratory (VRL), Jefferson Medical College, Philadelphia, Pa., where serological testing was done.

In a recent study, a total of 13 isolates (all of which were isolated from the alkaline peptone broth medium) were identified as *V. cholerae* by the API computer profile index. Serological examination of the isolates by the CDC[5] confirmed identification of the organisms as NAG *V. cholerae*, that is, *V. cholerae* strains not agglutinating in cholerae O group 1 antiserum. Selected biochemical characteristics of the 13 isolates are listed in Table 3. Serotyping provided by the VRL[6] is summarized in Table 4.

NAG vibrios are divided into six groups, following the Heiberg scheme. These groups are based upon ability to ferment arabinose, mannose, and sucrose. As shown in Table 3, strains V11 through V29 were sucrose negative, arabinose negative, and mannose positive, thereby identified as Heiberg group V and strains V3, V4, and V5, sucrose and mannose positive, and arabinose negative, as Heiberg group I. Strains V2 and V10 were sucrose positive, arabinose negative, and mannose negative as were Heiberg group II (see Table 3).

Pathogenicity of NAG vibrios has been recognized only recently. These organisms can cause cholera-like diarrhea and have been isolated from many outbreaks of diarrheal disease [*8*]. Microbial taxonomists include NAG vibrios within the taxospecies and genospecies *V. cholerae* [*3*], based on numerical taxonomy, overall G + C composition of the DNA, and DNA homology [*9-11*]. Lack of an antigen appears to be the only criterion distinguishing NAG vibrios from agglutinable strains of *V. cholerae*.

Discovery of NAG vibrios in brackish water, estuaries, and coastal areas of the United States may be quite unexpected and perhaps difficult to accept by epidemiologists and ecologists. However, upon reflection, this discovery should not be surprising. As just stated, NAG vibrios can survive in water for weeks, given proper environmental conditions. A shallow, slightly brackish (about 4 parts per thousand salinity) area of an estuary that receives a constant influx of domestic sewage effluent, as well as pollution from an international armada of oceangoing liners, tankers, and freighters should be a likely site for isolation of these organisms. The Jones Falls area in Baltimore Harbor is such a site, and the water at Jones Falls, which contains total hetero-

[5]Feeley, J. and DeWitt, W., personal communication.
[6]Smith, H. L., Jr., personal communication.

TABLE 3—*Biochemical characteristics of NAG vibrios isolated from estuarine water samples.*

Strain No.	ONPG	Arginine	Lysine	Ornithine	Citrate	H_2S	Urea	Tryptophan Deaminase	Indole	VP	Gelatin	Glucose	Mannitol	Inositol	Sorbitol	Rhamnose	Sucrose	Melibiose	Amygdalin	Arabinose	Mannose	Oxidase	0% NaCl	6% NaCl	String Test	Heiberg Group
V2	+	−	+	+	+	−	−	−	+	−	+	+	+	(+)	−	−	+	−	(+)	−	−	+	+	−	+	II
V3	+	−	+	+	+	−	−	−	+	−	+	+	+	−	+	−	+	−	−	−	+	+	+	−	+	I
V4	+	−	+	+	+	−	−	−	+	−	+	+	+	−	+	−	+	−	−	−	+	+	+	−	+	I
V5	+	−	+	+	+	−	−	−	+	−	+	+	+	−	−	−	+	−	−	−	+	+	+	−	+	I
V10	+	−	+	+	+	−	−	−	+	−	+	+	+	−	−	−	+	−	−	−	−	+	+	−	+	II
V11	+	−	+	+	+	−	−	−	+	−	+	+	+	−	−	−	−	−	−	−	+	+	+	−	+	V
V15	+	−	+	+	+	−	−	−	+	−	+	+	+	−	−	−	−	−	−	−	+	+	+	−	+	V
V19	+	−	+	+	+	−	−	−	+	−	+	+	−	−	−	−	−	−	−	−	+	+	+	−	+	V
V20	+	−	+	+	+	−	−	−	+	−	+	+	−	−	−	−	−	−	−	−	+	+	+	−	+	V
V24	+	−	+	+	+	−	−	−	+	−	+	+	+	−	−	−	−	−	−	−	+	+	+	−	+	V
V25	+	−	+	+	+	−	−	−	+	−	+	+	+	−	−	−	−	−	−	−	+	+	+	−	+	V
V26	+	−	+	+	+	−	−	−	+	−	+	+	+	−	−	−	−	−	−	−	+	+	+	−	+	V
V29	+	−	+	+	+	−	−	−	+	−	+	+	+	−	−	−	−	−	−	−	+	+	+	−	+	V

TABLE 4—*Serotype of* V. cholerae *isolated from Chesapeake Bay.*[a]

University of Maryland Strain No.	Vibrio Reference Laboratory No.	Serotype[b]
V3	9144	a-I
V4	9145	a-I-24
V5	9146	a-I-23
V10	9147	a-II-14
V11	9148	a-v
V15	9149	a-v
V19	9150	a-v
V20	9151	a-v
V24	9152	a-v
V25	9153	a-v
V26	9154	a-v
V29	9155	a-v

[a]Serotyping reported here was courtesy of Dr. H. L. Smith, Jr., Vibrio Reference Laboratory, Jefferson Medical College, Philadelphia, Pa.

[b]a = Gelatin positive, indole positive, nitrate positive; roman numerals indicate the Heiberg fermentation pattern.

trophic aerobic viable bacterial counts of 10^5/ml, fecal coliform most probable number (MPN) greater than 2400/100 ml, and *Salmonella* species [*12*, *7*] now has been found to contain NAG vibrios. The serotypes isolated from the Chesapeake Bay samples are of widespread geographical occurrence (Table 5).

It is interesting to note that only the alkaline peptone broth, a nonselective enrichment medium, yielded NAG vibrio isolates. The medium contains no selective agents and has an alkaline pH. However, the alkalinity is not great enough to be excessively inhibitory. The taurocholate tellurite peptone medium proved unsatisfactory for isolation of NAG vibrios. Successful recovery of *Salmonella* species has been obtained when a nonselective enrichment medium was employed, the nonselective enrichment acting, most probably, to "resuscitate" environmentally stressed cells [*12*]. Mossell et al [*13*] emphasized the importance of resuscitating environmentally stressed strains

TABLE 5—*Source of other strains received at the Vibrio Reference Laboratory (VRL) of serotypes similar to the Chesapeake Bay strains.*[a]

Classification According to Serotype	Number of Strains Received at VRL	Geographical Source of the Strains	Type of Specimen from Which Strains were Isolated
a-II-14	45	Maryland; Bulgaria; Iraq; Czechoslovakia; Japan; Philippines; Bangladesh; Thailand	brackish water; diarrhea; night soil; freshwater; animal feces; patient contact
a-I-23	10	Philippines; Sudan; Oman; Bangladesh	shellfish; diarrhea; animal feces
a-I-24	8	Philippines; Bangladesh; Thailand	freshwater; animal feces; diarrhea

[a]Courtesy of Dr. H. L. Smith, Jr., Vibrio Reference Laboratory, Jefferson Medical College, Philadelphia, Pa.

of Enterobacteriaceae before transfer to a selective medium. Additional evidence supporting this view was provided by Bissonnette et al [*14*], who demonstrated that environmentally stressed strains of *Escherichia coli* could be revived by inoculation into a nonselective medium. It is clear, from the data available in the literature and from that reported here, that methods for isolation of *Vibrio* and related species from the estuarine and marine system must be designed carefully to take into account adaptations, or lack of adaptations, made by the microorganisms to the environment. Thus, either new techniques or modifications of methods routinely used in clinical laboratories must be developed. Nonselective enrichment by preincubation of samples at environmental temperatures is far more satisfactory for isolation of vibrios than stringent conditions of selective media and elevated temperature in the case of samples collected from the natural environment.

Other *Vibrio* species besides *V. cholerae* and NAG vibrios should be considered when samples collected from the aquatic environment are examined. *V. parahaemolyticus* is one such species that is regarded as a potential human pathogen. However, the persistence of this organism in the estuarine environment is not dependent upon proximity of a human host, since the organism multiplies in nature and appears to perform an ecologically significant role [*10*]. Identifying characteristics of *V. parahaemolyticus* and related species are given in Table 1.

Enumeration

Accurate enumeration of *Vibrio* species in samples collected in the natural environment is complicated by the need for enrichment prior to enumeration. The requirement for enrichment necessitates the use of the MPN procedure, similar to that used in the enumeration of coliforms. Several MPN enrichment broths have been developed and applied in field studies. However, it is unfortunate that there is no standard procedure or uniformity of methodology yet agreed upon, in contrast to the standard methods applied to the procedures used for enumeration of coliforms and fecal coliforms. In the recent study of the Upper Chesapeake Bay detailed previously, a MPN of vibrios procedure was utilized to enumerate *V. cholerae* in Baltimore Harbor and other estuarine water samples. A three-dilution, five-tube replication series was used, in which 1 litre and 100-ml volumes of water were filtered. Procedures outlined previously for incubation of the samples were followed, with the filters placed in alkaline peptone broth after filtration was completed. A third dilution was obtained by inoculation of a 10-ml sample into 10-ml double-strength alkaline peptone broth. All samples were incubated and pure cultures isolated and characterized as described previously.

The MPN of vibrios index for *V. cholerae* in water collected at the Jones Falls area of Baltimore Harbor was 3.3 organisms per litre (95 percent confidence limits of 1.1 to 9.3). The MPN for *Salmonella* species in samples collected in the same area were reported to be 110 organisms/100 g of sediment and fecal coliforms, 24 000/100 ml [*7*].

Research on alternative approaches to the MPN procedure has been underway in several laboratories, and, recently, a new method was reported [*15*]. A membrane filtration procedure for the enumeration of *V. parahaemolyticus* was developed and is very similar to the membrane filtration procedure routinely employed for coliform enumeration. A newly devised, selective primary medium, coupled with elevated temperature of incubation and a series of *in situ* biochemical tests selective for *V. parahaemolyticus*, has proved successful for examination of environmental samples. The procedure of Watkins et al [*15*] can be carried out within 30 h with 95 percent accuracy. When compared with other methods for enumeration, the membrane filter procedure yielded consistently higher recoveries of *V. parahaemolyticus*. Thus, enumeration of *V. parahaemolyticus* may be efficiently and accurately accomplished at the present time with the methods available.

Significance

The significance of *Vibrio* species as bacterial indicators of potential health hazards associated with water is becoming increasingly evident. The two biotypes of *V. parahaemolyticus* (parahaemolyticus and alginolyticus) have been isolated from artesian waters in Florida [*16*] and estuarine waters within 500 miles of the Arctic circle [*17*]. There is an increasing interest in *V. parahaemolyticus* and related vibrios, not only in the United States but also in the Netherlands, India, Japan, and other countries since the organism is present in significant numbers in nutrient enriched waters. Since the ecology of *Vibrio* species, with the exception of *V. parahaemolyticus*, is poorly understood, their significance as indicators of water quality similarly remains to be more fully clarified. The primary indicator organism of water quality for decades has been *E. coli*, the presence of which denotes fecal contamination and the presence of potentially pathogenic bacteria. In contrast to the established role of *E. coli* as an indicator species, vibrios, including *V. cholerae*, *V. parahaemolyticus* (both biotypes), and *V. anguillarum*, have been demonstrated to be pathogens and are associated with food-borne disease and wound, ear, and eye infections of individuals in contact with marine or estuarine waters [*18*]. However, the significance of large numbers of *V. parahaemolyticus*, for example, in an embayment or estuary, has yet to be determined a public health hazard. Fortunately, the occurrence of large numbers of *V. parahaemolyticus* in shellfish destined for human consumption is now unequivocably recognized as a public health threat. It is yet to be proven that there is a correlation between standard pollution indexes and numbers of *Vibrio* species in water or food. In fact, it has been shown that no relationship exists between the incidence of *V. parahaemolyticus* and high counts of that species and *E. coli* counts in estuaries [*1*, *19*]. Biotype 2 of *V. parahaemolyticus* may prove to be a more useful standard for water quality, since it generally occurs in greater numbers in natural waters than *V. parahaemolyticus*. A proportional relationship between the numbers of both biotypes of *V. parahaemolyticus* has been suggested.[7] In summary, it remains to be deter-

[7]Baross, J., personal communication.

mined whether *Vibrio* species can serve as a water quality indicator, although the available evidence clearly points in that direction.

Conclusion

In conclusion, vibrios are a natural component of the microbial ecology of freshwater and saltwater systems, and the presence of specific human pathogenic species of *Vibrio* can serve as an indicator of the public health safety of water or of food destined for human consumption. However, methods for enumeration, isolation, identification, and classification of *Vibrio* species remain relatively cumbersome compared to tests for *E. coli*, type I, for example. Sampling methods used for isolation of *Vibrio* species include membrane filtration, gauze pad "traps," and collection with grab devices. But none of these has been perfected yet to the point of statistical reliability. Work presently underway in our laboratory and in the laboratories of other investigators is aimed at the development of a "minimum plexus" of unit characters for rapid identification of *Vibrio* species in order that the major hurdle in the employment of vibrios as an indicator group and of *Vibrio* species selected as indicator organisms may be overcome. If successful employment of *Vibrio* species as indicators of water quality can be achieved, evaluations of brackish, estuarine, and coastal aquatic environments thereby will be greatly improved.

References

[*1*] Kaneko, T. and Colwell, R. R., *Applied Microbiology*, Vol. 30, 1975, pp. 251-257.
[*2*] Rubin, S. J. and Tilton, R. C., *Journal of Clinical Microbiology*, Vol 2, 1975, pp. 556-558.
[*3*] Buchanan, R. E. and Gibbons, N. E., Eds., *Bergey's Manual of Determinative Bacteriology*, 8th ed., Williams and Wilkins, Baltimore, Md., 1974.
[*4*] Sebald, M. and Véron M., *Annales De L'Institut Pasteur De Lille*, Vol. 105, 1963, p. 897.
[*5*] Sakazaki, R., Kazunichi, T., Gomez, C. Z., and Sen, R., *Japanese Journal of Medical Science and Biology*, Vol. 23, 1970, p. 23.
[*6*] Baumann, P., Baumann, L., and Mandel, M., *Journal of Bacteriology*, Vol. 107, 1971, pp. 268–294.
[*7*] Kaper, J., Sayler, G. S., Baldini, M. M., and Colwell, R. R., *Applied and Environmental Microbiology*, Vol. 33, 1977, pp. 829–835.
[*8*] Finkelstein, R. A., *Chemical Rubber Company Critical Reviews in Microbiology*, Vol. 2, 1973, pp. 553–623.
[*9*] Citarella, R. V. and Colwell, R. R., *Journal of Bacteriology*, Vol. 104, 1970, pp. 434–442.
[*10*] Colwell, R. R., *Journal of Bacteriology*, Vol. 104, 1970, pp. 410–433.
[*11*] Sakazaki, R., Gomez, C. Z., and Sebald, M., *Japanese Journal of Medical Science and Biology*, Vol. 20, 1967, pp. 265–280.
[*12*] Sayler, G. S., Nelson, J. D., Jr., Justice, A., and Colwell, R. R., *Applied and Environmental Microbiology*, Vol. 31, 1976, pp. 723–730.
[*13*] Mossel, D. A. A., Harrewijn, G. A., and Nesselrooy-Yan-Zadelhoff, C. F. M., *Public Health Laboratory*, Vol. 11, 1974, pp. 260–267.
[*14*] Bissonnette, G. K., Jezeski, J. J., McFeters, G. A., and Stuart, D. G., *Applied Microbiology*, Vol. 29, 1975, pp. 186–194.
[*15*] Watkins, W. D., Thomas, C. D., and Cabelli, V. J., *Applied and Environmental Microbiology*, Vol. 32, 1976, pp. 679–684.
[*16*] Koburger, J. A. and Lazarus, C. R., *Applied Microbiology*, Vol. 27, 1974, pp. 435–436.
[*17*] Vasconcelos, G. J., Stang, W. J., and Laidlaw, R. H., *Applied Microbiology*, Vol. 29, 1975, pp. 557–559.
[*18*] Golten, C. and Scheffers, W. A., *Netherlands Journal of Sea Research*, Vol. 9, 1975, pp. 351-364.
[*19*] Thompson, C. A. and Vanderzant, C., *Journal of Food Science*, Vol. 41, 1976, pp. 117–122.

J. B. Evans[1]

Coagulase Positive Staphylococci as Indicators of Potential Health Hazards From Water

REFERENCE: Evans, J. B., **"Coagulase Positive Staphylococci as Indicators of Potential Health Hazards From Water,"** *Bacterial Indicators/Health Hazards Associated With Water, ASTM STP 635*, A. W. Hoadley and B. J. Dutka, Eds., American Society for Testing and Materials, 1977, pp. 126–130.

ABSTRACT: Coagulase-positive staphylococci are synonymous with the species *Staphylococcus aureus*. They are normal inhabitants of the nose, skin, and intestinal tract of healthy individuals as well as a cause of purulent infections of the skin, eyes, and ears. They are a major component of the bacterial flora of swimming pools and recreational waters that have a high density of swimmers and are potentially a good index of the health hazard associated with bathing in such waters. The major obstacle to their use as an indicator organism is the lack of a medium and method that is sufficiently selective and reliable for the quantitative recovery of these bacteria from aquatic habitats.

KEY WORDS: bacteria, water, coliform bacteria, staphylococci

The coagulase positive staphylococci are essentially synonymous with the well-defined and extensively studied species *Staphylococcus aureus*, a more precise and less cumbersome name for these bacteria. As members of the genus *Staphylococcus*, they are gram-positive, catalase positive cocci that ferment glucose and can grow both aerobically and anaerobically. Other criteria, of course, are included in a more complete taxonomic description. The species *S. aureus* is diagnostically identified by its ability to coagulate rabbit plasma in the presence of the usual anticoagulating agents. Other important diagnostic criteria include the anaerobic fermentation of mannitol and the production of a heat stable nuclease.

S. aureus is a major pathogen, and all strains are considered to be potentially pathogenic. Other species of staphylococci only rarely cause disease or infections. *S. aureus* produces a wide array of toxins and is responsible for most purulent cutaneous infections such as boils and infected cuts and scratches. Some particularly virulent strains cause eruptions on the unbroken skin.

[1]Head, Department of Microbiology, North Carolina State University, Raleigh, N. C. 27607.

Others cause deep infections such as osteomyelitis and postoperative infections. One of the toxins produced by many strains of *S. aureus* is responsible for the majority of cases of bacterial food poisoning. Paradoxically, the extensive disease producing capability of this species is accompanied by its ability to coexist harmlessly on our skin and in our environment. It is present, often in fairly large populations, in many of the foods that we consume, our nose and intestinal tract, our skin, and the air of occupied rooms.

Fortunately, *S. aureus* is not found in large numbers in water and is unlikely to multiply in most waters. It requires a considerable array of organic nutrients and relatively warm temperatures, growing slowly at temperatures below 20°C and not at all at temperatures below 10°C. However, it is well adapted to survival outside the human host, being quite resistant to sunlight, drying, or salinity. Its ability to grow readily in media with up to 10 percent salt concentration provides one of the major selective factors in media designed for its selective enumeration.

The major significance of *S. aureus* in water is under circumstances where it might infect cuts and scratches on the skin, infect the ears or the eyes of bathers, or be used in the preparation of foods in which it might multiply extensively. Therefore, it is an indicator of potential importance in recreational waters, swimming pools, water that might be added to foods, and special problem areas such as hydrotherapy pools.

In waters where we are concerned with the possible infection of the skin, eyes, and ears by *S. aureus* from other users of the same body of water, it is logical to assume that the degree of risk is directly correlated with the population density of *S. aureus* in the water. However, there is a lack of convincing data to substantiate this supposition. Robinton and Mood [*1*][2] using the membrane filter technique and staphylococcus medium 110, found staphylococci to be the most numerous bacteria shed by swimmers, and *S. aureus* was consistently present. They suggested staphylococci as a good choice for an index of body contamination but presented insufficient data to correlate the numbers found with bathing load. Favero and Drake [*2*] found staphylococci and bacilli to be the predominant bacteria in chlorinated swimming pools, and about half of the staphylococci were *S. aureus*. In iodinated water, about the same number of staphylococci were present, but much larger numbers of pseudomonads were found and became the primary flora. Keirn and Putnam [*3*] used M-staphylococcus broth with membrane filters, and their results closely paralleled with those of Favero and Drake. Neither group established a correlation between bacterial population and bathing load. Palmquist and Jankow [*4*] reported that some pools that were considered safe on the bases of coliform counts and chlorine levels were causing a high incidence of otitis externa and suggested *S. aureus* and *Pseudomonas* species as indicators of proper disinfection, but supporting data were not reported. Boccia and Montanaro [*5*] also supported the value of staphylococci as indicators of

[2]The italic numbers in brackets refer to the list of references appended to this paper.

the hygienic quality of swimming pool water. Perhaps the best study of this problem is summarized in the report by Crone and Tee [*6*]. They examined 1192 samples from 89 different swimming pools using the membrane filter technique and staphylococcus medium 110 and found no correlation between numbers of staphylococci and either bathing load or concentration of free chlorine. However, staphylococci were found more frequently than were coliform bacteria, and approximately two thirds of the staphylococci proved to be *S. aureus*. They concluded that the presence of any staphylococci in a 100-ml sample from a pool that had been out of use for at least 10 h was an indication of inadequate filtration and chlorination. The staphylococci in these waters are shed from the skin, nose, and intestinal tract by bathers in the water. It should be noted that most of the foregoing reports have stressed total staphylococci rather than *S. aureus*, and there is reason to expect that total staphylococci may be a better indicator of human contamination. Kloos and Musselwhite [*7*] have shown that *S. aureus* and *S. epidermidis* are the predominant staphylococci in the nose and *S. epidermidis* is the predominant species on the skin.

Perhaps the biggest problem in the evaluation or use of either *S. aureus* or of total staphylococci as indicators of a health hazard in water is the lack of information on the effectiveness of media and methods employed for their selective enumeration. As noted previously, only a very limited number of studies on staphylococci in water have been published, and these studies were not designed to evaluate media or methods. Most recent investigators have used the membrane filter technique with M-staphylococcus broth [*8*]. This medium which is staphylococcus medium 110 without the gelatin or agar is neither sufficiently differential nor selective for quantitative recovery of staphylococci, especially *S. aureus* [*9*].[3] The most recent edition of *Standard Methods* [*10*] states that enumeration of total staphylococci and of *S. aureus* is useful in evaluating water quality in swimming pools. It states, however, that no method for *S. aureus* has been adequately tested, and they recommend the use of Chapman-Stone agar with a membrane filter technique for determining total staphylococci. To my knowledge, there are no published reports of a critical evaluation of this medium for this purpose. It is recommended that mucoid and rough colonies are not to be counted because they are presumed to be bacilli. Experience with this medium suggests that such a differentiation on the basis of colony morphology is likely to be inaccurate, and the differential value of gelatin hydrolysis is questionable, at best, since bacilli are more likely to hydrolyze gelatin than are staphylococci.

In a paper presented by Olivieri and Riggio at an Environmental Protection Agency (EPA) workshop on microorganisms in stormwater, these workers reported that less than 10 percent of the typical colonies on M-staphylococcus medium proved to be *S. aureus*, and most such colonies proved to be gram-positive or gram-negative rod-shaped bacteria. They recommend the addition

[3]Brodsky, M. H., personal communication.

of 0.75 m*M* sodium azide to M-staphylococcus broth and use of it in a multiple tube dilution procedure. Addition of this level of sodium azide to staphylococcus medium 110 was originally recommended by Smuckler and Appleman [*11*] to inhibit growth of rod-shaped bacteria.

Food microbiologists have carried out extensive comparative studies of media for the enumeration of *S. aureus* from foods. Invariably they seem to conclude that Baird-Parker's medium or other media containing both egg yolk and tellurite (for example, Baird-Parker's ETGPA medium or Crisley's TPEY medium) are superior to other media for the quantitative recovery of *S. aureus* from food [*12–14*]. Indicative of the general lack of interest in staphylococci in water is the absence of published reports of the use of these media for water samples. However, Michael Brodsky has indicated (personal communication) that his laboratory has been using both Baird-Parker medium and TPEY medium to enumerate *S. aureus* in hydrotherapy pools with results that are much superior to those with M-staphylococcus medium. It is probable that other laboratories also may have pertinent unpublished data from use of these media for water samples. It would be highly desirable to conduct and publish the results of carefully controlled experiments evaluating these and other media for selective enumeration of *S. aureus* and total staphylococci in appropriate water habitats.

We also should not lose sight of the possibility of the development of new and improved methods or media. Schleifer and Kloos [*15*] have reported a simple test system for separation of staphylococci from micrococci based on the ability of staphylococci to produce acid from glycerol in the presence of erythromycin (0.4 μg/ml) and on their sensitivity to lysozyme. Perhaps the use of glycerol and erythromycin in the presence of other selective agents might lead to an improved medium for total staphylococci.

In conclusion, it seems apparent that staphylococci are a major bacterial contaminant of swimming pools and other recreational waters that have a high density of swimmers, and a significant portion of these staphylococci are *S. aureus*. Hence, enumeration of either total staphylococci or of *S. aureus* logically would seem to provide a good index of the level of contamination of recreational waters by swimmers and the concurrent effectiveness of filtration and chlorination procedures. Unfortunately, data presently available do not clearly establish a quantitative correlation among these factors or their relationship to the incidence of microbial infection of bathers. A primary problem in obtaining suitable data for the determination of these correlations is the need for a medium and method that is sufficiently selective, accurate, and reliable. It seems probable that total staphylococci may be a more suitable indicator than *S. aureus*, but only through carefully planned and controlled studies and analyses of the data will we be able to establish the value of these indicator organisms.

References

[*1*] Robinton, E. D. and Mood, E. W., *Journal of Hygiene*, Vol. 64, 1966, pp. 489–499.
[*2*] Favero, M. S. and Drake, C. H., *Applied Microbiology*, Vol. 14, 1966, pp. 627–635.

[3] Keirn, M. A. and Putnam, H. D., *Health Laboratory Science*, Vol, 5, 1968, pp. 180–193.
[4] Palmquist, A. F. and Jankow, D., *Journal of Environmental Health*, Vol. 36, 1973, pp. 230–232.
[5] Boccia, A. and Montanaro, D., *Igiene Moderna*, Vol. 67, 1974, pp. 43–47.
[6] Crone, P. B. and Tee, G. H., *Journal of Hygiene*, Vol. 73, 1974, pp. 213–220.
[7] Kloos, W. E. and Musselwhite, M. S., *Applied Microbiology*, Vol. 30, 1975, pp. 381–395.
[8] *Standard Methods for the Examination of Water and Wastewater*, 13th ed., American Public Health Association, Inc., New York, 1971.
[9] Alico, R. K. and Palenchar, C. A., *Health Laboratory Science*, Vol. 12, 1975, pp. 341–346.
[10] *Standard Methods for the Examination of Water and Wastewater*, 14th ed., American Public Health Association, Inc., New York, 1976.
[11] Smuckler, S. A. and Appleman, M. D., *Applied Microbiology*, Vol. 12, 1964, pp. 355–359.
[12] Collins-Thompson, D. L., Hurst, A., and Aris, B., *Canadian Journal of Microbiology*, Vol. 20, 1974, pp. 1072–1075.
[13] Gray, R. J., Gaske, M. A., and Ordal, Z. J., *Journal of Food Science*, Vol. 39, 1974, pp. 844–846.
[14] Stiles, M. E., *Canadian Journal of Microbiology*, Vol. 20, 1974, pp. 1735–1744.
[15] Schleifer, K. H. and Kloos, W, E., *Journal of Clinical Microbiology*, Vol. 1, 1975, pp. 337–338.

M. A. Levin[1]

Bifidobacteria as Water Quality Indicators

REFERENCE: Levin, M. A., "**Bifidobacteria as Water Quality Indicators,**" *Bacterial Indicators / Health Hazards Associated With Water, ASTM STP 635*, **A. W. Hoadley and B. J. Dutka, Eds., American Society for Testing and Materials, 1977, pp. 131–138.**

ABSTRACT: The potential value of bifidobacteria as indicators of fecal pollution in water has not been fully explored. Nevertheless, a number of workers have suggested that bifidobacteria may be useful as indicator organisms because (*a*) they do not multiply in nature, (*b*) they exhibit survival characteristics which compare favorably to *Escherichia coli*, and (*c*) they may be useful as specific indicator organisms to differentiate animal from human fecal pollution.

An identification scheme for the most common species and which may be applicable in determining the type of pollution has been developed, and current techniques of speciation and quantitation are discussed.

KEY WORDS: bacteria, water, health, indicator bacteria, pathogenic bacteria, bifidobacteria

The role of indicator organisms in sanitary microbiology is well established. The use of coliform organisms, streptococci, and clostridia have been proposed and investigated in depth. These groups of organisms have been studied in relation to the presence of fecal pollution in potable water, surface water supplies, recreational waters, and shellfish growing waters. Differentiation between animal and human sources of fecal pollution has been attempted by comparing densities of fecal streptococci to fecal coliform densities.

The rationale for the use of indicator organisms—that is, the logistical and cost problems involved in examining samples for all possible pathogenic microorganisms—is well understood. However, each group of indicator organisms has been found to be less than perfectly satisfactory, and there has been a trend to attempt to select a more specific group or member of a group. Thus, emphasis on coliforms has shifted to examination of fecal coliforms, and the use of "thermotolerant *Escherichia coli*" has been proposed.

[1]Microbiologist, Marine Field Station, Health Effects Research Laboratory-Cin., U. S. Environmental Protection Agency, West Kingston, R.I. 02892.

The ideal indicator of fecal pollution should be present in sufficient density to permit sensitive quantitation, should not multiply in natural situations, and should persist (survive in nature) to the same degree as do the pathogenic organisms in sewage. A technique for quantitation which is precise, sensitive, and not overly expensive or time consuming is necessary. There is also a requirement for differentiation of the source (animal versus human) of the pollution.

None of the indicators currently in use meet all of the foregoing requirements. The possibility of multiplication under natural conditions (coliforms), low density (streptococci), or extreme persistence (clostridia) has been pointed up by a number of workers. To obviate some of these problems, bifidobacteria have been proposed as indicators of fecal pollution. These organisms have been demonstrated in human and animal feces at high densities, are strictly anaerobic and, hence, cannot multiply in the presence of air, have been shown to possess survival characteristics under laboratory conditions which are similar to *E. coli*, and may permit differentiation between animal and human sources of pollution. This group of organisms has not been well examined, possibly because of the difficulties associated with culturing anaerobic bacteria, although techniques now available permit quantitative recovery and speciation and should permit the evaluation of these microorganisms as indicator organisms to aid in evaluating the health hazards associated with a particular water source.

Bifidobacteria have been the least studied of all the actual and potential indicators of fecal pollution. There is little available ecological, taxonomical, and physiological information about these microorganisms, despite the fact that they were first described in 1899 [*1*].[2] The biology of the genus was reviewed in 1973 by Poupard et al [*2*], and the classification has been updated in the most recent edition of *Bergey's Manual of Determinative Bacteriology* [*3*] in which eleven species have been listed.

The potential significance of bifidobacteria as indicator organisms in water and food has been pointed up by several authors over the past 20 years, beginning in 1958 with Mossel [*4*] and continuing through 1974 with the work of Evison and James [*5*]. The impetus for considering the use of bifidobacterium as indicator organisms can be traced to the numerous reports documenting their presence in the feces of warm-blooded animals [*2, 6–9*]. There is general agreement that levels of 10^9 to 10^{11} organisms/g of feces can be expected.

Speciation of fecal isolates has indicated that four species are frequently isolated from human feces. They are *Bifidobacterium bifidum, B. adolescentis, B. infantis*, and *B. breve*. The species *B. longum* (also isolated from human feces), *B. pseudolongum, B. thermophilum*, and *B. suis* are associated generally with lower animal feces, while *B. asteroides, B. indicum*, and *B. coryneforme* are insect associated [*3*]. While *B. longum* is associated with

[2]The italic numbers in brackets refer to the list of references appended to this paper.

both human and animal feces, data to date indicate that it constitutes only a small fraction of the total bifidobacteria population in human feces [*6*]. The postulated subspecies (*B. longum* subspecies *animalis*) has not yet been differentiated from *B. longum* and in fact may be "a synonym of *B. longum*" [*3*].

The impetus for considering the use of bifidobacteria as a water quality indicator was the work of Gylenberg et al [*10*] demonstrating the survival characteristics of these organisms in nature. He found that there was no difference between the survival of bifidobacteria and coliforms at room temperature. This fact, coupled with the fact that bifidobacteria are strict anaerobes and, hence, incapable of multiplication in the presence of air, strongly suggested the use of this group of organisms as an indicator of fecal pollution.

Identification

Bifidobacteria are defined as strictly anaerobic, nonspore forming, nonmotile, gram-positive, morphologically thick pleomorphic rods. They may exhibit branching bulbs, clubs, coryneforms, buds, spheroids, and bifurcated Y and V forms when freshly isolated from fecal sources (the morphology is influenced by nutritional conditions and may appear different when isolated from an aquatic environment). They are catalase negative. Acetic and lactic are the minor acids produced from glucose fermentation; therefore, tests to identify the fermentation products by gas-liquid chromotography are most useful for identification to the generic level. A scheme for identification to this level is shown in Fig. 1. The inoculation of brain-heart-infusion (BHI) agar for aerobic incubation and peptone yeast glucose (PYG) agar [*11*] for anaerobic incubation is required to exclude the possibility of a facultative anaerobe. A second anaerobic isolation on PYG agar and an aerobic one on BHI is needed since there are reports of *Lactobacillus* species which appear as strict anaerobes on initial isolation. Gas chromotography, with quantitation of the acetic and lactic acids produced, will eliminate eubacteria. Gram's stain characteristics, motility (–), catalase reaction (–), and production of acid without gas when grown in glucose broth confirm the identity of the isolate as a member of the genus *Bifidobacterium*.

As can be seen from Fig. 2, use of the two carbohydrates, arabinose and gluconate, will divide the eleven species of the genus *Bifidobacterium* into four groups, with *B. indicum* alone in group 4. Groups 1, 2, and 3 can be subdivided using lactose, ribose, xylose, mannitol, and starch (Fig. 3). However, for use as indicators of fecal pollution, group 2 need not be further divided since all of the members originate in warm-blooded animals. Group 1 isolates can be divided into human and insect species by examining utilization of lactose. Subdivision of group 3 requires the use of ribose, xylose, mannitol, and starch for complete identification although the ribose and starch tests would suffice if one is simply interested in identifying organisms primarily from feces of warm-blooded animals. The eight tests will permit

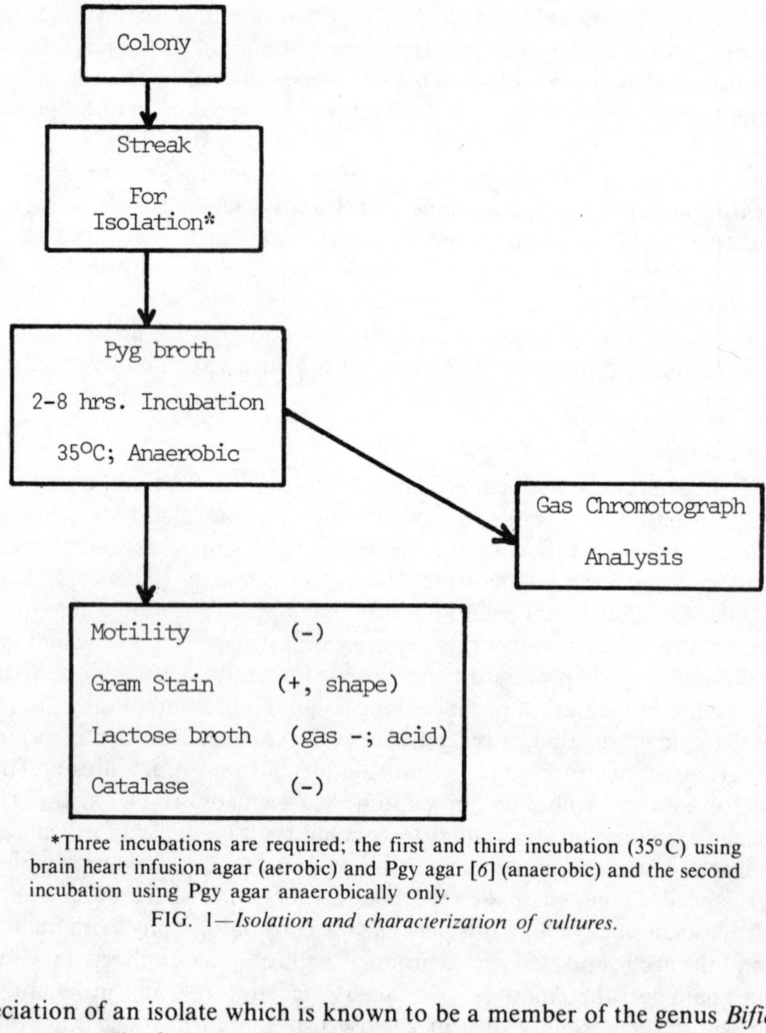

*Three incubations are required; the first and third incubation (35°C) using brain heart infusion agar (aerobic) and Pgy agar [*6*] (anaerobic) and the second incubation using Pgy agar anaerobically only.

FIG. 1—*Isolation and characterization of cultures.*

speciation of an isolate which is known to be a member of the genus *Bifidobacterium*. All of these tests are reliable within 90 percent of the isolates tested [*12,3*].

Importance

The importance of bifidobacteria as indicators of fecal pollution has been pointed up by Mossel [*4*], Gylenberg [*10,13*], and Evison [*5,14*]. These organisms meet all the criteria of an indicator of fecal pollution. They are associated with feces, do not multiply in nature, and their survival characteristics appear to be similar to other indicator organisms.

However, there have been only a limited number of studies in which attempts have been made to quantitate these organisms, specifically those of Evison and James [*14*]. Working in the United Kingdom and Kenya they

Isolate

Arabinose	Gluconate →	Group
+	+	1
+	-	2
-	-	3
-	+	4

NOTE—Group 1—*B. adolescentis, B. asteroides, B. coryneforme;*
Group 2—*B. longum, B. pseudolongum, B. suis;*
Group 3—*B. bifidum, B. brevis, B. infantis, B. thermophilum;* and
Group 4—*B. indicum.*

FIG. 2—*Speciation of Isolate 1.*

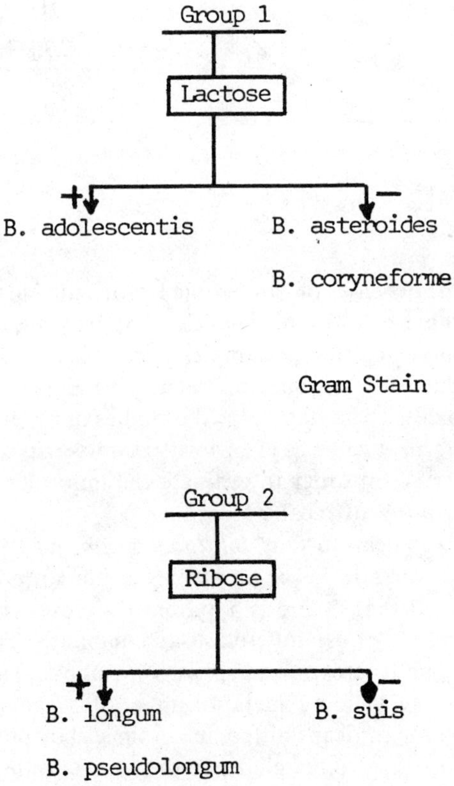

FIG. 3—*Speciation of Isolate 2.*

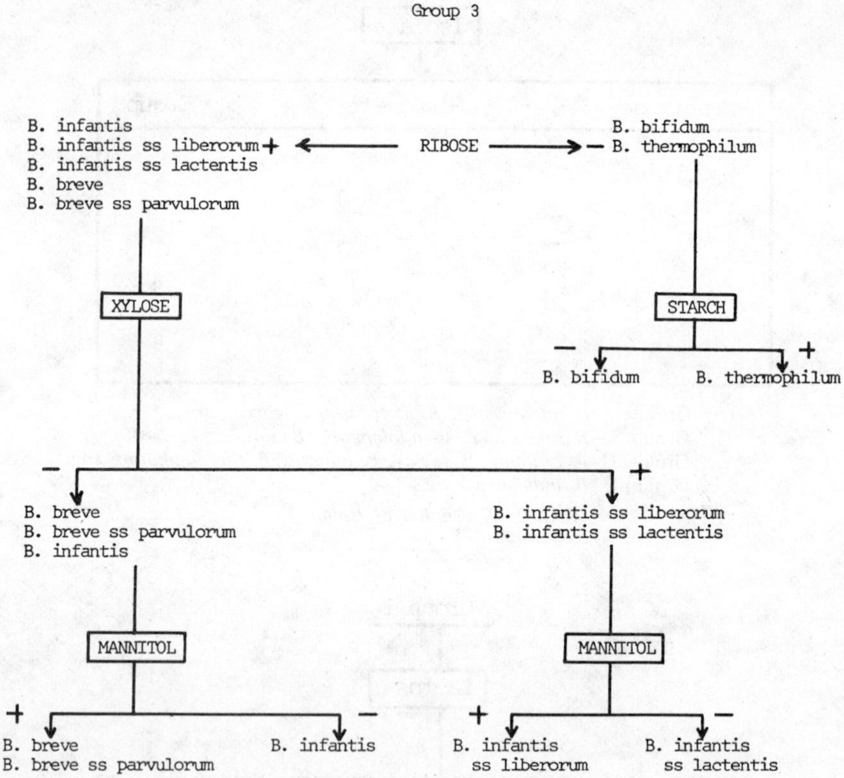

FIG. 4—*Speciation of Isolate 3.*

have shown that the densities of "anaerobic lactobacilli" in water samples are generally comparable to the *E. coli* levels of approximately $10^2/100$ ml. A similar comparison, using sewage samples, indicates ten fold higher concentrations of the "anaerobic lactobacilli" relative to *E. coli* densities ($10^6/100$ ml versus $10^5/100$ ml). These data, plus the studies of Gylenberg [*10*] demonstrating the long-term (twelve day) survival characteristics, strongly suggest that further studies are in order to evaluate the importance of these microorganisms as indicators of fecal pollution.

There is also an indication that *Bifidobacterium* may provide a specific assay which could separate lower animal from human pollution. Scardovi et al [*15*] have stated that "there is apparently a close relatedness between electrophoretic type of phosphoketolase and habitat." This was based on studies of the enzyme fructose-6-phosphate phosphoketolase which is a key enzyme in the energy yielding metabolism of this group. These authors found that there is a significant difference in the electrophoretic mobility of the enzyme which seemed to be related to the source of the organism. Thus, avian, cattle, and human isolates had distinctly different migration distances.

To date, the only information concerning the presence and densities of bifidobacteria in the environment is the work of Evison and James [*5,14*] using a modification of the medium of Gylenberg [*13*]. The medium and the technique used (gas pack jars) have not been sufficiently evaluated. The medium is used with membrane filters which permits concentration of the sample. The presence of triphenyltetrazolium chloride in the medium causes the colonies to assume a dark red color which makes them easy to distinguish. However, all of the dark red colonies are not bifidobacteria. Work in this laboratory has shown that, depending on the source of the sample, the percentage of false positives ranges from 20 to 55 percent, most of which were identified as lactobacilli. In addition, the use of arabinose as the sole carbohydrate limits the medium to organisms in groups 1 and 2. Thus, *B. bifidum, B. infantis*, and *B. breve*—three of the five species known to be associated with human feces—would not be recovered.

It is clear that a great deal of work is necessary to develop the techniques needed to determine the true potential of bifidobacteria as indicators of fecal pollution. The genus has, in theory, all of the characteristics of an excellent, if not ideal, indicator. The organisms do not multiply in nature, the sole source is fecal material, the survival characteristics appear similar to currently used indicator organisms, and, finally and perhaps most significantly, the possibility of developing a reliable test to determine the source of pollution (human or animal) exists. However, epidemiological data are not available.

The need for a quantitative isolation medium for *Bifidobacterium* species has been pointed up by a committee in 1973 [*16*]. Our laboratory has been attempting to develop a medium using the work of Nakamura et al [*17*] as a point of departure. The medium has been developed to the point where it can recover quantitatively all strains of bifidobacteria tested to date. However, a sufficient number of strains have not been tested to permit any definite statements as to its overall suitability. The specificity of the medium is such that no gram-negative organisms or cocci have been recovered using water samples from polluted areas and sewage. However, a significant number of facultative, anaerobic, gram-positive rods do multiply on the filters and have in many cases overgrown the bifidobacteria. Work is in progress to improve its selective and differential properties.

References

[*1*] Tissier, M. H., *Compte Rendus of Academy of Science*, Vol. 51, 1889, pp. 943–945.

[*2*] Poupard, A., Husain, Intisar, and Norris, R. F., *Bacteriological* Reviews, Vol. 2, 1973, pp. 136–165.

[*3*] Bergey's Manual of Determinative Bacteriology, 8th ed., Williams & Wilkins, Baltimore, Md., 1974.

[*4*] Mossel, D. A. A., *Abstracts of Papers*, VIIth International Congress of Microbiology, 1958, p. 440.

[*5*] Evison, L. M. and James, A., "Bifidobacterium as an Indicator of Faecal Pollution in Water," *Proceedings*, 7th International Conference on Water Pollution Research, Pergamon Press Ltd., New York, 1974.

[6] Holdeman, I. V., Good, I. J., and Moore, W. E. C., *Applied Microbiology*, Vol. 31, 1976, pp. 359–375.
[7] Mata, L. J., Carrillor, C., and Villatoro, E., *Applied Microbiology*, Vol. 17, 1969, pp. 596–602.
[8] Mata, L. J. and Urrutia, J. J., *Annals of the New York Academy of Science*, Vol. 176, 1971, pp. 93–108.
[9] Zani, G., Crociani, B., and Matteuzzi, D., *Journal of Applied Bacteriology*, Vol. 37, 1974, pp. 537–547.
[10] Gyllenberg, H., Niemela, S., and Sormunen, T., *Applied Microbiology*, Vol. 8, 1960, p. 20.
[11] Holdeman, I. V. and Moore, W. E. C., *Anaerobic Laboratory Manual*, Virginia Polytechnic Institute and State University, Blacksburg, Va., 1972.
[12] Scardovi, U., Trovatelli, L. A., Zani, G., Crociani, F., and Matteuzzi, D., *International Journal of Systematic Bacteriology*, Vol. 21, 1971, pp. 276–294.
[13] Gyllenberg, H. and Niemela, S., *Maatalouts, Aikakausk*, Vol. 31, 1959, p. 94.
[14] Evison, L. M. and James, A., *Journal of Applied Bacteriology*, Vol. 36, 1973, pp. 109–118.
[15] Scardovi, V., Sgorbati, B., and Zani, G., *Journal of Bacteriology*, Vol. 106, 1971, pp. 1036–1039.
[16] Sykes, G. and Skinner, F. A., Eds., *Actinomycetales: Characteristics and Practical Importance*, Academic Press, London and New York, 1973.
[17] Nakamura, H., Samejima, K., and Tamura, Zenzo., *Japanese Journal of Microbiology*, Vol. 17, No. 4, 1973, pp. 283–289.

J. D. Buck[1]

Candida albicans

REFERENCE: Buck, J. D., ***"Candida albicans,"*** *Bacterial Indicators / Health Hazards Associated With Water, ASTM STP 635*, A. W. Hoadley and B. J. Dutka, Eds., American Society for Testing and Materials, 1977, pp. 139–147.

ABSTRACT: This paper presents current knowledge regarding the potential of the human pathogenic yeast *Candida albicans* to serve as an indicator of health hazards associated with the use of natural waters. In lieu of defined procedures for the specific detection of the organism from natural habitats, the discussion emphasizes what is known concerning distribution of the yeast in nature and possible sources for subsequent occurrence in water. The necessity is stressed for the development of selective media to allow comparative quantitative and qualitative assessment of yeast densities, particularly in recreation waters.

KEY WORDS: bacteria, water quality, yeasts, health hazards, recreational waters, survival, selective media

In January of 1975, the ASTM Task Group D19.01.04.01 on Microorganisms of Health and Sanitary Significance was established under Section D.19.01.04 on Microbiology of Subcommittee D19.01 on Biological Monitoring of Committee D-19 on Water to examine *Candida albicans* as a potential indicator of water quality. This review will consider the rationale of this approach as expressed by contributing members of the ASTM Task Group and the existing literature. The current status of quantitative and qualitative aspects of *C. albicans* occurrence in natural waters is presented.

Description of *C. albicans*

Functional Definition of the Species

C. (then *Oidium*) *albicans* was described originally in 1853. Since then, the organism has been subjected to a variety of taxonomic treatments (see reviews of Winner and Hurley [*1*],[2] Lodder and Kreger-van Rij [*2*], and Skinner and Fletcher [*3*]). The current standard working definition of the species is given by van Uden and Buckley [*4*], supplemented technically by

[1]Associate professor, Biological Sciences Group and Institute of Marine Sciences, Marine Research Laboratory, University of Connecticut, Noank, Conn., 06340.

[2]The italic numbers in brackets refer to the list of references appended to this paper.

Barnett and Pankhurst [*5*]. Essentially, *C. albicans* is a "typical" yeast in culture; that is, cells are generally ovoid and not routinely of a characteristic shape, asexual reproduction occurs by multipolar budding, there is no visibly unique pigment, and fermentation occurs in carbohydrate media. Detailed laboratory studies including a pattern of the fermentation and assimilation of a variety of carbon compounds and demonstration of pseudomycelial production can be used to identify accurately the species. Fortunately, these tedious and time-consuming procedures can be eliminated, since under appropriate cultural conditions *C. albicans* produces characteristic morphological structures called germ tubes. Details of the procedure for their detection are given by Ahearn [*6*] and elsewhere. Briefly, cells from young colonies on plates or agar slant growth are suspended in 0.5 ml of serum, incubated at 37° C for 2 to 3 h, and examined microscopically by the wet mount technique for the presence of germ tubes. *C. albicans* produces thin cellular extensions 3 to 4 μm wide and up to 20 μm in length with no constriction at the point of origin with the producing yeast cell. Cultures which exhibit this characteristic feature can be presumed to be *C. albicans*.

Potential Difficulties Associated with the Definition

C. stellatoidea is a species closely related to *C. albicans* [*1,3,5,7*] and also produces germ tubes [*6,8*]. Although the organism is associated with humans [*9–11*], it is evidently rare in occurrence in marine waters [*12,13*]. Its presence would assume the same significance as that of *C. albicans*.

Importance of C. albicans

Disease in Man

It is likely that Hippocrates described cases of oral and vaginal candidosis. The role of *C. albicans* in human disease has been reviewed amply [*1,8,9,10, 14*]. However, in addition to the "classical" involvement of *C. albicans* in oral, vaginal, and cutaneous mycoses, the organism has been implicated recently in a variety of other clinical situations. Examples include veneral disease [*15*], heart surgery [*16*], denture stomatitis [*17*], epithelial hyperplasia [*18*], endophthalmitis [*19*], cystic fibrosis [*20*], and mucocutaneous candidiasis [*21*]. In fact, *C. albicans* is becoming more recognized as an infectious agent in hospital patients [*22*]. In one study, *C. albicans* was the most common species (56 percent of the total number of yeasts) recovered from individuals at 14 hospitals [*23*]. Also, skin carrier rates among hospital patients are high [*24*].

Disease in Animals

A general treatise on the role of *C. albicans* as a pathogen of domestic and farm animals is provided by Jungerman and Schwartzman [*25*]. The occurrence of the yeast in wild animals is described by Winner and Hurley [*1*] and van Uden [*26*]. The sensitivity or resistence of an animal to infection may be dependent upon genetic background and sex of the host as well as on the previous history of the yeast and its genetic information [*27*]. *C. albicans* has been found in droppings from 30 percent of the pigeons examined with up to 2.8×10^6 organisms/g noted [*28*]. Curiously, in another study of pigeons, the yeast was relatively rare [*29*]. In seagulls, 21.7 percent of fresh feces showed the presence of *C. albicans*, but only 1.6 percent of dry feces were positive [*30*]. Both types of birds are frequent in aquatic habitats and could well serve as a constant source of local inoculum.

Characteristics of Population in Water

Population Densities

Several reports have noted the presence of *C. albicans* in fresh, estuarine, and marine waters [*31–38*]. While some of these studies include the abundance of the yeast relative to other species or as frequency of isolation, there are few quantitative data. Fell and van Uden [*39*] reported up to 320 cells per litre in a semitropical marine bathing lagoon. Recently, we have found up to 1000 *C. albicans* per litre during peak summer usage of a large municipal marine beach in the northeast (Buck, unpublished data). The paucity of accurate enumeration figures is a consequence of the lack of a *C. albicans* selective medium which is efficient in sampling natural waters.

Biotypes

Meyer and Roth [*31*] found no significant metabolic or physiologic differences between *C. albicans* isolated from a marine lagoon and those from human sources. Fell and Meyer [*11*] noted marine distributional patterns of animal-associated yeasts, including *C. albicans*, but did not comment on any specific differences between human and marine isolates.

Role of Water in the Spread of *C. albicans*

There are no definitive epidemiological studies linking *C. albicans* infections in man with the occurrence of the pathogen in natural waters. The fact that the organism has been found in recreational and adjacent waters [*31,32, 34–39*] is only suggestive of infective potential in lieu of quantitative data and clinical case histories. Similarly, the occurrence of *C. albicans* in public freshwater swimming pools and associated shower tiles and floors [*40*] does not establish that these areas are hazardous to users of these facilities. Dabrowa et al [*41*] recovered *C. albicans* from California crabs (no species given), and we have isolated the organism on several occasions from both liquor and meat of the quahog (*Mercenaria mercenaria*) and oyster (*Crassostrea virginica*) in the Long Island Sound area (Buck, data in press). Again, there are no substantiated epidemiological implications from these observations. Winner and Hurley [*1*] cite a reference to a condition called "surfer's foot" in Australia in which *C. albicans* infects macerated skin.

Factors Affecting the Use of *C. albicans* as an Indicator

Source of the Organism

The general inability of *C. albicans* to exist without a host [*1,42,43*] indicates that there are no truly autochthonous populations of the yeast in natural waters. Lacking data to the contrary, it must be presumed that no significant reproduction occurs in either water or aquatic animals (for example, molluscs). Thus, the origin of *C. albicans* in aquatic habitats is allochthonous. The yeast is not usually a major component of the soil mycoflora, although it does occur [*1,44*]. Its presence in bird droppings [*28–30,41*] and, hence, soil may provide a source of the yeast in water via runoff. In fact, Fell [*33*] remarked that birds were prevalent in the only sampling area in the Indian Ocean where *C. albicans* was found.

Yeasts, in general, occur in relatively high numbers in both raw [*45*] and treated [*46*] sewage. Although the numbers are low in some chlorinated effluent mixing areas [*47*], Englebrecht et al [*45*] have found several unidentified chlorine resistent yeasts. Specifically, *C. albicans* is a common component of human urine [*48*] and feces [*49*] and, thus, is found in settling ponds and privies [*34*] and raw municipal sewage [*34,42*] but not in treated effluents [*42*]. It is apparent that untreated human and animal wastes are the major source of *C. albicans* in natural waters.

There are no direct temperature data relating to the abundance (or survival) of *C. albicans* in natural waters; however, some general comments are appropriate. Van Uden and Castelo Branco [*50*] indicated that recent pollution and a high water temperature were necessary for the marine occurrence of intestinal yeasts. Van Uden [*51*] and Fell and Meyer [*11*] noted the differential maximal temperature optima for marine occurring and intestinal yeasts and reinforced the role of water temperature in establishing population densities. Ahearn [*42*] speculated that water temperatures lower than that of the human host were probably responsible for poor survival of *C. albicans* in water. While no correlation of numbers with temperature has been attempted, we have recovered 80 cells of *C. albicans* per litre from estuarine water at an ambient temperature of 8° C; the organism has been found also in water at a temperature of 3° C (Buck, unpublished data).

Meyer and Roth [*31*] observed that cells of *C. albicans* isolated from a marine bathing lagoon survived for at least 10 days when inoculated back into lagoon water. Dzawachiszwili et al [*52*] reported the viability of clinical isolates of *C. albicans* for over six weeks in distilled water, autoclaved and filtered natural seawater, and sodium chloride (NaCl) concentrations up to 6.8 percent (double the total solute content of seawater). They concluded that the ocean may represent a reservoir of human pathogenic fungi. Madri, Claus, and Moss [*53*] indicated that human strains of *C. albicans* not only survived in filtered natural seawater for eight weeks but also retained pathogenicity for mice. Madri [*54*] showed that the organism was capable of reproducing by budding even after several transfers in seawater. Ahearn

[*42*] reported that, in pure culture, *C. albicans* suspended in lake and coastal waters at 22 to 25°C showed restricted budding for 4 h and the water demonstrated increased absorbance at 260 to 280 nm and a positive ninhydrin reaction. After 10 to 12 h, budding was initiated. After 12 days, cell numbers fell to approximately 20 percent of the original (lake water), but 77 percent survived after the same period in coastal water. Jamieson [*55*] demonstrated that *C. albicans*, while undergoing an initial decrease in concentration, will stabilize its population at a more or less defined level after seven days of exposure to temperatures of 4, 25, and 37°C and salinities of 5, 20, and 35 parts per thousand. Di Menna and Parle [*56*] recovered *C. albicans* after nine months from artificially contaminated turf.

While these data are suggestive of the ability of *C. albicans* to not only survive but possibly to reproduce in water, we must consider that *in vitro* conditions do not duplicate the natural environment. The experiments cited [*31,42,52–55*] were all conducted with seawater or other menstrum in laboratory glassware. Preliminary studies in our laboratory with a variety of strains of *C. albicans* have confirmed the survival of the organism under similar *in vitro* conditions. However, the same cultures placed in dialysis bags in the intertidal zone were reduced to undetectable numbers within 24 h. These observations were made at or near 20°C; further studies are in progress at different temperatures and with other human-associated species of *Candida*. In the lagoon studies noted previously, Meyer and Roth [*31*] found a marked reduction in the number of *C. albicans* survivors during the first 24 h of incubation with water obtained during periods of rich phytoplankton development. Fell and van Uden [*39*] were unable to culture *C. albicans* in either sterile macroalgal extracts or nutritionally supplemented extracts lacking only a carbon source. This may reflect an antibiotic effect of marine algae toward *C. albicans* [*57–61*]. Seshadri and Sieburth [*62*] reported that 95 percent of yeasts isolated from inshore seaweeds were members of *Candida*; no *C. albicans* were found.

Ahearn [*42*] noted that when *C. albicans* and *Rhodotorula rubra* (a common aquatic occurring yeast) were inoculated together into lake and coastal water the survival of *C. albicans* was reduced dramatically. In lake water, only 1 percent survived after 12 days while 28 percent remained in coastal water (compare these with previous data for pure cultures). Ahearn feels that *C. albicans* may undergo a "shock excretion" of nitrogen compounds in water and, hence, is unable to compete with other microorganisms for essential nutrients in a natural situation.

Coleman, Cook, and Ahearn [*63*] have shown the selective inhibition of *C. albicans* in primary sewage settling tank effluents by a *Bacillus subtilis*-like bacterium and proposed this antagonism as effecting the limited survival of the yeast in a host-free system. Marine bacteria have shown anti-*C. albicans* properties also [*64–67*]. The marine sponge *Microciona prolifera* (redbeard sponge) produces a broad spectrum antimicrobial which is active against *C. albicans* [*68*]. Kunen et al [*69*] have demonstrated that the sponge can ingest and digest *C. albicans* cells, an activity not ascribed to antibiotic production. Other sponges [*70*], bivalve molluscs [*71*], and a wide variety of marine invertebrates [*72*] are inhibitory to *C. albicans* when extracts were assayed by the disc procedure under laboratory conditions. *C. albicans*, in fact, may produce self-inhibitory compounds [*73*].

Meyer and Roth [*31*] were unable to recover any yeasts from chlorinated freshwater pools while *C. albicans* has been found by others [*40*].

The literature reports are thus equivocal on the matter of *C. albicans* survival in the aquatic habitat and range from the ability of the organism to reproduce and maintain infectivity over a several week period to a rapid reduction of numbers in particular environments in a short period of time.

Methods for Enumeration

There are no recognized "standard" procedures for the isolation of *C. albicans* from aquatic habitats or, in fact, for the recovery of yeasts in general, although a wide variety of techniques are available [*74–77*]. Population densities cited previously represented primarily chance isolations of the organism. The membrane filter method has been used extensively in mycological studies of water. Unless, however, all yeast colonies are picked for the time-consuming identification process or confirmed as *C. albicans* more expediently by the germ tube procedure, quantitative estimates of the organism will continue to be sporadic in the literature. Some studies have utilized the broth enrichment technique; the difficulty encountered is rapid overgrowth of *C. albicans* by other yeasts, especially if the incubation temperature is too low. Recovery of human-associated yeasts from water is enhanced at 37° C [*37*].

Roth and Rogoff [*78*] developed a selective medium for *C. albicans* using clinical material and marine waters or sediments. Subsequent extensive marine field testing was disappointing, however, with many "false positive" yeasts developing.[3] We have experienced similar frustrations in our laboratory in early attempts to culture *C. albicans* selectively for recreational waters. However, studies with a newly developed medium for detection and enumeration of the organism have been most encouraging, and we anticipate reporting our observations in the near future.

Conclusions

C. albicans is a well-defined, easily identified yeast associated with humans and animals. Despite an obvious lack of quantitative information on distribution in natural waters, the occurrence of the organism in association with human-urban activities has stimulated support for its use as an indicator of potential health hazard [*34,35*].

[3]Personal communication with Dr. F. J. Roth, Jr., University of Miami, Medical School, Department of Microbiology, Miami, Fla.

A reliable enumeration procedure must be developed so that numbers can be correlated with other parameters; only then can epidemiological studies be meaningful. Supplementary evaluation of survival rates as a function of salinity, temperature, organic matter, and especially distance from inoculum source will reveal the utility of the organism to serve as an indicator of the recentness of its environmental introduction. Of additional interest will be the comparative abundance, quantitatively and qualitatively, of *C. albicans* and traditional indicators. The yeast is easily able to contact body surfaces (skin, nails) and can gain entry to other susceptible tissues (for example, vagina, oral cavity) through bathing and other activities connected with recreational usage of natural waters.

C. albicans is, thus, a most reasonable candidate for further study of health hazards in recreational areas which are subject currently to considerable discussion regarding microbiological quality [*79–82*].

Acknowledgments

This represents Contribution No. 120 from the University of Connecticut, Marine Sciences Institute, Marine Research Laboratory, Noank, Ct. Portions of the research cited herein were supported by Grant No. R802827–01–0 from the U.S. Environmental Protection Agency, Health Effects Research Laboratory, West Kingston, R.I. The author is indebted to Dr. J. W. Fell for aid in manuscript preparation and P. M. Bubucis for providing help in collecting certain laboratory data included.

References

[*1*] Winner, H. I. and Hurley, R., *Candida albicans*, Little, Brown and Co., Boston, 1964.
[*2*] Lodder, J. and Kreger-van Rij, N. J. W., *The Yeasts*, North-Holland, Amsterdam, 1952.
[*3*] Skinner, C. E. and Fletcher, D. W., *Bacteriological Reviews*, Vol. 24, 1960, pp. 397–416.
[*4*] van Uden, N. and Buckley, H. in *The Yeasts*, J. Lodder, Ed., North Holland, Amsterdam, 1970, pp. 893–1087.
[*5*] Barnett, J. A. and Pankhurst, R. J., *A New Key to the Yeasts*, North-Holland, Amsterdam, 1974.
[*6*] Ahearn, D. G. in *Opportunistic Pathogens*, J. E. Prier and H. Friedman, Eds., University Park Press, Baltimore, 1974, pp. 129–146.
[*7*] Campbell, I., *Journal of General Microbiology*, Vol. 90, 1975, pp. 125–132.
[*8*] Buckley, H. R. in *Methods in Microbiology*, Vol. 4, C. Booth, Ed., Academic Press, 1971, pp. 461–478.
[*9*] Gentles, J. C. and LaTouche, C. J. in *The Yeasts*, Vol. 1, A. H. Rose and J. S. Harrison, Eds., Academic Press, 1969, pp. 107–182.
[*10*] Silva-Hutner, M. in *Manual of Clinical Microbiology*, J. E. Blair, E. H. Lennette, and J. P. Truant, Eds., American Society for Microbiology, Washington, 1970, pp. 352–363.
[*11*] Fell, J. W. and Meyer, S. A., *Mycopathologia et Mycologia Applicata*, Vol. 32, 1967, pp. 177–193.
[*12*] Morris, E. O. in *Annual Review of Oceanography and Marine Biology*, Vol. 6, H. Barnes, Ed., Hafner Publ. Co., New York, 1968, pp. 201–230.
[*13*] Fell, J. W. in *Recent Advances in Aquatic Mycology*, E. B. G. Jones, Ed., Paul Elek Books, Ltd., London, 1976, pp. 93–124.
[*14*] Stockdale, P. M. in *Methods in Microbiology*, Vol. 4, C. Booth, Ed., Academic Press, 1971, pp. 429–460.

[15] Eriksson, G. and Wanger, L., *British Journal of Veneral Diseases*, Vol. 51, 1975, pp. 192–197.
[16] Parsons, E. R. and Nassau, E., *Journal of Medical Microbiology*, Vol. 7, 1974, pp. 415–423.
[17] Budtz-Jorgensen, E., *Sabouraudia*, Vol. 12, 1971, pp. 266–271.
[18] Cawson, R. A., *British Journal of Dermatology*, Vol. 89, 1973, pp. 497–503.
[19] Weinstein, A. J., Johnson, E. H., and R. C. Moellering, Jr., *Archives of Internal Medicine*, Vol. 132, 1973, pp. 749–752.
[20] Hughes, W. T. and Kim, H. K., *Mycopathologica et Mycologia Applicata*, Vol. 50, 1973, pp. 261–269.
[21] Kroll, J. J., Einbinder, J. M., and Merz, W. G., *Archives of Dermatology*, Vol. 108, 1973, pp. 259–262.
[22] Anonymous, *Clinical Laboratory Forum*, Vol. 7, 1975, p. 8.
[23] Mackenzie, D. W. R., *Sabouraudia*, Vol. 1, 1961, pp. 8–15.
[24] Somerville, D. A., *Journal of Hygiene*, Cambridge, Vol. 70, 1972, pp. 667–675.
[25] Jungerman, P. F. and Schwartman, R. M., *Veterinary Medical Mycology*, Lea and Febiger, Philadelphia, 1972.
[26] van Uden, N., *Annals of the New York Academy of Sciences*, Vol. 89, 1960, pp. 59–68.
[27] Saltarelli, C. G., Gentile, K. A., and Mancuso, S. C., *Canadian Journal of Microbiology*, Vol. 21, 1975, pp. 648–654.
[28] Hasenclever, H. F. and Kocan, R. M., *Sabouraudia*, Vol. 13, 1975, pp. 116–120.
[29] Ramirez, R., Robertstad, G. W., Hutchinson, L. R., and Chavez, J., *Journal of Wildlife Diseases*, Vol. 12, 1976, pp. 83–85.
[30] Cragg, J. and Clayton, Y. M., *Journal of Clinical Pathology*, Vol. 24, 1971, pp. 317–319.
[31] Meyer, S. A. and Roth, F. J., Jr., *Bacteriological Proceedings*, 1961, p. 71.
[32] Capriotti, A., *Archiv fuer Mikrobiologie*, Vol. 42, 1962, pp. 407–414.
[33] Fell, J. W., *Bulletin of Marine Science*, Vol. 17, 1967, pp. 454–470.
[34] Ahearn, D. G., Roth, F. J., Jr., and Meyers, S. P., *Marine Biology*, Vol. 1., 1968, pp. 291–308.
[35] Cook, W. L. in *Recent Trends in Yeast Research*, D. G. Ahearn, Ed., Spectrum, Vol. 1, School of Arts and Sciences, Georgia State University, Atlanta, 1970, pp. 107–112.
[36] Combs, T. J., Murchelano, R. A., and Jurgen, F., *Mycologia*, Vol. 63, 1971, pp. 178–181.
[37] Buck, J. D., *Mycopathologia*, Vol. 56, 1975, pp. 73–79.
[38] Messineva, M. A. and Skadooskii, S. N., *Mikrobiologiya*, Vol. 16, 1947, pp. 43–49.
[39] Fell, J. W. and van Uden, N. in *Symposium on Marine Microbiology*, C. H. Oppenheimer, Ed., C. C. Thomas, Springfield, Ill., 1963, pp. 329–340.
[40] Kraus, H. and Tiefenbrunner, F., *Zentralblatt fuer Bakteriologie, Parasitenkunde, Infektionskrankheit und Hygiene*, I. Abteilung Originale, Vol. 160, 1975, pp. 286–291.
[41] Dabrowa, N., Landau, J. W., Newcomer, V. D., and Plunkett, O. A., *Mycopathologia et Mycologia Applicata*, Vol. 24, 1964, pp. 137–150.
[42] Ahearn, D. G. in *Estuarine Microbial Ecology*, Vol. 1, L. H. Stevenson and R. R. Colwell, Eds., Belle W. Baruch Library in Marine Science, University of South Carolina Press, 1973, pp. 433–439.
[43] van Uden, N. and Do Carmo-Sousa, L., *Journal of General Microbiology*, Vol. 16, 1957, pp. 385–395.
[44] McDonough, E. S., Ajello, L., Ausherman, R. J., Balows, A., McClellan, J. T., and Brinkman, S., *American Journal of Hygiene*, Vol. 73, 1961, pp. 75–83.
[45] Englebrecht, R. S., Foster, D. H., Greening, E. O., and Lee, S. H. EPA-670/2-73-082, Environmental Protection Technology Series, U.S. Environmental Protection Agency, Office of Research and Development, Washington, 1974.
[46] Cook, W. B., *Mycologia*, Vol. 57, 1965, pp. 696–703.
[47] Buck, J. D., *Florida Scientist* (*Quarterly Journal of the Florida Academy of Sciences*), Vol. 32, 1976, pp. 111–120.
[48] Ahearn, D. G., Jannach, J. R., and Roth, F. J., Jr., *Sabouraudia*, Vol. 5, 1966, pp. 110–119.
[49] Cohen, R., Roth, F. J. Jr., Delgado, E., Ahearn, D. G., and Kalser, M. H., *New England Journal of Medicine*, Vol. 280, 1969, pp. 638–641.
[50] van Uden, N. and Castelo Branco, R., *Limnology and Oceanography*, Vol. 8, 1963, pp. 323–329.

[*51*] van Uden, N. in *Estuaries*, G. H. Lauff, Ed., American Association for the Advancement of Science, Washington, 1967, pp. 306–310.
[*52*] Dzawachiszwili, N., Landau, J. W., Newcomer, V. D., and Plunkett, O. A., *Journal of Investigative Dermatology*, Vol. 43, 1964, pp. 103–109.
[*53*] Madri, P. P., Claus, G., and Moss, E. E., *Revista de Biologia* (Portugal), Vol. 5, 1966, pp. 371–381.
[*54*] Madri, P., *Botanica Marina*, Vol. II, 1968, pp. 31–35.
[*55*] Jamieson, W., Ph.D. dissertation, New York University, 1974.
[*56*] Di Menna, M. E. and Parle, J. N., *Proceedings of the University of Otago Medical School*, Vol. 32, 1954, p. 2.
[*57*] Burkholder, P. R., Burkholder, L. M., and Almodovar, L. R., *Botanica Marina*, Vol. 2, 1960, pp. 149–156.
[*58*] Welch, A. M., *Journal of Bacteriology*, Vol. 83, 1962, pp. 97–99.
[*59*] Nadal, N. G. M., Rodriguez, L. V., and Liburd, E. M., *Caribbean Journal of Science*, Vol. 4, 1964, pp. 347–349.
[*60*] Nadal, N. G. M., Rodriguez, L. V., and Casillas, C. in *Antimicrobial Agents and Chemotherapy-1964*, J. C. Sylvester, Ed., American Society for Microbiology, Washington, 1965, pp. 131–134.
[*61*] Sims, J. J., Donnell, M. S., Leary, J. V. and Lacy, G. H., *Antimicrobial Agents and Chemotherapy*, Vol. 7, 1975, pp. 320–321.
[*62*] Seshadri, R. and Sieburth, J. M., *Marine Biology*, Vol. 30, 1975, pp. 105–117.
[*63*] Coleman, A., Cook, W. L. and Ahearn, D. G. in *Abstracts of the Annual Meeting*, American Society for Microbiology, Washington, 1975, p. 187.
[*64*] Krasil'nikova, E. N., *Mikrobiologiya* (English traslation), Vol. 30, 1962, pp. 545–550.
[*65*] Buck, J. D., Meyers, S. P. and Kamp, K. M., *Science*, Vol. 138, 1962, pp. 1339–1340.
[*66*] Buck, J. D., Ahearn, D. G., Roth, F. K., Jr. and Meyers, S. P., *Journal of Bacteriology*, Vol. 85, 1963, pp. 1132–1135.
[*67*] Lebedeva, M. N. and Markianovic, E. M., *Revue Internationale d'Oceanographie Medicale*, Vol. 24, 1971, pp. 136–137.
[*68*] Nigrelli, R. F., Jakowska, S. and Calventi, I., *Zoologica*, Vol. 44, 1959, pp. 173–177.
[*69*] Kunen, S., Claus, G., Madri, P., and Peyser, L., *Hydrobiologia*, Vol. 38, 1971, pp. 565–576.
[*70*] Burkholder, P. R. and Ruetzler, K., *Nature*, Vol. 222, 1969, pp. 983–984.
[*71*] de Calventi, I. B., Perez, J., and Jakowska, S., *American Zoologist*, Vol. 7, 1967, p. 722.
[*72*] Constantine, G. H., Jr., Catalfomo, P., and Chou, C., *Aquaculture*, Vol. 5, 1975, pp. 299–304.
[*73*] Lingappa, B. T., Prasad, M., and Lingappa, Y., *Science*, Vol. 163, 1969, pp. 192–193.
[*74*] Beech, F. W. and Davenport, R. R. in *Isolation Methods for Microbiologists*, D. A. Shapton and G. W. Gould, Eds., Technical Series No. 3, The Society for Applied Bacteriology, Academic Press, 1969, pp. 71–88.
[*75*] van der Walt, J. P. in *The Yeasts*, J. Lodder, Ed., North Holland, Amsterdam, 1970, pp. 34–113.
[*76*] Beech, F. W. and Davenport, R. R. in *Methods in Microbiology*, Vol. 4, C. Booth, Ed., Academic Press, 1971, pp. 153–182.
[*77*] Morris, E. O., *Journal of Applied Bacteriology*, Vol. 38, 1975, pp. 211–223.
[*78*] Roth, F. J., Jr. and Rogoff, G. in *Abstracts of the Annual Meeting*, American Society for Microbiology, Washington, 1974, p. 143.
[*79*] Bott, T. L. in *Biological Methods for the Assessment of Water Quality*, J. Cairns, Jr. and K. L. Dickson, Eds., American Society for Testing and Materials, Philadelphia, 1973, pp. 61–75.
[*80*] Geldreich, E. E., *Ocean Management*, Vol. 3, 1974, pp. 225–248.
[*81*] Cabelli, V. J., Levin, M. A., Dufour, A. P., and McCabe, L. J. in *Discharge of Sewage from Sea Outfalls*, Pergamon Press, 1975, pp. 63–73.
[*82*] Fuhs, G. W., *The Science of the Total Environment*, Vol. 4, 1975, pp. 165–175.

E. H. Kampelmacher[1]

Spread and Significance of Salmonellae in Surface Waters in The Netherlands

REFERENCE: Kampelmacher, E. H., **"The Spread and Significance of Salmonellae in Surface Waters in The Netherlands,"** *Bacterial Indicators / Health Hazards Associated With Water, ASTM STP 635*, A. W. Hoadley and B. J. Dutka, Eds., American Society for Testing and Materials, 1977, pp. 148–158.

ABSTRACT: The spread of salmonellae in surface water in The Netherlands was studied during recent years. Epidemiological studies revealed that infection cycles exist probably in the following way: food of animal origin, humans, sick persons, discharge of contaminated feces into the sewage system, contamination of surface water, carry-over via insects, rodents, and birds, into stables, and on feed or directly on food. Data are given in order to substantiate this hypothesis. *Salmonella* contamination to surface water is discussed with regard to the significance in *Salmonella* epidemiology. Also the effect of the chlorination of effluents in order to break possible contamination infection cycles is summarized. Moreover, attention is given to the fact that also other infectious diseases may be maintained by infection cycles in the environment.

KEY WORDS: bacteria, water, environmental pollution, water pollution, infection cycles, bacterial water control, epidemiology

The Netherlands is a relatively small country of approximately 16 000 square miles, situated in Northwestern Europe. The country lies in a delta of two polluted rivers, the Rhine and the Meuse, and roughly one fifth of its area consists of surface water (Fig. 1). These two rivers run through densely populated areas. The Rhine starts in Switzerland and goes through Germany, and the Meuse comes from France through Belgium.

The Netherlands have 13.5 million inhabitants, 7.5 million pigs, 5 million cattle, 70 million poultry, 2 million dogs, 1 million cats, and 1 million pets other than dogs and cats. About 40 percent of the wastewater derived from inhabitants as well as from industry undergoes full biological treatment in various types of purification plants; the remaining 60 percent is discharged without treatment to surface water. We have collected much information in

[1]Laboratory for Zoonosis and Food Microbiology, National Institute of Public Health, Bilthoven, The Netherlands.

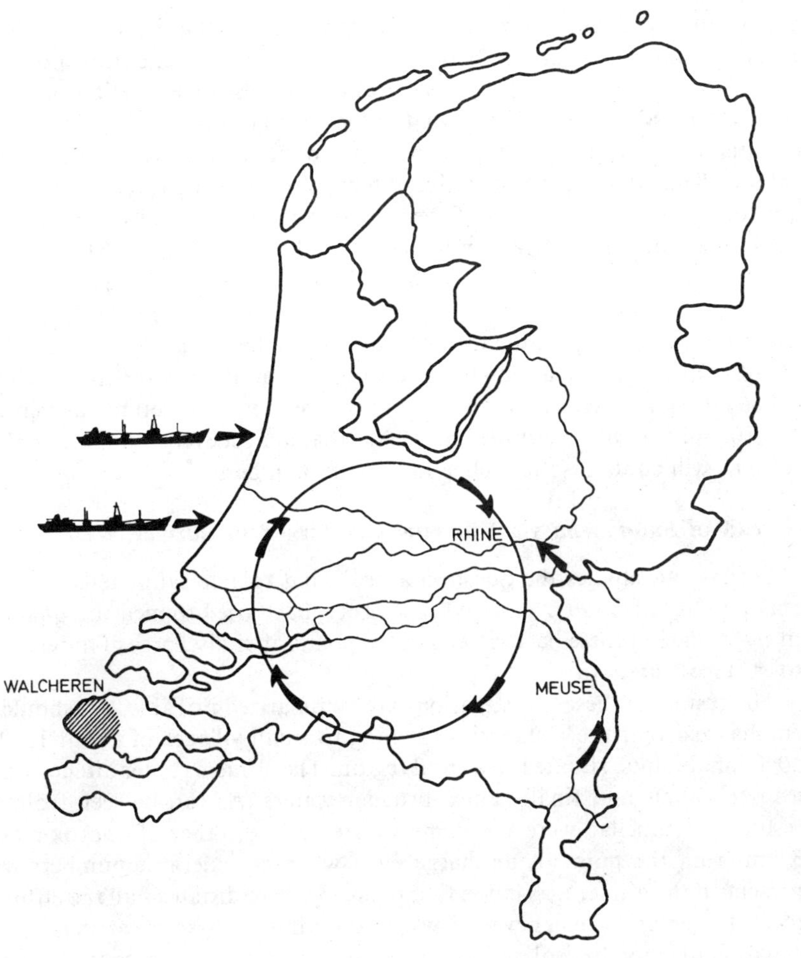

FIG. 1—*The Netherlands, import of foods and feeds by ship; delta of two large rivers (Rhine and Meuse); existence of contamination cycles built up in the country.*

recent years on the microbiological pollution of surface water and especially to the genus *Salmonella*. This paper summarizes our research work with regard to the reduction in the number of *Salmonella* during the purification of sewage, the spread of these bacteria via effluents or purification plants, the discharge of salmonellae to rivers, the effect of chlorination of effluents, and finally the significance of contaminated surface water in the epidemiology of salmonellosis.

Reduction of *Salmonella* During Wastewater Purification

In order to obtain data about their numerical reduction, we have measured quantitatively the natural *Salmonella* content of both the influent and effluents of different types of sewage purification plants, that is, plants with primary settling, installations with trickling filters, and oxidation ponds or extended aeration plants [*1–4*].[2] The results of these investigations are summarized in Table 1. This shows that in the installations with primary settling and plants with trickling filters there is little or no reduction in the numbers of salmonellae, while in the oxidation ditches there is a slight reduction, probably of the order of 1 to 2 logs. Several years ago, we artificially contaminated the influent of the purification plant of Utrecht (population 325 000; sewage volume 48 000 m^3/day) with large numbers (up to 10^{11}) of *S. utrecht* in order to study its reduction in this trickling filter installation. The results showed that there was only a reduction of about 1 to 2 logs [*1*].

Subsequent investigations have confirmed that at least 10^2 to 10^4 salmonellae/100 ml are constantly present in the influents. Sometimes as many as 10^6/100 ml may be observed. This means that almost every 100-ml sample of effluent will contain salmonellae in varying numbers.

Spread of *Salmonella* via Effluents Discharged to Surface Waters

During our studies the question arose: what role do salmonellae in effluents play in infection cycles? We, therefore, have tried to measure quantitatively the fate of these bacteria at various points after discharge of the effluents to surface waters [*3*].

The results of these investigations are summarized in Table 2. It should be emphasized that the installations belong to small villages of about 1000 to 5000 inhabitants, situated in a rural region. The effluents enter ditches which lead to canals and finally enter broader waters. As can be seen from the results, salmonellae were not demonstrated at a distance of approximately 3 km from the point of discharge even when considerable numbers were present in the effluent, as in the first plant. At these distances all the effluents reach (or enter) broader waters where the dilution is so great that salmonellae could not be isolated from 100-ml samples. In the context of our studies on infection cycles, this could mean that only the installations and

TABLE 1—*Reduction of* Salmonella *in sewage treatment plants.*

Treatment	Number of Samples Examined	Number of Samples Containing *Salmonella*/100 ml			
		<2	2 to 10^2	10^2 to 10^4	>10^4
Primary settling					
influent	48	1	13	32	2
effluent	48	. . .	14	33	1
Tickling filters					
influent	96	. . .	10	58	28
effluent	96	3	26	66	1
Oxidation ditches					
influent	39	. . .	12	25	2
effluent	39	8	24	7	. . .

[2]The italic numbers in brackets refer to the list of references appended to this paper.

TABLE 2—*MPN of* Salmonella / *100 ml in effluents and ditchwater 1973.*

Place	Effluent	Ditchwater Distance of Sampling Point from Discharge Point				
		±15 m	±150 m	±600 m	±1.5 km	±3 km
Aagtekerke	1800	1800	1800	3	...	...
Domburg	11	14	5	7	...	...
Grijpskerke	250	900	70	3	5	...
Kouderkerke	50	45	35	...	...	...
Meliskerke	1	13	17	1	...	...
Oost-Kapelle	50	120	14	20	4	...

their immediate surroundings are important in relation to the spread of salmonellae. However, it must be emphasized that in order to survey the situation, we deliberately chose very small installations which discharge only small volumes of effluent. Where larger plants discharge a large effluent stream continuously, salmonellae may be present in receiving water at greater distances from points of discharge.

Since the greater part of all effluents finally reach the great rivers, 100-ml samples from these rivers ought to contain salmonellae. For several years, we have investigated such samples from both the rivers Rhine and Meuse at the point where they enter The Netherlands [*5*]. Samples were obtained once a month from the Meuse and, in recent years, approximately twice a month from the Rhine. In Tables 3 and 4, the frequency of *Salmonella* isolations from both rivers is summarized, and, in Tables 5 and 6, the serotypes that were isolated from each river are shown. From the results, it can be seen that salmonellae were isolated from the majority of 100-ml samples and that a variety of serotypes were isolated. When salmonellae were demonstrated, most probable numbers (MPN) generally were between 1 and 10 salmonellae/100 ml. Most serotypes isolated are important in human salmonellosis in The Netherlands as well as in Germany, which contributes by far the greatest volume of effluents into the Rhine. The same is true for the Meuse, which flows into The Netherlands from Belgium. By knowing the average water flow, we can say theoretically that 32, 85, and 158 million salmonellae were entering The Netherlands in the Rhine every second during 1972, 1974, and

TABLE 3—Salmonella *isolations from River Rhine.*

	1971–1972	1974	1975
Number of			
samples	12	16	21
cultures	50	56	102
serotypes	17	18	23
MPN/100 ml			
<1	3	2	...
1 to 10	8	11	14
10 to 30	1	3	5
30 to 100	...	...	2

TABLE 4—Salmonella *isolations from River Meuse*.

	1971×1972	1974	1975
Number of			
samples	11	11	13
cultures	41	31	41
serotypes	18	13	21
MPN / 100 ml			
<1	3	3	3
1 to 10	6	8	8
10 to 30	2	. . .	2

1975, respectively. The Meuse, which is a much smaller river, contributed 3.5, 4.4, and 4.2 million salmonellae/s, respectively. By measuring the fecal contamination in general using *Escherichia coli* as an indicator organism, the conclusion was reached that the water of both rivers were equivalent to sewage effluents diluted 10 to 100 times.

Significance of Contaminated Surface Water in the Epidemiology of Salmonellosis

In the foregoing paragraphs, it has been concluded that relatively large numbers of salmonellae are spread continuously by effluents and that big rivers entering from neighboring populated areas contribute to the total *Salmonella* load. Since about 1959, salmonellosis has been one of the most important infectious diseases in this country. A greater number of investigations in The Netherlands and many other countries have shown that food of animal origin, especially meat, plays a significant role [*1*]. Pork, veal [*6-9*], and poultry [*10*] are important sources of infection [*11,12*]. These animals, which are usually clinically healthy carriers of salmonellae, are mainly infected by contaminated feed. However, research especially in an experimental farm at the National Institute of Public Health has shown also that the environment contributes to spread of the infection among the animals [*13*].

Even when *Salmonella*-free piglets were selected and fed with pellets which either contained no *Salmonella* or such small numbers that infection did not occur, up to 75 percent of the animals proved to be carriers when they were slaughtered [*6,15,16*]. Only after rigorous isolation in order to prevent access to the stables of insects, rodents, and birds which contaminate

TABLE 5—*Frequency of isolation of* Salmonella *serotypes in River Rhine.*

Serotype	1971–1972	1974	1975
S. brandenburg	3	7	11
S. enteritidis	4	2	. . .
S. infantis	7	6	16
S. panama	8	10	14
S. paratyphi B	5	2	1
S. typhimurium	6	8	11
Other types	17	21	49

TABLE 6—*Frequency of isolation of* Salmonella *serotypes in River Meuse.*

Serotype	1971–1972	1974	1975
S. brandenburg	4	4	3
S. enteritidis	1	. . .	1
S. infantis	4	3	6
S. panama	7	5	6
S. paratyphi B	7	3	. . .
S. typhimurium	3	5	4
Other types	15	11	21

feeds and water, could *Salmonella*-free animals be reared. It was suggested that rodents and birds (but possibly insects originally) carry salmonellae acquired from contaminated grassland or ditches which surround farms. In order to learn more about the ecology of *Salmonella*, investigations have been carried out for several years on the Dutch island of Walcheren; these experiments became known as the "Walcheren Project" (Fig. 2). The island contains a limited number of farms and a few small communities in addition to two larger towns, two slaughterhouses, and a total of 14 sewage purification plants. We were able to control the slaughter pigs and investigate insects, chopping block shavings, various foods of animal origin, effluents, and feces from human patients and droppings from seagulls [*17*].

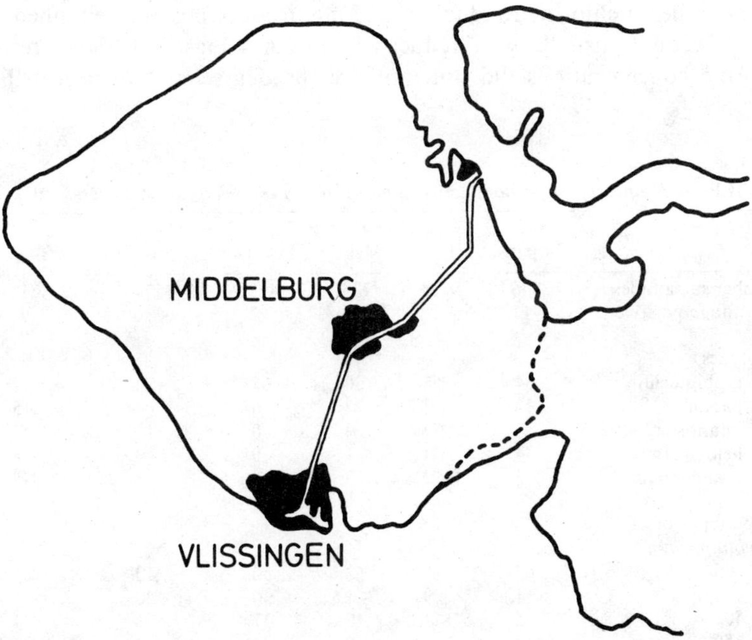

FIG. 2—*The former Island Walcheren, where* Salmonella *epidemiology has been studied in recent years.*

Moreover, the physicians on the island gave special attention to human patients suffering from diarrhea. In Table 7, some of the results of these investigations which began in 1971 are summarized. It is quite clear that only a small number of *Salmonella* serotypes is responsible for infections in man and animal and contamination of foods. The predominant serotype was *S. typhimurium*, which causes about 60 percent of human salmonellosis in The Netherlands. Initially, phage type II 505 was most important. The way in which insects were collected and investigated and the results obtained are shown in Figs. 3 and 4. If contaminated effluents do play a role in the spread of infection, then the cycles might be broken by disinfection of effluents in order to reduce the amount of contamination from animals and man.

Effect of Chlorination of Effluents on the Isolation of *Salmonella*

For several years, effluents, which are discharged into the "randmeren"—the remaining water areas between "old land" and new reclaimed inland sea areas—have been chlorinated because these lakes are used extensively for recreational purposes. The effects of chlorination on the reduction of salmonellae and other fecal bacteria at two treatment plants are summarized in Tables 8 and 9. When a total residual chlorine concentration of not less than 0.10 mg/litre (as measured by the orthotolidine method) remained after 15 min contact time according to the "Ten States Standards" [*18*], numbers of coliform organisms and fecal streptococci were reduced by approximately 3 logs. When a chlorine residual of 0.25 mg or more per litre remained after 15 min, counts usually were reduced by about 4 logs, but higher residual chlorine concentrations did not improve the degree of reduction [*4*].

TABLE 7—*Frequency of predominants salmonellae on the Island of Walcheren 1971–1975.*

	Man Patients	Pigs	Minced Meat	Insects	Feces of Seagulls	Effluents
Number of samples	846	9135	725	412	174	311
Percentage positive	11	22	23	2	25	94
Serotypes						
S. typhimurium	64[a]	58	30	67	23	29
S. panama	14	1	27	0	4	15
S. infantis	3	7	3	0	12	7
S. heidelberg	4	<1	<1	0	0	8
S. brandenburg	3	<1	13	0	4	3
Phage types of *S. typhimurium*						
II 505	63	62	63	50	10	63
	<1	9	<1	0	20	<1
II 502	3	<1	10	17	<1	3
I 650	7	6	0	33	0	7

[a]percent.

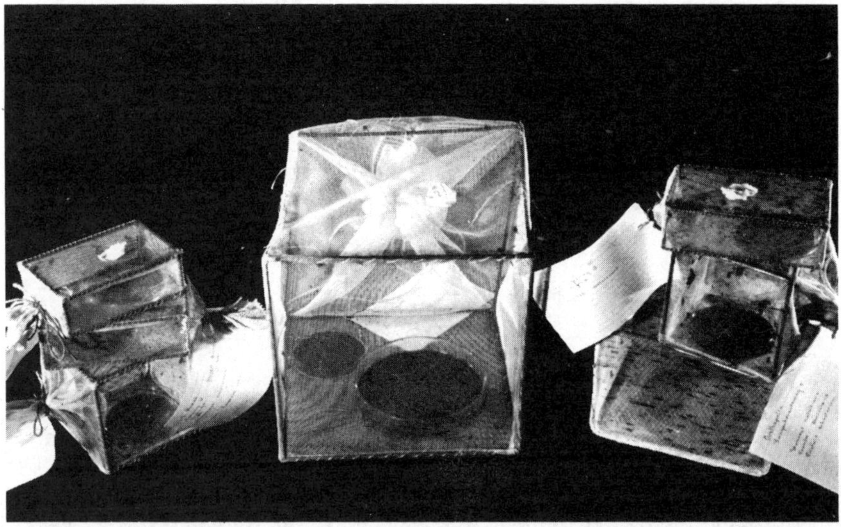

FIG. 3—*Insects are caught in special cages with media which are opened after catching.*

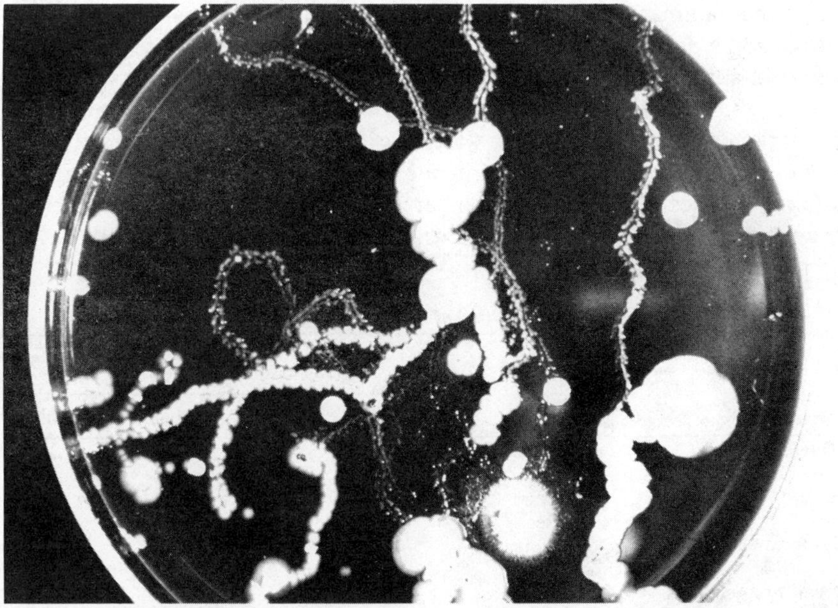

FIG. 4—*Bacterial growth, mainly salmonellae, from insects which moved over the medium.*

TABLE 8—*Effect of chlorination of effluents on fecal bacteria at plant I.*

Organism	Number of Samples	Number of Organisms/100 ml			
		<2	2 to 10^2	10^2 to 10^4	>10^4
Salmonella	32 before	...	...	32	...
	21 after	19	1	1	...
E. coli	32 before	...	...	...	32
	21 after	...	6	14	1
Coliforms	32 before	...	...	...	32
	21 after	...	3	17	1
Fecal Streptococcus	32 before	...	...	...	32
	21 after	4	16	...	1

Discussion

As a result of investigations carried out over several years in The Netherlands, it is concluded that *Salmonella* infections are maintained by cycles in which the polluted environment, and especially contaminated surface water, plays a major role (Fig. 5). This may be the case not only with salmonellae, but also with other microorganisms. As far as *Salmonella* is concerned, some parts of the cycle are quite clear. Animals are infected by contaminated feeds and become healthy carriers. Products from these animals, especially meat, frequently are contaminated and, thus, cause salmonellosis in man. However, even if animal feeds are *Salmonella* free, or if salmonellae in such feeds are reduced to such small numbers that infection in animals does not occur, contamination may still be introduced into the animal quarters from surrounding polluted environments. It is our belief that it is here that salmonellae in surface waters polluted by untreated sewage enter the cycle. Such waters often surround farms in The Netherlands. Moreover, because The Netherlands is a small country, water purification plants and animal quarters are never far from sewage effluents or polluted surface waters, perpetuating the cycles of infection. Such carry-over from contaminated waters is probably accomplished by man, birds, rodents, and insects. The latter espe-

TABLE 9—*Effect of chlorination of effluents on fecal bacteria at plant II.*

Organism	Number of Samples	Number of Organisms/100 ml			
		<2	2 to 10^2	10^2 to 10^4	>10^4
Salmonella	32 before	3	17	12	...
	21 after	18	2	1	...
E. coli	32 before	...	...	1	31
	21 after	1	8	8	4
Coliforms	32 before	...	...	...	32
	21 after	1	2	11	7
Fecal Streptococcus	32 before	...	1	9	22
	21 after	5	10	4	2

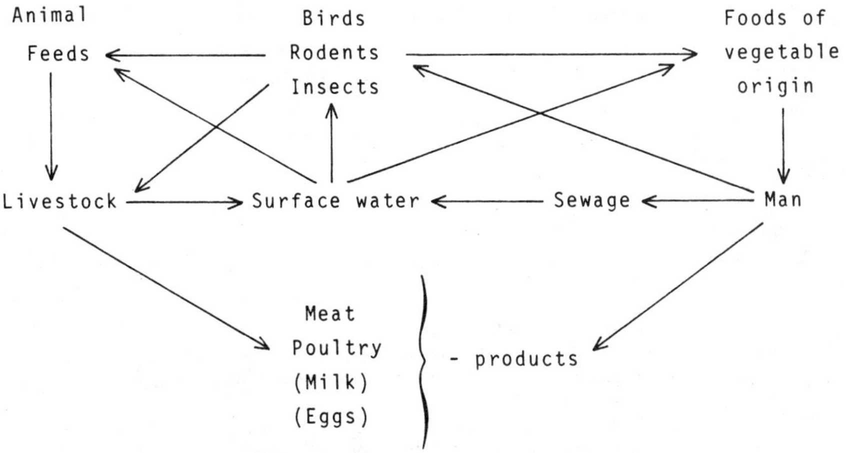

FIG. 5—Salmonella *contamination cycles.*

cially seem to play an important role, and further studies are in progress with entomologists to investigate this point. In addition to the carry-over from contaminated water to animal quarters, there also may be direct contamination of food.

We believe that our theory is supported by the striking results of the Walcheren Project, where a limited number of predominant serotypes, such as *S. typhimurium* and *S. panama*, are common to animals, food, human patients, effluent waters, insects, and birds. But we are still puzzled by the fact that only certain serotypes and phage types occur constantly in man, animal, and the environment, whereas many others are only found sporadically. The predominance of *S. typhimurium* and some of its phage types found not only in our country but in other parts of the world as well is an observation for which no explanation so far has been found.

Another puzzling observation is that of the continuous isolation of salmonellae from the sewage effluents of small villages where clinical salmonellosis has not been observed in humans. Farms are not connected to these sewage systems, and no other potential sources can be found to explain the frequent presence of salmonellae in the effluents. Perhaps salmonellae can grow in the sewage system or in the treatment plant; nutrients, temperatures, and perhaps other conditions are sometimes suitable. Research now is being carried out to investigate this possibility. Although chlorination of effluents is partially effective in reducing the numbers of salmonellae, it may, if applied on a wide scale, encounter valid opposition both by toxicologists and ecologists. Moreover, it may well be asked whether the vast amount of untreated wastewater and the sewage purification plants themselves are not even more significant than treated effluents in the overall context of *Salmonella* pollution. If this view is correct, it is perhaps necessary to realize that increasing pollution of our environment may promote outbreaks of certain infectious diseases, such as salmonellosis, cholera, and hepatitis, which have become evident in recent years. We must await results of future research to determine whether our hypothesis with regard to the significance of polluted surface waters and effluents within the *Salmonella* cycle is valid.

References

[*1*] Kampelmacher, E. H. and van Noorle Jansen, L. M., *Journal of the Water Pollution Control Federation*, Vol. 42, 1970, pp. 2069–2073.

[*2*] Kampelmacher, E. H. and van Noorle Jansen, L. M., *Journal of the Water Pollution Control Federation*, Vol. 45, 1973, pp. 348–352.

[*3*] Kampelmacher, E. H. and van Noorle Jansen, L. M., *Zentralblatt für Bakteriologie und Hygiene*, I Abstract Original, Vol. 162, 1976, pp. 307–319.

[*4*] Kampelmacher, E. H., Fonds, A. W., and van Noorle Jansen, L. M., "Reduction of Salmonella, E. coli, Coliforms and Fecal Streptococci by Chlorination of Effluents," *Water Research*, Vol. 2, 1977, pp. 545–550.

[*5*] Kampelmacher, E. H., and van Noorle Jansen, L. M., *Zentralblatt für Bakteriologie und Hygiene*, I Abstract Original, Vol. 158, 1973, pp. 177–182.

[*6*] Edel, W. and Kampelmacher, E. H., *Zentralblatt für Veterinärmedizin*, Vol. 17, 1970, pp. 875–879.

[*7*] Guinée, P. A. M., Edel, W., and Kampelmacher, E. H., *Zentralblatt für Veterinärmedizin*, Vol. 14, 1967, pp. 163–169.

[*8*] Edel, W., Guinee, P. A. M., and Kampelmacher, E. H., *Zentralblatt für Veterinärmedizin*, Vol. 17, 1970, pp. 479–484.

[*9*] Edel, W. and Kampelmacher, E. H., *Zentralblatt für Veterinärmedizin*, Vol. 18, 1971, pp. 619–621.

[*10*] van Schothorst, M., van Leusden, F. M., Edel, W., and Kampelmacher, E. H., *Zentralblatt für Veterinärmedizin*, Vol. 21, 1974, pp. 723–728.

[*11*] van Schothorst, M., Edel, W., and Kampelmacher, E. H., *Tijdschrift Voor Diergeneeskunde*, Vol. 95, 1970, pp. 279–282.

[*12*] den Dulk-Gerritsen, M. A., Klaarenbeek, T., Tamminga, S. K., and Kampelmacher, E. H., *Voeding*, Vol. 10, 1974, pp. 506–509.

[*13*] Edel, W., Guinée, P. A. M., van Schothorst, M., and Kampelmacher, E. H., *Proceedings*, Symposium on Microbiological Foodborne Infections and Intoxication, Ottawa, Canada, 1973, pp. 64–67.

[*14*] Edel, W., van Schothorst, M., Guinée, P. A. M., and Kampelmacher, E. H., *Zentralblatt* für Veterinärmedizin, Vol. 17, 1970, pp. 730–738.

[*15*] Edel, W., van Schothorst, M., Guinée, P. A. M., and Kampelmacher, E. H. in *The Microbiological Safety of Food*, Academic Press, England, 1972.

[*16*] Edel, W., van Schothorst, M., Guinée, P. A. M., and Kampelmacher, E. H., *Zentralblatt für Bakteriologie und Hygiene*, I Abstract Original, Vol. 158, 1974, pp. 568–577.

[*17*] Edel, W., Guinée, P. A. M., van Schothorst, M., and Kampelmacher, E. H., *Zentralblatt für Bakteriologie und Hygiene*, I Abstract Original, Vol. 221, 1972, pp. 547–549.

[*18*] Juliano, F. E., *Water and Sewage Works*, Vol. 28, 1968, Table IV.

Gertrud Müller[1]

Bacterial Indicators and Standards for Water Quality in the Federal Republic of Germany

REFERENCE: Müller, Gertrud, "**Bacterial Indicators and Standards for Water Quality in the Federal Republic of Germany**," *Bacterial Indicators / Health Hazards Associated With Water, ASTM STP 635*, A. W. Hoadley and B. J. Dutka, Eds., American Society for Testing and Materials, 1977, pp. 159–167.

ABSTRACT: In the Federal Republic of Germany, the choice of the indicator differs with the water that has to be examined. For drinking water, the classical parameters are the total colony count and the *Escherichia coli* test.

While the colony count seems to be considered of less importance in other European countries, the German legislation contains mandatory limits. Furthermore, *E. coli* must not be present in 100 ml of drinking water. Fecal streptococci and *Pseudomonas aeruginosa* are good indicators for the examination of swimming water in pools which have been treated by flocculation, filtration, and chlorination.

KEY WORDS: bacteria, water, coliform bacteria, colony count, cholera, typhoid fever, incubation conditions, sewage, swimming pools

Since the discovery of bacteria at the end of the 19th century and since the application of slow sand filtration to treat surface waters, the colony count has been employed in Germany to control the function and efficiency of treatment of public water supplies. It was Robert Koch who concluded that, if the effluent from a slow sand filter has a colony count of less than 100 bacteria/ml, it should be suitable for drinking purposes and present no hazard to human beings of acquiring cholera or typhoid fever [*1,2*].[2]

I think this is the reason why in Germany total plate counts still constitute a part of the conventional bacteriological examination of drinking water. For the same reason, 100 colony forming units/ml became something of a classical standard [*3*]. Today, this standard is no longer in use for control of slow sand filter effluents only. It has become a standard for drinking waters and swimming pools. Thus, the total plate count became part of the German

[1]Professor, Bundesgesundheitsamt, Institut for Wasser, Boden u. Luft Hygiene, Corrensplatz, West Germany.

[2]The italic numbers in brackets refer to the list of references appended to this paper.

drinking water law, after having been a standard method for nearly 35 years [*4*]. The standard method for performing colony counts was integrated into the text of the law, so there does exist the possibility of comparing the results of different laboratories and evaluating them.

The colony count is defined as "that number of visible colonies which are able to develop from those bacteria contained in 1 ml of water by using pour plate cultures of a medium enriched with meat extract and peptone, a fixed incubation temperature, and a fixed time of incubation. The counting must be done using a microscope with a magnification of eight times."

By using media which are enriched with respect to peptone and meat extract and by limiting time and temperature of incubation, this method will not support the growth of all bacteria in a water sample, but it will provide an optimal count of those species which are, in the broadest sense of the word, epidemiologically relevant—not by reproducing germs of waterborne infections directly, but by providing an indication of the level of contamination by heterotrophic species [*5*]. The purpose of the temperature, incubation time, and enriched medium is the prevention of formation of colonies by bacteria indigenous to water, that is, the autochtonous flora.

In contrast to the German standard methods for drinking water which offer incubation temperatures of 37 ± 0.5°C and 20 ± 2°C, the drinking water law allows only one temperature of incubation, that is, the temperature of 20 ± 2°C. This was necessary as the drinking water law has mandatory and recommended values, and, after the experience of the last 100 years, all values are related to this temperature. According to the law, the recommended and mandatory limits on colony counts have to be tested only for water which leaves the waterworks or its reservoirs. The force of the law ends at the water meter prior to the connection to the consumer's system. Nevertheless, there are other laws included in the public health regulations to control the quality of tap water, and the mandatory and recommended values of the drinking water law apply accordingly. In addition to those tests which the waterworks and the public health authorities are obliged to perform, it may be desirable to control individual steps of the water treatment process periodically at certain times of the year and to do so also at different points of the distribution system. Only when applied in this way is the colony count an effective method of quality control. Variations of normally good results may suggest disturbances in the treatment or distribution systems. Sudden increases in colony counts which had been low for several years have indicated the sources of waterborne outbreaks in Germany. Such an increase in the total count occurred prior to the typhoid fever outbreak in Hannover in 1926 [*6*], the cause of which was flooding by highly contaminated river waters of wells which were the source of raw water for the drinking water supply (Fig. 1). The most interesting observation was that before the water in the distribution system was positive for *Escherichia coli* or coliforms, high colony counts were observed, and an outbreak of nonspecific gastrointestinal illness (named the "water disease") consisting of 40 000 cases occurred

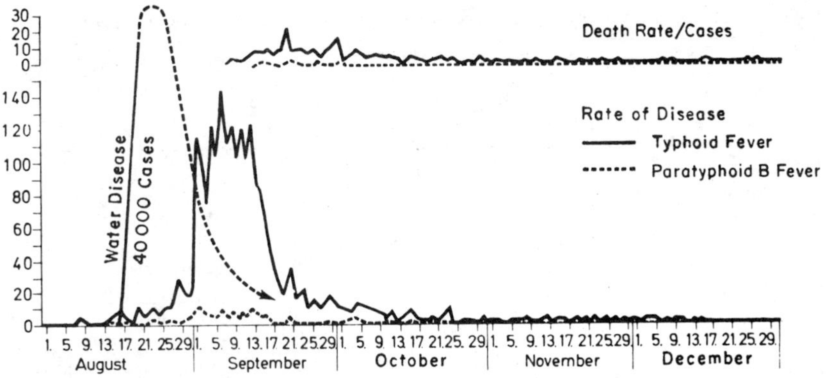

FIG. 1—*Epidemic outbreaks of typhoid and paratyphoid B fever in Hanover in 1926.*

two weeks before the first cases of typhoid fever were reported. The same observations were made during the typhoid fever outbreaks at Pforzheim in 1919 and at Gelsenkirchen in 1889. Therefore, we read with great interest that the rules and regulations for the U.S. Safe Drinking Water Act say that "the administrator has had evidence that the standard plate count does have health significance and in addition is a valid indicator of bacteriological quality of drinking water" [*7*].

The method which is part of the German drinking water law, as well as the German standard method for performing the colony count, requires that colonies be counted after two days of incubation at 20°C and with a microscope at ×8 magnification. Deviations from this procedure result in "variations" which invalidate comparisons of results [*8*]. As many investigations have shown, there are no differences in the results whether one uses agar, gelatin, or silacagel for consolidating the medium [*9*].

Looking for recommended or mandatory values for colony counts in Europe (Table 1), one can make the following observation: the farther the State is situated on the eastern part of the globe, the lower are the recommended or mandatory values. For example, Poland has a standard and a mandatory limit of 25 colonies/ml, Czechoslovakia has regulation and mandatory limits of 20 and 100 colonies/ml (incubation temperatures 37 and 20°C) for public water supply and 100 and 500 colonies/ml (37 and 20°C) for well water. Yugoslavia has established a mandatory limit of 10 colonies/ml (37°C) for treated water, 100 colonies/ml (37°C) for raw ground waters, and 300 colonies/ml (37°C) for raw surface waters. Romania has a standard including mandatory limits of 20 colonies/ml for public water supplies for more than 70 000 consumers and of 100 to 300 colonies/ml for other supplies. On the other hand, Switzerland requires that raw waters entering treatment works contain no more than 100 colonies/ml; untreated waters for consumption, no more than 300 colonies/ml; treated water immediately after treatment, no more than 20 colonies/ml; and treated water in

TABLE 1—*Colony count; recommended or mandatory values in Europe.*

Country	Recommended Value/ml	Mandatory Value/ml	Temperature	Comments
Poland	...	25	...	
Czechoslovakia	...	20	37°C	
	...	100	20°C	public supply
	...	100	37°C	
	...	500	20°C	well water
Yugoslavia	...	10	37°C	for treated water
		100	37°C	raw underground water
		300	37°C	raw surface water
Romania	...	20	...	public water supply for 70 000 consumers
		100 to 300	...	other supplies
Switzerland	...	100	...	raw water entering works
		300	...	raw water during distribution
		20	...	treated water immediately after treatment
		300	...	treated water in distribution system
Spain	...	50 to 65	37°C	good quality water
		100	37°C	tolerable quality water
Netherlands, Sweden, GDR	...	100	20°C	
Federal Republic of Germany				
Disinfected water	20	...	20 ± 2°C	
All other kinds of drinking water	100	...	20 ± 2°C	
Water tanks	1000	...	20 ± 2°C	
Bottled waters	...	1000	20 ± 2°C	
France, Austria, Great Britain	...	...	...	no guide or imperative values

the distribution system, no more than 300 colonies/ml. Spain has a food code which distinguishes between good and tolerable numbers. Water of good quality has a mandatory limit of 50 to 65 colonies/ml (37°C), and water of tolerable quality has a limit of 100 colonies/ml (37°C). While France, Austria, and Great Britain have no mandatory limits for the colony count, The Netherlands, Sweden, and the German Democratic Republic (GDR) have a norm or a rule which prescribes a limit of 100 colonies/ml (20°C) [*9*]. The Federal Republic of Germany distinguishes between recommended and mandatory limits. Drinking water after chlorination has a recommended limit of 20 colonies/ml, undisinfected water of 100 colonies/ml, and recommended limits for water on ships, trains, or aircraft, or reservoirs of rainwater are 1000 colonies/ml [*10,11*]. The mandatory limit for bottled and packed water is 1000 colonies/ml. This mandatory limit for bottled water gave rise to many discussions. In Germany, the bacteriological quality of bottled water, which is free of carbon dioxide, often is very bad. In more than 50 percent of bottled waters sampled, colony counts reached

1 or 10 million/ml. A great part of the bacterial populations responsible for such high counts are gelatin liquifying fluorescent pseudomonads [*12*]. Furthermore, up to 30 percent of bottled waters contain clostridia and *Pseudomanas aeruginosa*. If these organisms have an opportunity to multiply in food, outbreaks of disease may occur [*12*]. Since advertisements of the producers of such bottled waters are very persistent in recommending it as more suitable for preparing infant food, ice cubes, and long drinks than tap water is, it is felt that adherance to standards is of utmost importance. Therefore, this mandatory limit of 1000 colonies/ml for bottled water was necessary on behalf of epidemiological considerations.

High colony counts occur also following installation of filters, ion exchange equipment, and phosphorus dosage equipment in households for the softening of water [*14*]. More than 150 000 such devices have been installed already in Germany and more are being installed at a rate of about 20 000/year [*15*]. It is well known that filter media (that is, charcoal, polysterol, asbestos, etc.) provide bacteria with the opportunity to multiply, and the effluent then is highly contaminated with bacteria. Ion exchange resins not only have the capacity to adsorb bacteria but, depending on the make, frequently contain defects which afford protection to bacteria during disinfection [*16*]. Resin granules consist of organic matter which promotes bacterial growth [*17*].

The second parameter which is regulated by law in Germany consists of the *E. coli* and coliform content of drinking waters. The limit for *E. coli* is mandatory and that for coliforms is a recommended limit. The difference between the definitions of "mandatory" and "recommended" limits in the drinking water law is established by the consequences which arise when the limits are exceeded. If *E. coli* or coliforms or both are present in the drinking water, waterworks have to report to public health authorities as soon as possible.

The presence of *E. coli* has the consequence that the waterworks are closed, and there will be no delivery of drinking water up to clearance (mandatory limit). Presence of coliforms give rise to considerations about the source of this contamination and measures have to be taken in order to keep the water supply safe (recommended limit). As *E. coli* is an enteric bacterium that under normal circumstances is not able to multiply in water or soil, it is considered of definite fecal origin, and, therefore, its presence is considered of greater significance. *E. coli* must not be present in 100 ml of drinking water. If *E. coli* is present in 100 ml of drinking water, the water must be suspected of being fecally polluted. The relevant paragraph of the drinking water law states that "drinking water and water used in the food industry must be free of bacterial pathogens. This condition is realized if 100 ml of water do not contain any *E. coli*." Furthermore, 100 ml of drinking water should be free of coliforms. Although the presence of coliforms in drinking water is not permitted, the interpretation of their demonstration is different. In contrast to *E. coli*, coliforms are able to multiply outside of the intestinal tract of warm-blooded animals. They may be of fecal origin, but they may occur also in water and soil. Nevertheless, the presence of both *E. coli* and coliforms has to be reported immediately to the local board of public health.

Methods for the detection of *E. coli* and coliforms in water, like the total plate count, are part of the law. This means that laboratories of waterworks and state laboratories are obliged to use prescribed methods. Only then is a comparison of the results possible. To demonstrate *E. coli* or coliforms in 100 ml of drinking water, there are two procedures offered.

1. Membrane filtration followed by incubation of the filter on Endo agar for 20 to 44 h at 37 ± 0.5°C.

2. Inoculation of 100 ml of water into 100 ml of double strength lactose broth followed by incubation for 20 to 44 h at 37 ± 0.5°C.

The procedures are described in Fig. 2.

Red colonies on the membrane filters may be *E. coli*, coliforms or *Aeromonas* species, and they must be differentiated as described in Fig. 1. In Germany, we have great difficulties in preventing the small, poorly equipped

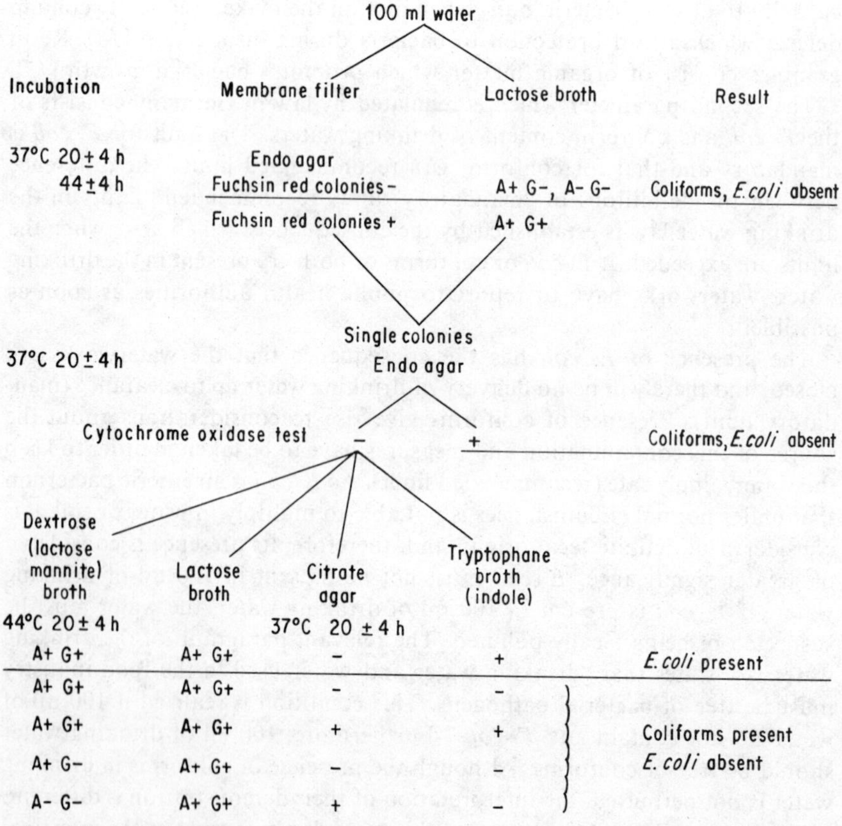

FIG. 2—E. coli *coliform detection in drinking water (production of A* = acid and G = gas).

laboratories from basing identification upon colonial appearance or gas formation in lactose broth alone. Coliforms may grow in dark red colonies, with or without a spot on the backside of the filter and with or without a white margin. Coliforms never grow as pink or colorless colonies, for one of the main criteria of the group is the fermentation of lactose which is always combined with a positive fuchsin red reaction [*18,19,20*].

Additional fecal indicators, such as fecal streptococci or sulfite-reducing spore forming anaerobes, are used only in special cases during the bacteriological examination of drinking water. Clostridia organisms have been of value for the detection of remote contamination, particulary in wells or reservoirs. The detection of clostridia has become a German standard method for the examination of water, sewage, and sludge. Besides, detection of clostridia is useful for the layout of protection areas for water catchment equipments.

Water for swimming purposes may be natural water (rivers, ponds, sea), or it may be treated and chlorinated in pools. For surface water, including seawater, the test for *E. coli* is most commonly employed. Salmonellae are used as an additional fecal indicator, and if the *E. coli* count or the load of *Salmonella* is high, or both, salmonellae are regarded as a potential health hazard. Laws, rules, or regulations do not exist for the regulation of swimming waters, and the responsibility for quality of surface swimming waters is still left to local boards of public health.

Water in swimming pools by regulation has to have drinking water quality from the bacterial point of view. This means that no *E. coli* can be present in 100 ml of water, and a colony count must be below 100 colonies/ml. This does not mean that other indicators such as fecal streptococci, staphylococci, or *P. aeruginosa* or both are not equally well suited for the examination of artificial pool water. These organisms are interesting, as the contamination of pool water may occur not only as a result of fecal pollution, but may arise from urine, the human skin, and the nose and throat region. Therefore, *P. aeruginosa* and perhaps staphylococci are good additional test organisms, especially if a pool is associated with a hospital and is used for ill people. In order to prevent infections in patients, effective disinfection of this water is very important. As gram-positive bacteria generally are more resistant to chlorination than gram-negative ones, the examination of chlorinated swimming pool water for fecal streptococci is very effective. On the other hand, since numbers of fecal streptococci in feces are low in relation to *E. coli*, only the presence of fecal streptococci in the absence of *E. coli* indicates poor conditions [*21*].

The method of detecting fecal streptococci is not yet standardized in Germany, but the method of detection of *P. aeruginosa* was standardized last year and is part of the German standard for examination of drinking water, sewage, and sludge [*22*]. The test method for fecal streptococci is membrane filtration and incubation of the membrane filter on sodium azide-tetrazolium chloride agar (incubation for 20–44 h at 37° C). *P. aeruginosa* is tested by membrane filtration and transfer of the membrane filter into malachite green broth, incubating at 42° C for 20 h, plating on nutriant agar, and confirmation by plating single colonies on King's A and King's B medium. If the examination of less than 100 ml of water is wanted, the water may be added to an equivalent amount of double-strength malachite green broth.

Summary

In the Federal Republic of Germany, the choice of the indicator differs with the water that has to be examined. For drinking water the classical parameters are the total colony count and the *E. coli* test. The colony count was the first test applied to drinking waters, having been developed in the years after 1880 to examine the effluents of slow sand filters. If the count was less than 100 counts/ml, it was considered that no typhoid bacteria or cholera vibrios were present in the effluent.

While the colony count seems to be considered of less importance in other European countries, the German legislation contains mandatory limits. Furthermore, *E. coli* must not be present in 100 ml of drinking water. In drinking water examination, coliforms are regarded as a recommended parameter which should be absent in drinking water and, if present, suggests that the treatment process is not functioning properly.

Additional parameters, such as clostridia, fecal streptococci, *Salmonella* or *P. aeruginosa*, are used under special circumstances. Salmonellae in surface water or seawater used for swimming purposes are looked on as an additional indicator of fecal pollution and only in special cases as a pathogen. Fecal streptococci and *P. aeruginosa* are good indicators for the examination of swimming water in pools which have been treated by flocculation, filtration, and chlorination.

References

[*1*] Koch, Robert, *Wasserfiltration und Cholera*, Gesammelte Werke Bd.2, Teil 1, Georg Thieme Verlag, Leipzig, 1912.

[*2*] Wallichs, G., *Deutsche Medizinische Wochenschrift*, Vol. 15, 1891.

[*3*] "Leitsatz für die zentrale Trinkwasserversorgung," DIN 2000, Deutsches Institut fur Normung, Berlin, 1973.

[*4*] "Deutsche Einheitsverfahren für Wasser-, Abwasser- und Schlamm- untersuchungen," Absch. K 5, 6.Lieferung, Verlag Chemie, Weinheim/Bergstraße, 1972.

[*5*] Habs, H., Müller, Gertrud, Schädlich, Vera, Schubert, R., and Selenka, F., Die neuen Einheitsverfahren für bakteriologische Wasseruntersuchung. *Städtehygiene*, Vol. 23, 1972, pp. 131–134.

[*6*] Mohrmann, R., *Veröffentl. Gebiete Med. Verw.*, Richard Schoetz, Berlin, Vol. 24, 1927, p. 5.

[*7*] "National Interim Primary Drinking Water Regulations," Part IV, No. 141.15, Environmental Protection Agency.

[*7a*] *Federal Register*, Vol. 40, No. 248, Rules and Regulations, Washington, D.C., Dec. 1975.

[*8*] Müller, Gertrud, *Koloniezahlbestimmungen im Trinkwasswer*, Vol. 113, Gas- u. Wasserfach (GWF), 1972, pp. 53–100.

[*9*] Müller, Gertrud, *Zbl. Bakt. Hyg. I.Abt. Orig. B*, Vol. 157, 1970, pp. 376–386.
[*10*] Müller, Gertrud, *Städtehygiene*, Vol. 12, 1961, pp. 113–114.
[*10a*] Müller, Gertrud, *Städtehygiene*, Vol. 16, 1965, pp. 240–241.
[*11*] Soenke, H., "Untersuchungen über bakterielle Indikatoren und deren Verhalten in Tafelwässern," Dissertation, Freie Universität Berlin, Fachbereich Hygiene, 1975.
[*12*] Müller, Gertrud, *Zbl. Bakt. Abt. I, Orig. A*, Vol. 227, pp. 50–55.
[*13*] Müller, Gertrud, *Hygiene der Trinkwassernachaufbereitung*, Internationale wasserfachliche Aussprachetagung des DVGW, Basel, 1975.
[*14*] *Z. Stiftung Warentest*, Vol. 2 1972, p. 77.
[*15*] Müller, Gertrud and Herzel, F., *Zb. Bakt. Hyg. Abt. I., Orig. B*, Vol. 156 1973, p. 524.
[*16*] Müller, Gertrud, *Zbl. Bakt. Hyg. Abt. I, Orig. B*, Vol. 158, 1974, p. 507.
[*17*] "Deutsche Einheitsverfahren für Wasser-, Abwasser- und Schlamm- untersuchungen," Absch. K 6, Verlag Chemie, Weinheim/Bergstraβe, 1972.
[*18*] Schubert, R. W. und Deutsch, E., *Archiv fuer Hygiene*, Vol. 153, 1969, p. 205.
[*19*] Schubert, R. W., *Archiv fuer Hygiene*, Vol. 154, 1971, p. 500.
[*20*] Müller, Gertrud, *Archiv fuer Hygiene*, Vol. 151, 1967, p. 403.
[*21*] Schubert, R. W. und Blum, U., *Zbl. Bakt. Hyg. Abt. I, Orig. B*, Vol. 158, 1974, p. 583.

W. O. K. Grabow[1]

South African Experience on Indicator Bacteria, *Pseudomonas aeruginosa*, and R+ Coliforms in Water Quality Control

---

REFERENCE: Grabow, W. O. K., "**South African Experience on Indicator Bacteria, *Pseudomonas aeruginosa*, and R$^+$ Coliforms in Water Quality Control,**" *Bacterial Indicators/Health Hazards Associated With Water, ASTM STP 635*, A. W. Hoadley and B. J. Dutka, Eds., American Society for Testing and Materials, 1977, pp. 168–181.

ABSTRACT: Drinking waters obtained from conventional sources and reclaimed wastewater in South Africa have been analyzed for standard bacterial plate count, total coliform bacteria, fecal coliforms, confirmed *Escherichia coli*, type I, *Pseudomonas aeruginosa*, fecal streptococci, *Clostridium perfringens*, *Staphylococcus aureus*, enteric viruses, and parasite ova. Apart from the standard plate count, waters which were free of coliforms rarely contained any of the other organisms. *P. aeruginosa* occasionally was isolated from waters with a zero coliform count, and it is suggested that this organism should be included in some routine quality tests. Epidemiological studies on an isolated community, which included periods when drinking water obtained by conventional methods was supplemented with supplies reclaimed from wastewater, indicated that water which conforms to the criteria of nil coliforms and *P. aeruginosa*/100 ml, a standard plate count of less than 100/1 ml, and no detectable enteric viruses/10 litre will not transmit microbial diseases. Hospital and city sewage, as well as a river and reservoir polluted with secondary treated sewage, contained large numbers of coliforms with transferable (R factor) resistance to one or more of five commonly used antimicrobial drugs. The survival of R$^+$ coliforms and the transfer of resistance during treatment of sewage and in polluted waters indicated that advanced treatment of wastewater will be necessary to protect water resources against these organisms which have potential health significance. It is suggested that specifications limiting R$^+$ bacteria should be included in water quality standards.

KEY WORDS: bacteria, water, coliform bacteria, health aspects, reclaimed water, resistant bacteria, R factors, hospital effluents, epidemiology, water treatment, coliforms, clostridia, enterococci, staphylococci, fecal streptococci, enteric viruses, intestinal parasites

[1]National Institute for Water Research, South African Council for Scientific and Industrial Research, Pretoria, South Africa.

Southern Africa's water resources are not sufficient to meet the increasing demand for conventionally produced potable water for more than another 25 years [*1*].[2] Extensive research is therefore in progress to protect available resources and to augment existing supplies by means such as the reuse of wastewater. Health considerations are of primary concern in this work, and certain findings, some of which form part of a national ten-year project on water quality [*1*], are presented here.

In order to establish microbiological quality standards for the prevention of pollution of water resources and protection of the health of consumers of reclaimed wastewater, the following lines of research are being followed. Conventionally derived drinking waters are analyzed for as many microbiological parameters as possible, and the findings are correlated with epidemiologic data. The study includes research on the microbiology of water supplies and on disease among consumers in Windhoek, Southwest Africa, where directly reclaimed wastewater has been used occasionally since 1968 to supplement drinking water derived by conventional methods from surface and bore-hole waters [*1,2*]. This city represents a relatively isolated community and offers an ideal model for the study of possible adverse health effects caused by the consumption of reclaimed water. The study also includes investigations of the behavior of indicator and potentially harmful organisms in conventional wastewater treatment processes, reclamation processes, and conventional processes for the treatment of drinking waters.

In addition to conventional microbiological parameters of water quality [*3,4*], investigations of coliform bacteria resistant to antimicrobial drugs have been undertaken. Special attention has been given to coliforms carrying resistance transfer (R) factors which are rapidly transferable by conjugation among many bacteria. R factors are extrachromosomal deoxyribonucleic acid elements (plasmids) which may confer on their hosts simultaneous resistance to a wide spectrum of antibacterial agents which include drugs, heavy metals, phages, bacteriocins, and ultraviolet light [*5*]. They may code for the synthesis of enzymes which catalyze the repair of damaged deoxyribonucleic acid, change the composition of the cell envelope and morphology, and control properties such as swarming. They may carry determinants coding for the excretion of proteolytic enzymes, hydrogen sulfide (H_2S) production, sugar fermentation, enteropathogenicity, enterotoxin production, lethality, hemolysin and bacteriocin production, cell surface charge, and the ability to grow at elevated temperatures or on selective media [*6*]. These alterations in the physiology and composition of bacteria may affect their survival during treatment of water. Under the application of selective pressures, which may be expected when water is continually reused, the incidence of bacteria resistant to water treatment processes may increase [*5,6*]. A condition similar to that in human and animal medicine, where the uncontrolled use of drugs rapidly selected for resistant strains which seri-

[2]The italic numbers in brackets refer to the list of references appended to this paper.

ously reduce the value of antimicrobial therapy, may develop [*5,6*]. In this study, R+ coliforms were used as a model to investigate the role of water in the spread of drug resistant bacteria as well as to investigate the general behavior of R+ bacteria in the water environment.

The results obtained support a proposed standard which specifies that drinking water should contain no coliforms or *P. aeruginosa*/100 ml, a standard plate count of less than 100/1 ml, and no detectable enteric viruses/10 litre.

Materials and Methods

Water Supplies, Wastewater, and Treatment Plants

The study deals primarily with Windhoek dam, bore-hole and reclaimed waters [*1,2*], Rand Water Board drinking water [*7*], bore-hole, reservoir, and river waters in the Pretoria area [*6,8,9*], city and hospital wastewaters in Worcester [*10*], Pietermaritzburg [*10*], Pretoria [*6*] and Windhoek [*8*], the 4.5 thousand litre/day experimental Stander [*1*] and pilot [*11*] reclamation plants in Pretoria, and conventional wastewater purification plants in Windhoek [*2*] and Pretoria [*6,12*].

Microbiological Analyses

Standard Bacterial Plate Count—Yeast extract agar pour plate cultures were incubated at 37°C for 48 h [*10*]. All colonies were counted.

Total Coliform Count—Membrane filtered specimens were incubated at 37°C for 24 h on a modified MacConkey medium [*10*]. Since March of 1976, duplicate tests for all coliform bacteria were done on this medium using a resuscitation top layer [*13*]. Sartorius SM11406 or Gelman GN-6 membranes, pore size 0.45 μm, were used throughout. All dark yellow colonies of lactose fermenters were counted.

Fecal Coliform Count (Presumptive Escherichia coli, *Type I)*— Membrane filtered specimens were incubated at 44.5°C for 24 h on the medium used for total coliforms.

Confirmed E. coli, *Type I*—Representative colonies were picked from fecal coliform plates and *E. coli*, type I was confirmed by means of the indole test [*10*].

Fecal Streptococci—Membrane filtered specimens were incubated at 44.5°C for 48 h on Difco M-enterococcus agar [*10*]. All dark red colonies were counted.

Clostridium perfringens—Membrane filtered specimens were incubated at 44.5°C for 14 h on Wilson and Blair glucose sulfite iron medium [*10*]. Black colonies with a dark halo were counted.

Staphylococcus aureus—Membrane filtered specimens were incubated at 37°C for 48 h on Difco staphylococcus medium No. 110 followed by purification on Difco mannitol salt agar [*10*] and confirmation by a DNase test [*14*].

P. aeruginosa—One or more of the following methods were used: the most probable number (MPN) method of Drake followed by confirmation on acetamide medium [*10*], the membrane filtration (MF) method of Levin and Cabelli [*15*], or incubation in Drake's medium of membranes through which 100 ml of water had been filtered, followed by confirmation as just mentioned in order to obtain a qualitative positive or negative result.

Virus Counts—Water samples were concentrated by means of ultrafiltration (Amicon Model 2000, Amicon Corporation, Lexington, Mass.) using 150-mm diameter type XM50 membranes followed by either evaluation of the 50 percent tissue culture infectious dose ($TCID_{50}$) in roller tubes or inoculation of 16-ml concentrate into one litre Roux flasks to obtain a qualitative positive or negative result. Primary monkey kidney tissue cultures were used throughout [*16*].

Counts of Parasite Ova—Counts of parasite ova were obtained by direct microscopic examination of 10 litre quantities of water concentrated by means of a nylon-gauze filter [*17*].

Drug Resistant Coliforms—The enumeration and isolation of resistant coliforms and the demonstration of resistance transfer to *E. coli* E25 and *Salmonella typhi* N have been described previously [*18*]. The following drugs were used in enumeration media: ampicillin, chloramphenicol, streptomycin, sulfonamide, and tetracycline. In later studies, sulfonamide was replaced by kanamycin.

Epidemiologic Studies

Epidemiologic studies were coordinated by the South African Institute for Medical Research. Detailed studies have been conducted in Windhoek where as many data as possible on communicable diseases caused by salmonella, shigella, enteropathogenic *E. coli*, vibrio, enterovirus, and schistosoma infections as well as viral hepatitis, meningitis, and encephalitis, have been recorded. Private medical practitioners, all hospitals and clinics, and health authorities participated in the study [*19*].

Results

Counts of Indicator Organisms

Although details are given for analyses which were conducted during the past two years only, the results are representative of routine analyses which in many cases cover ten years or more. Among conventional drinking waters, supplies derived by the Rand Water Board from surface waters had the highest quality judged by all the parameters, while supplies derived from boreholes had the poorest quality.

Standard Plate Count—Table 1 shows that 51.0 percent of conventional

TABLE 1—*Conformation of drinking water samples to proposed microbiological quality standard.*

Parameter	Standard	Samples Failing to Conform to Proposed Limits			
		Conventional Supplies		Reclaimed Water	
		Ratio[a]	Percentage	Ratio[a]	Percentage
Standard plate count	100/ml	147/288	51.0	31/83	37.3
Total coliforms	0/100 ml	75/317	23.7	12/84	14.3
Fecal coliforms	0/100 ml	42/318	13.2	7/84	8.3
Virus count	0/10 litre	1/318	0.3	0/84	0.0

[a]Number failing to meet standard/total number of samples.

drinking water samples did not conform to the proposed standard of 100 colonies/ml and that this criterion is the one which was most frequently exceeded. Among samples of reclaimed water from the Stander reclamation plant, 37.3 percent failed to meet the proposed standards during experimental runs only, and, in all cases, failure to meet the standards was associated with experimental stressing of the system or intentional technical defects. Failure to meet fecal and total coliform limits was less frequent. These studies, and similar experience on the Windhoek reclamation plant, proved the standard plate count a most useful tool in the general control of advanced treatment plants, since it was always the first parameter to indicate anomalies such as microbial contamination of activated carbon units or insufficient disinfection. Table 2 shows that among 300 samples which conformed to a standard plate count of 100 colony forming units/ml, numbers which exceeded maximum levels for other parameters (100/ml for the standard plate count, 0/10 litre for the virus count, and 0/100 ml for all other parameters, throughout) ranged from 0 (0.0 percent) to 55 (18.3 percent). Counts of fecal streptococci and viruses rarely were isolated when the standard plate count was less than 100/ml and rarely contributed information on the quality of drinking water. *C. perfringens* which was isolated frequently may be considered a potentially useful additional indicator.

Total Coliform Count—The proposed limit of 0 coliforms/100 ml was exceeded in 23.7 percent of conventionally derived drinking water samples and in 14.3 percent of samples taken from the final product during experimental runs on the Stander reclamation plant (Table 1). Among 574 samples from which coliforms were not isolated, the number of samples which exceeded maximum permissible levels for other microorganisms ranged from 0.0 and 0.6 percent for fecal streptococci and viruses, respectively, to 29.1 percent for the standard plate count (Table 2). Consequently, the latter would be a profitable indicator to use in addition to the total coliform count. *Aeromonas, Pseudomonas,* and *Acinetobacter* species, and related bacteria often overgrew membranes, interfering with counting of total coliforms.

TABLE 2—*Comparison of frequencies with which samples satisfying individual limits fail to satisfy other limits.*[a]

Number of Samples not Exceeding Maximum/ Permissible Counts	Frequency of Samples Exceeding Maximum Permissible Counts for Alternate Parameters															
	Plate Count		Total Coliforms		Fecal Coliforms		*P. aeruginosa*		Fecal Streptococci		*S. aureus*		*C. perfringens*		Virus Count	
	No.	%	No.	%	No.	%	No.	%	No.	%	No.	%	No.	%	No.	%
Plate count 300	...	...	42	14.0	33	11.0	9	3.0	0	0.0	9	3.0	55	18.3	2	0.7
Total coliforms 547	159	29.1	...	...	4	0.7	49	9.0	0	0.0	21	3.8	116	21.2	3	0.6
Fecal coliforms 613	221	36.1	76	12.4	...	...	71	11.6	0	0.0	26	4.2	131	21.4	2	0.3
P. aeruginosa 413	185	44.8	77	18.6	56	13.6	...	...	0	0.0	23	5.6	98	23.7	9	2.2
Fecal streptococci 10	1	10.0	2	20.0	1	10.0	1	10.0	...	...	0	0.0	0	0.0	0	0.0
S. aureus 480	192	40.0	155	32.3	110	22.9	136	28.3	0	0.0	...	...	165	34.4	52	10.8
C. perfringens 521	204	39.2	128	24.6	91	17.5	58	11.1	0	0.0	28	5.4	...	...	36	6.9
Virus count 782	384	49.1	267	34.1	188	24.0	176	22.5	0	0.0	81	10.4	268	34.3	...	...

[a] Maximum permissible counts: standard plate count 100/ml, virus count 0/10 litre, all other parameters 0/100 ml.

Fecal Coliform Count—The proposed limit of zero fecal coliforms/100 ml was exceeded by 13.2 percent of conventional drinking water samples and by 8.3 percent of experimentally reclaimed water samples (Table 1). More than 36 percent of 613 samples from which fecal coliforms were not isolated exceeded the maximum permissible level for the standard plate count while the corresponding figures for fecal streptococci and virus counts were only 0.0 and 0.3 percent, respectively. The standard plate count, therefore, should be considered seriously as an additional parameter to fecal coliform counts. The contaminants which interfered with total coliform counts were no problem here, since they were inhibited by the high incubation temperature.

E. coli, *Type I*—Details on analyses are not presented since confirmation was not done regularly. In general, *E. coli*, Type I counts were between 25 and 50 percent lower than fecal coliform counts, and, consequently, the percentage of drinking waters which exceeded 0/100 ml was lower than that for fecal coliforms.

Fecal Streptococci—Counts of fecal streptococci were obtained only rarely. This should be taken into account when evaluating results in Table 2. Fecal streptococci counts generally were omitted since previous experience has shown that they contribute little to coliform and standard plate counts in evaluating the waters concerned. Ten samples of reclaimed wastewater, two of which contained coliforms (Table 2), were analyzed, but all were negative for fecal streptococci. In secondary treated effluent supplying the reclamation plant, counts of fecal streptococci were usually about 50 percent lower than those of total coliforms.

P. aeruginosa—This bacterium was isolated from 54 (20.8 percent) of 260 samples of drinking water derived from conventional sources and from four (9.5 percent) of the samples taken from the final product during experimental runs on the Stander reclamation plant. According to these results, the waters concerned were more frequently positive for total coliforms but less frequently for fecal coliforms (Table 1) than for *P. aeruginosa*. Among 413 samples which contained no *P. aeruginosa*, the number of samples which did not conform to maximum levels for other parameters ranged from zero for fecal streptococci to 185 (44.8 percent) for the standard plate count. Using the standard plate count in addition to tests for *P. aeruginosa* is, therefore, highly desirable. The MPN method consistently yielded higher *P. aeruginosa* counts than the MF method, and, at present, all drinking waters are tested by incubating membranes through which 100 ml has been filtered in Drake's medium.

S. aureus—*S. aureus* was present in only 7 (3.0 percent) of 231 conventional drinking water samples and not in 42 samples of reclaimed water. Many samples negative for this organism were positive for others (Table 2), which indicates that *S. aureus* would be a poor indicator of microbiological contamination of water supplies.

C. perfringens—*C. perfringens* was demonstrated in 77 (26.6 percent) of 289 samples of drinking water produced by conventional methods and in ten

(11.9 percent) of 84 samples of experimentally reclaimed water. Among 521 samples which contained no *C. perfringens*, a relatively large number exhibited counts of other indicators, especially virus counts, exceeding maximum permissible levels (Table 2). These findings indicate that *C. perfringens* should be used only in combination with other indicators. Recent investigations have revealed that the *C. perfringens* test used may yield false positive results. This implies that counts of typical *C. perfringens* were lower than those recorded and that the relative value of the organism for indicator purposes is lower than that reflected in Table 2.

Virus Counts—Only 1 (0.3 percent) of 318 samples of conventional drinking water contained virus/10 litre, and reclaimed water samples were always free of virus (Table 1). Among 782 samples which conformed to a virus limit of zero virus particles/10 litre, the number of samples which exceeded maximum levels for other parameters was as high as 384 (49.1 percent) for the standard plate count (Table 2). This leaves no doubt that the virus count should never be used without bacteriological counts for evaluation of water quality. Only 2 (0.4 percent) out of 473 drinking water samples yielded positive virus tests while satisfying maximum permissible levels for all bacteriological counts. One of the samples contained an enterovirus and the other a reovirus.

Parasite Ova—Parasite ova never were isolated from drinking waters. In the treatment processes concerned, including those for the direct reclamation of wastewater, parasite ova were removed efficiently by sedimentation and filtration processes.

Drug Resistant Coliforms—The incidence in various water environments of coliforms carrying transferable (R^+) or nontransferable resistance to at least one of five commonly used drugs is listed in Table 3. Some findings on the incidence and behavior of these organisms in waters may be summarized as follows. Large numbers occur in wastewater, particularly that from hospitals, and in water polluted with these effluents (Table 3). Survival in the water environment compares favorably with that of sensitive bacteria (Table 3). Removal is optimal in purification processes which involve rapid passage over stony surfaces such as biofilters and sand filtration [*6*] while percentage incidence may increase in stagnant waters such as sedimentation units and maturation ponds (Table 3). R factor transfer may occur in water [*9*], and R factors with wide resistance spectra survive best [*5*]. R factors are stable, and segregation of markers occurs rarely [*5*].

Epidemiologic Studies—Transmission of disease has never been associated with any of the water supplies concerned. Detailed epidemiologic studies failed to reveal any adverse effect of directly reclaimed wastewater on the health of consumers in Windhoek.

Discussion

The findings reported here indicate that water which conforms to a proposed drinking water standard specifying a standard plate count of 100/ml,

TABLE 3—*Incidence in various water environments of coliforms carrying transferable or nontransferable resistance to at least one drug.*[a]

Water Environment	Transferable Resistance Count/100 ml	Transferable Resistance Total Coliforms, %	NonTransferable Resistance Count/100 ml	NonTransferable Resistance Total Coliforms, %
Hospital wastewater Pietermaritzburg	129×10^5	26.0	151×10^5	30.0
City wastewater Pietermaritzburg	40×10^5	4.0	212×10^5	21.0
River before discharge of sewage works effluent	7×10^3	1.4	45×10^3	9.1
River after discharge of sewage works effluent	56×10^3	2.2	316×10^3	12.6
River 10 km downstream from sewage works discharge	3×10^3	1.5	22×10^3	13.2
Effluent of Bon Accord dam	14	1.4	143	14.3
Maturation pond system				
influent	69×10^3	0.9	632×10^3	7.9
effluent	18×10^3	2.5	49×10^3	6.7
Daspoort sewage works				
primary sedimentation	33×10^5	4.0	88×10^5	10.7
biofiltration	11×10^4	2.9	41×10^4	11.1
secondary sedimentation	50×10^3	2.8	201×10^3	11.1
chlorination	60×10^2	2.5	213×10^2	8.9
sand filtration	3×10^2	2.4	14×10^2	10.7

[a]Abstracted from Refs *6,9,12*, and *18*; drugs: ampicillin, chloramphenicol, kanamycin, streptomycin, sulfonamide, or tetracycline.

total coliform and *P. aeruginosa* counts of 0/100 ml, and a virus count of 0/10 litre will not transmit microbial diseases. This proposed standard holds both for waters derived by conventional methods or by direct reclamation of wastewater. The standard plate count often is not regarded as an important parameter to measure drinking water quality [*20*]. It is included in the standard since it has proved to be a highly sensitive tool for the general control and surveillance of water treatment processes, which is particularly important in the advanced reclamation of wastewater. Similar views have been expressed previously [*20,21*]. Although the limit proved low enough generally to cover most other microbiological parameters (Table 2), it is not unreasonably stringent since large-scale suppliers such as the Rand Water Board as well as plants which directly reclaim potable water from wastewater, produce water of a quality well within this limit. The Rand Water Board supplies about 1645 thousand litres/day to an area of 4548 km^2 which accommodates 4 300 000 people and 60 percent of the country's manufacturing production [*7*]. The raw water sources are river and impounded waters which are subject to heavy pollution from densely populated and industrialized areas [*1*]. This study yielded no evidence in support of a standard plate count limit of 0/379 litre (100 gal) [*22*].

The value of the total coliform count in standards for drinking water is generally recognized [4,20] and has been confirmed in Table 2. The *E. coli*, type I count is included in the proposed standard since it provides vital information on the possible fecal origin [4] of accidental contamination of drinking water. It also proved valuable when total coliform counts were obscured by contaminant overgrowth.

Although viruses rarely were isolated from waters which conform to the proposed bacteriological criteria (0.4 percent of 473 samples), the observation that this is indeed possible is one reason for including a virus count in water quality standards. In addition, the method used for the detection of viruses, which is the best available at present, is suitable only for enteric viruses such as polio, Coxsackie B, echo and reoviruses, while it cannot detect most Coxsackie A [23], hepatitis [24,25], rota [26], adeno [27], and other viruses [28] which may be transmissible by the water route. Viruses are also more resistant to certain treatment processes than many conventional bacteriological parameters [29]. The proposed standard of zero virus/10 litre is based on comparisons of virus and bacteriological counts, epidemiologic data, and practical considerations regarding virus concentration procedures. The findings reported here contain no evidence supporting suggested standards of no virus per 379 litre (100 gal) [22] or 3785 litre (1000 gal) [30] based on laboratory experiments [31] which do not necessarily reflect practical conditions. Worldwide epidemiologic data provide no evidence for waterborne transmission of viruses other than the type A hepatitis virus [24,25], and this virus, as well as others suspected of waterborne transmission [27,28], never have been shown to be transmitted by water which conforms to established bacteriological criteria such as the total coliform count [3,24,25,32]. The hepatitis A virus is probably an enterovirus sharing many physical and chemical properties with detectable viruses such as polio and Coxsackie, which suggests that the latter are valid indicators for the hepatitis A virus in water [24]. Finally, it should be kept in mind that waterborne transmission of any microbial disease agent has never been attributed to shortcomings of microbiological quality standards or technological know-how but always to the choice of treatment processes or their control [33–35].

Evaluation of parameters indicates that among those not included in the proposed standards, *P. aeruginosa* would be the most meritorious candidate for inclusion in water quality criteria. The organism should prove most useful in areas where it occurs frequently in water supplies [36]. In view of its intrinsic resistance to antibacterial agents such as drugs and disinfectants [10] and its frequent carriage of R factors [5], *P. aeruginosa* should prove an excellent indicator for waters like those in swimming pools [37] which are continually treated under conditions which favor selection of resistant bacteria or waters associated with hospital wastewater which contains exceptionally high numbers of *P. aeruginosa* [10]. The methods used to enumerate *P. aeruginosa* counts in this study were cumbersome, time-consuming, and of suspect sensitivity, and research on this disadvantage compared to coliform counts is necessary. *C. perfringens* and *S. aureus* counts had similar disadvantages in addition to their relatively low indicator value for the waters concerned. Both these organisms, as well as fecal streptococci which are more resistant to many treatment processes than coliforms [*11,38*], may however prove useful for particular purposes [*4*].

The standard for drinking water which is proposed here is based on the evaluation of various microbiological parameters and epidemiologic studies on the consumers of water derived by conventional methods or reclamation of wastewater, and it is largely in agreement with the standard recommended by the South African Bureau of Standards (Table 4) [*39*] and the European Standards of the World Health Organization (WHO) which include a similar virus count [*40*]. South Africa has no statutory standards for potable water [*1*].

Counts of R^+ coliforms similar to and higher than those in South African waters (Table 3) have been reported in Britain [*41*], Europe [*41*], and the United States [*43,44*]. These figures present only a portion of the total number of R^+ bacteria in polluted water which includes many organisms other than coliforms, such as *P. aeruginosa* [*5*]. These organisms may find their way back to man or animals [*5*]. R^+ coliforms alone may be ingested in quantities of two per day by drinking from many approved potable supplies or 20 by normal accidental intake at many bathing beaches [*5*]. Shellfish harvested from approved waters may contain 50 or more R^+ coliforms/100 g, and crops may be spray irrigated with water containing 100 R^+ coliforms/100 ml [*5*]. Ingested R^+ bacteria may either colonize the digestive tract or transfer the R factors to members of the permanent intestinal flora. These organisms may eventually transfer the R factors to pathogens which are then resistant to antimicrobial therapy [*5*]. Since R factor transfer occurs in the water environment [*9*], the natural flora of streams, rivers, and impoundments may serve as reservoirs of resistance plasmids. Apart from the health hazards associated with drug-resistant pathogens, these observations indicate that the possibility of similar selection and spread of R factors enhancing resistance to water treatment processes grows as the continuous reuse of water increases [*1,5*].

In view of the foregoing considerations, coliforms should no longer be

TABLE 4—*Microbiological drinking water standards recommended by the South African Bureau of Standards.*[a]

Organism	Recommended Limit	Maximum Allowable Limit
Total coliform count/100 ml	. . .[b]	10
E. coli, type I count/100 ml	0	0
Total viable organisms/1 ml	100	not specified

[a]Abstracted from Ref *39*.

[b]If any coliform organisms are found in a sample, a second sample shall be taken immediately after the tests on the first sample have been completed and shall be free from coliform organisms, and not more than 5 percent of the total number of water samples from any one reticulation system tested per year may contain coliform organisms.

regarded as harmless indicators of fecal pollution but rather as allies of pathogens providing them with means to resist antimicrobial therapy or water treatment processes. Water quality standards should be reevaluated accordingly [*5*], and it is recommended that whenever such standards allow the presence of coliforms, they should specify that these coliforms should not carry R factor resistance to at least five commonly used and therapeutically important drugs such as ampicillin, chloramphenicol, kanamycin, streptomycin, sulfonamides, or tetracycline. With the knowledge of microbiological methodology required from workers in public health and water research laboratories today, it should be possible to perform these elementary transfer experiments to a standard recipient on a routine basis in a central or even a local laboratory.

Since drinking waters should be free of all coliforms, the limitations on R factors proposed here are primarily aimed at effluents discharged into the environment and waters used for purposes such as recreation or irrigation. In South Africa, this would relate to recommended standards such as the 1000 *E. coli* type I/100 ml for irrigation of crops unlikely to be consumed raw, fruit trees, sports fields and grazing pasturages, and for discharge into streams, rivers, and the wave zone near bathing beaches [*45*]. In other countries, it would include recommendations for overhead irrigation of crops (maximum coliform count of 20 000/100 ml) [*46*], natural surface waters (2400 coliforms/100 ml) [*47*], shellfish waters (70 coliforms/100 ml) [*46*], recreational waters for bathing purposes (1000 coliforms/100 ml, log mean of 200 *E. coli*, type I/100 ml, or absence of undisintegrated feces) [*5,46–48*], and recreational waters for purposes other than primary contact (log mean of 1000 *E. coli* type I/100 ml) [*46,47*].

The experience recorded here unfortunately indicates that many quality standards will have to be reviewed and tightened up in order to protect existing water resources and ensure waters for drinking, recreational, agricultural, and industrial purposes which will not endanger the health of man or animals. Fortunately, however, the findings also prove that the technological knowledge and experience necessary to meet these demands is available.

Acknowledgments

The kind cooperation of the following is greatfully acknowledged: Department of Health, Water Research Commission, South African Institute for Medical Research, Rand Water Board, South West Africa Administration, and Municipalities of Pretoria and Windhoek. Sincere thanks are due to O. W. Prozesky for his advice and the many colleagues who contributed to this study. The paper is published with the permission of the Director of the National Institute for Water Research.

References

[*1*] Barnard, J. J. and Hattingh, W. H. J., "Health Aspects of the Re-Use of Wastewater for Human Consumption," presented at a Water Research Centre Colloquium entitled *Drinking Water Quality and Public Health*, High Wycombe, England, 4–6 Nov., 1975.

[2] Van Vuuren, L. R. J., Henzen, M. R., Stander, G. J., and Clayton, A. J. in *Advances in Water Pollution Research*, Vol. 1, S. H. Jenkins, Ed., Pergamon Press, Oxford, 1971, pp. I-32/1-I-32/9.

[3] Grabow, W. O. K., *Water Research*, Vol. 2, 1968, pp. 675–701.

[4] Grabow, W. O. K., "Literature Survey: The Use of Bacteria as Indicators of Faecal Pollution in Water," CSIR Special Report O/WAT 1, CSIR, Pretoria, 1970, pp. 1–27.

[5] Grabow, W. O. K., Prozesky, O. W., and Smith, L. S., *Water Research* Vol. 8, 1974, pp. 1–9.

[6] Grabow, W. O. K., Van Zyl, M., and Prozesky, O. W., *Water Research*, Vol. 10, 1976, pp. 717–723.

[7] Annual Report, Rand Water Board, Johannesburg, South Africa, 1975, pp. 5–21.

[8] Nupen, E. M. and Hattingh, W. H. J., "Health Aspects of Reusing Wastewater for Potable Purposes—South African Experience," presented at a Conference on *Wastewater Reclamation Research Needs*, University of Colorado, Boulder, 20–22 March, 1975.

[9] Grabow, W. O. K., Prozesky, O. W., and Burger, J. S., *Water Research*, Vol. 9, 1975, pp. 777–782.

[10] Grabow, W. O. K. and Nupen, E. M., *Water Research*, Vol. 6, 1972, pp. 1557–1563.

[11] Grabow, W. O. K., Grabow, N. A., and Burger, J. S., *Water Research*, Vol. 3, 1969, pp. 943–953.

[12] Grabow, W. O. K., Middendorff, I. G., and Prozesky, O. W., *Water Research*, Vol. 7, 1973, pp. 1589–1597.

[13] Rose, R. E., Geldreich,E. E., and Litsky, W., *Applied Microbiology*, Vol. 29, 1975, pp. 532–536.

[14] Cruickshank, R., Duguid, J. P., Marmion, B. P., and Swain, R. H. A., *Medical Microbiology*, 12th ed., Vol. II, Churchill Livingstone, New York, 1975, pp. 358–359.

[15] Levin, M. A. and Cabelli, V. J., *Applied Microbiology*, Vol. 24, 1972, pp. 864–870.

[16] Nupen, E. M., *Water Research*, Vol. 4, 1970, pp. 661–672.

[17] Visser, P. S. and Pitchford, R. J., *South African Medical Journal*, Vol. 46, 1972, pp. 1344–1346.

[18] Grabow, W. O. K. and Prozesky, O. W., *Antimicrobial Agents and Chemotherapy*, Vol. 3, 1973, pp. 175–180.

[19] Grové, S. S., "The Epidemiology of Reclaimed Water," presented at the 51st Conference of the Institute of Municipal Engineers of South Africa, Windhoek, June, 1974.

[20] Train, R. E., *Federal Register*, Vol. 40, 1975, pp. 59566–59588.

[21] Geldreich, E. E., "Is the Total Count Necessary," presented at the AWWA first Water Quality Technology Conference, Cincinnati, Ohio, Dec. 2–4, 1973.

[22] Berg, G. in *Progress in Water Technology*, Vol. 3, S. H. Jenkins, Ed., Pergamon Press, New York, 1973, pp. 87–94.

[23] Hoskins, J. M., *Virological Procedures*, Butterworths, London, 1967, pp. 271–300.

[24] Grabow, W. O. K., *Water S. A.*, Vol. 2, 1976, pp. 20–25.

[25] Mosley, J. W. in *Transmission of Viruses by the Water Route*, G. Berg, Ed., Wiley, New York, 1965, pp. 5–23.

[26] Editorial, *The Lancet*, Vol. I, 1975, pp. 257–259.

[27] Foy, H. M., Cooney, M. K. and Hatlen, J. B., *Archives of Environmental Health*, Vol. 17, 1968, pp. 795–802.

[28] Moore, B. in *Microbial Aspects of Pollution*, G. Sykes and F. A. Skinner, Ed., Academic Press, New York, 1971, pp. 11–32.

[29] Kott, Y., Nupen, E. M. and Ross, W. R., *Water Research*, Vol. 9, 1975, pp. 869–872.

[30] Melnick, J. L., "Recommendations," presented at an *International Conference on Viruses in Water*, Mexico City, 9–12 June, 1974.

[31] Katz, M. and Plotkin, S. A., *American Journal of Public Health*, Vol. 57, 1967, pp. 1837–1840.

[32] Committee Report, *Journal of the Sanitary Engineering Division, Proceedings of the American Society of Civil Engineers*, Vol. 96, SA1, 1970, pp. 111–161.

[33] Craun, G. F. and McCabe, L. J., *Journal of the American Water Works Association*, Vol. 65, 1973, pp. 74–84.

[34] McDermott, J. H., *Civil Engineering, American Society of Civil Engineers*, Vol. 43, 1973, pp. 100–102.

[35] Hughes, J. M., Merson, M. H., Craun, G. F. and McCabe, L. J., *Journal of Infectious Diseases*, Vol. 132, 1975, pp. 336–339.

[36] Schubert, R. and Scheiber, P., *GWF-Wasser / Abwasser*, Vol. 116, 1975, pp. 413–415.
[37] Hoadley, A. W., Ajello, G., and Masterson, N., *Applied Microbiology*, Vol. 29, 1975, pp. 527–531.
[38] Cohen, J. and Shuval, H. J., *Water, Air, and Soil Pollution*, Vol. 2, 1973, pp. 85–95.
[39] *Specification for Water for Domestic Supplies*, SABS 241-1971, South African Bureau of Standards, Pretoria, 1971, pp. 6–9.
[40] *European Standards for Drinking Water*, 2nd ed., World Health Organization, Geneva, 1970, pp. 28–29.
[41] Smith, H. W., *Nature, London*, Vol. 234, 1971, pp. 155–156.
[42] Janouskova, J., Kremery, V. and Kadlecova, O., *Zentralblatt für Bakteriologie, Parasitenkunde, Infektionskrankheiten und Hygiene, Originale*, Vol. A233, 1975, pp. 495–504.
[43] Sturtevant, A. B., Cassell, G. H., and Feary, T. W., *Applied Microbiology*, Vol. 21, 1971, pp. 487–491.
[44] Feary, T. W., Sturtevant, A. B., Jr., and Lankford, J., *Archives of Environmental Health*, Vol. 25, 1972, pp. 215–220.
[45] Smith, L. S., *Water Pollution Control*, Vol. 68, 1969, pp. 544–549.
[46] Wolf, H. W. in *Water Pollution Microbiology*, R. Mitchell, Ed., Wiley, New York, 1972, pp. 333–345.
[47] Geldreich, E. E. in *Water Pollution Microbiology*, R. Mitchell, Ed., Wiley, New York, 1972, pp. 207–241.
[48] Committee Report, *Journal of Hygiene, Cambridge*, Vol. 57, 1959, pp. 435–472.

M. J. Suess[1]

Bacterial Water Quality and Standards: The Role of the World Health Organization

REFERENCE: Suess, M. J., "**Bacterial Water Quality and Standards: The Role of the World Health Organization,**" *Bacterial Indicators/Health Hazards Associated With Water, ASTM STP 635*, A. W. Hoadley and B. J. Dutka, Eds., American Society for Testing and Materials, 1977, pp. 182–195.

ABSTRACT: One of the tasks of the World Health Organization (WHO) is the development of environmental health criteria. In this connection, it has been concerned with some problems related to the concept of indicator bacteria and the rapid and reliable identification of pathogenic ones. WHO requires the absence of any organisms of fecal origin in drinking water and refers to *Escherichia coli*, fecal streptococci, and *Clostridium perfringens* as the most representative of fecal pollution. Potable water supplies for aviation and shipping have to be of similar quality. To assist in the control of surface water pollution, a manual, which includes a chapter on bacteriological examination, has been prepared to encourage the use of approved analytical methods. Attention has also been given to the survey and examination of recreational and shellfish-bed waters.

KEY WORDS: bacteria, water, coliform bacteria, drinking water, recreational waters, shellfish-bed waters, health criteria, indicator bacteria, pathogenic bacteria, fecal streptococci

Since its Constitution was adopted on 22 July 1946, the World Health Organization (WHO) has been engaged in the improvement of man's health and safeguarding his environment. Thus, WHO always has been concerned with the development of safe supply of drinking water supplies, on the one hand, and the protection of environmental waters from unacceptable pollution on the other. Furthermore, in recent years, special attention has been given to the quality of recreational waters and beaches.

It is well recognized that in accordance with the principles of its establishment and organization, and following its declared mode of operation, WHO relies heavily on scientific work undertaken by Member States as well as specialist groups who review various issues related to current health prob-

[1]Regional officer for Environmental Pollution Control, World Health Organization, Regional Office for Europe, Copenhagen, Denmark.

lems, reach conclusions on the present state of affairs, and recommend courses of action. Therefore, while WHO routinely reports on the proceedings of the various meetings which have taken place under its auspices and submits these reports to its Member States, it should be recognized that such reports contain the collective views of groups of experts and do not necessarily represent the decisions or the stated policy of WHO.

It must be emphasized also that while it is very much concerned with the establishment of health criteria, WHO is not engaged in the development of standards of any kind. Whereas criteria represent principles for arriving at decisions (usually on the basis of a dose-response relationship) and are essentially determined by means of scientific research and experimentation, standards are quality requirements based on such criteria and established by a regulatory agency on the basis of cost-benefit and other considerations. Nevertheless, there continues to be a demand for water-quality standards—hence, the European and International Standards for Drinking Water (see Appendixes I and II)—although it is realized that specific limits for particular constituents will have to be looked at in the light of particular circumstances in each country.

Environmental Health Criteria

Thus, following the 1972 Stockholm United Nations (UN) Conference on the Human Environment, WHO was assigned the enormous task of developing environmental health criteria [*1*].[2] While industrial countries have achieved control of waterborne diseases, this is not the case in developing countries. The former, however, face severe problems related to industrial pollution, and much attention is being devoted to this. Yet, in April 1973, when reviewing current problems and priorities, a WHO scientific group identified a number of environmental healthrelated problems for which further information will be required before they can be resolved. These problems include the relative survival time of indicator bacteria in water; the concept of indicator bacteria, particularly the validity of using coliform bacteria as an indicator of the presence or absence of viruses; the use and value of bacterial indicators under different environmental conditions (for example, in hot climates); and the development of new, rapid, and reliable techniques for the isolation, identification, and quantitative determination of pathogenic bacteria in water [*2*].

Drinking Water

The first WHO document relating to standards for drinking water quality was issued in 1956 by the European Regional Office [*3*], and the first edition of the International Standards for Drinking-Water was published in 1958 [*4*]. However, the considerations involved in the development of such

[2]The italic numbers in brackets refer to the list of references appended to this paper.

standards had already been discussed by U. A. Corti of the Swiss Federal Institute for Water Resources and Water Pollution Control at the Second Seminar for European Sanitary Engineers which was organized by the WHO Regional Office for Europe in Rome in 1951. It was stated as an absolutely essential requirement that water not contain pathogenic organisms detrimental to the health of the consumer [*5*].

The unquestionable validity of this requirement (which, of course, had been established long before) is reflected again in a recent WHO contribution to the preparatory papers for the UN Water Conference, which was held in Mar Del Plata, Argentina, 14 to 25 March 1977.

However, the major contribution of WHO in determining the desired drinking water quality, including bacterial quality, is concentrated in the third and second editions, respectively, of the "International Standards for Drinking-Water" [*6*] and "European Standards for Drinking-Water" [*7*].[3] Originally, the publication of two versions, one an international and the other European, reflected their different scope. The first, published by WHO Headquarters, was directed primarily to developing countries in order to offer technical guidance to health and sanitation administrations wishing to revise their regulations on water-quality control, while the purpose of the European version, issued by the WHO Regional Office for Europe, was to encourage the European countries with advanced technological economies and intense industrial and agricultural development to attain higher and stricter standards than the ones specified in the international edition.

However, with the years, experts have recognized that there cannot be much of a divergence in requirements no matter where the consumer is located, when the subject concerns potable water. This, in turn, has led to very similar texts in the present editions of these two publications, with almost identical method for the bacteriological examination of drinking water (see Appendixes I and II). The major difference between the two lies in having two separate limits for *Escherichia coli* in the international version, one for treated and another for nontreated water supplies—a distinction which is not made in the European version. Consequently, when the two versions are next revised (probably in 1978), it might be possible to combine them into one.

The discussion of bacteriological examination in both versions proceeds along similar lines. The first part in the European version, dealing with the organisms as indicators of pollution, states that "the greatest danger associated with drinking-water is the possibility of its recent contamination by sewage or human excrement; and the danger of animal pollution must not be overlooked" [*7*]. Reference is then made to bacteria indicative of fecal

[3]The word "standards," while not referring to the present WHO concept of the term, as defined in the beginning of the paper, has been retained by the revising groups of experts in the title of the subsequent editions by reasons of tradition, as these texts have been identified worldwide for many years already by the titles given.

pollution, mentioning as the most common, representative, and characteristic organisms, *E. coli, Streptococcus faecalis*, and *Clostridium perfringens* (*C. welchii*), and it is recommended in both versions that "water circulating in the distribution system, whether treated or not, should not contain any organisms which may be of faecal origin" [*6,7*].

The second part describes recommended methods for the detection and estimation of organisms indicative of pollution, while the sampling procedures are given as a fourth part. The third part presents the standards of bacterial quality applicable to piped supplies[4] of drinking water.

Potable Water Supply for Aviation and Shipping

Bearing in mind its concern for safe conditions in the international transportation industry, WHO has published two guides concerning sanitary conditions in aviation and shipping in which a special section is devoted to potable water supply.

In the "Guide to Hygiene and Sanitation in Aviation" (which is presently being revised and updated), it is stated that all water for drinking and other personal uses should be free from microorganisms in amounts which might cause illness in any form [*8*]. Also, the "Guide to Ship Sanitation" [*9*] makes special reference to the bacteriological requirements, and each respectively maintains that the potable water provided must be of a quality not less than that described in the WHO International Standards for Drinking-Water.

Environmental Waters

The subject of the Fourth European Seminar for Sanitary Engineers, which took place in Yugoslavia in April 1954, was water pollution in Europe. In their presentation on the health aspects of polluted water, H. Emili and P. S. Tomašic of the Institute of Hygiene, Zagreb, Yugoslavia, stated that "the bacterial pollution of water is the commonest type of water pollution throughout the world," and added the observation that "the better the bacterial balance of stagnant water, the quicker the disappearance of pathogenic germs" [*10*].

The ever-growing concern of the WHO Regional Office for Europe about water pollution problems and its intensified activities in this field have finally been concentrated into sectorial programs on water pollution control management and ecology within its long-term program in environmental pollution control. To prepare the basis for these sectorial programs, a working group was convened in Copenhagen in September 1969 to discuss the trends and developments in water pollution control in Europe, propose priorities for action, and make detailed recommendations on suitable projects which

[4]The word "piped" is omitted in the international version, as this version discusses separately later in the text both piped supplies and individual or small community supplies.

the European office might sponsor [*11*]. Their recommendations have led eventually to the preparation of a manual on analysis for water pollution control. Its draft chapter on bacteriological examination was discussed by another working group in Mainz, Federal Republic of Germany, in April 1975 [*12*]. A part of the summary report of the meeting and the group's recommendations are given in Appendix III.

Another set of recommendations and proposed projects were made by the working group mentioned earlier in relation to biological agents, with particular reference to health risks from aquatic recreation. The group believed that the quantity of water ingested during bathing or other sports is too small to present any cumulative long-term hazard, but, normally, under natural conditions, there is no safeguard against pathogenic organisms [*11*].

Recreational Waters

Three groups of experts were convened by WHO to examine the quality requirements for recreational waters. The first two groups—one organized by WHO Headquarters in Ostend, Belgium, in March 1972, and the other organized by the Regional Office for Europe in Bilthoven, The Netherlands, in October 1974—concentrated on the proposal for quality criteria with special reference to coastal waters and beaches, whereas the third group, convened jointly by WHO and the United Nations Environmental Program (UNEP) in Geneva in December 1975, was assigned the task of preparing a water pollution monitoring and research program for the Mediterranean coastal-water, quality-control project.

In the report on the first meeting, it was suggested that from the standpoint of communicable disease hazards, the most common pathogenic and indicator bacteria for which bathing waters should be examined were *Salmonella, Shigella*, and *E. coli*. Yet, for various reasons, the recovery of pathogens from bathing waters does not necessarily indicate a health risk, and such findings must be critically evaluated pending further epidemiological studies. Recovery of pathogens does, however, serve as confirmatory evidence that sewage pollution has taken place [*13*]. In practice, routine monitoring tests for pathogens present various difficulties apart from the expertise required. Therefore, more practicable as a routine procedure is the use of tests for the presence of excretal indicator organisms, as in the routine examination of drinking water supplies [*13*]. With these various limitations in mind, the group suggested that "as a schematic guideline for classification and action, samples with faecal coli MPNs of greater than 2000 per 100 ml should be deemed heavily polluted and objectionable. Faecal counts of between 1000 and 2000 would indicate distinct pollution and be suspect. Counts of 50 to 200 per 100 ml would indicate slight pollution and counts of less than 50 be considered highly satisfactory. The higher the faecal coliforms count, of course, the greater is the likelihood that sewage-derived pathogens are present in the sample" [*13*].

The second group considered during its deliberations the report of the first meeting. It went on to emphasize that "it is important that coastal bathing waters be sufficiently free of pathogenic microbial life to preclude a significant risk to the health of bathers." The consensus view was that "it was generally feasible and desirable to set broad upper limits for the number of faecal indicator organisms in coastal bathing waters." It was recognized that "in certain countries local environmental conditions might complicate the implementation of this policy." Because of the high variability of bacterial counts in the marine environment, the group agreed that "the recommended upper limits for indicator organisms should be expressed in broad terms of orders of magnitude rather than as rigidly stated specific numbers. Highly satisfactory bathing areas should, however, show *E. coli* counts of consistently less than 100 per 100 ml, and, to be considered acceptable bathing, waters should not give counts consistently greater than 1000 *E. coli* per 100 ml." It was agreed that "this guideline was not at the present time backed by direct epidemiological data." Nevertheless, the group believed that "on general public health grounds the level of contamination of bathing waters with faecal microorganisms should be kept at as low a level as was reasonably feasible" [*14*].

The third group specifically proposed, with regard to the Mediterranean project, that an examination of water zones where body contact water sports take place should be carried out with respect to a number of parameters, including the bacteriological characteristics. The bacteriological tests are "limited to the most common indicators of faecal pollution. The density differences between faecal coliforms and faecal streptococci might indicate the possible sources of pollution. Pathogenic organisms are, of course, of particular importance and whereas it is not practicable to specify a generally applicable programme, it should be emphasized that assays of *Salmonella, Shigella, V. cholerae*, etc., should be carried out in a systematic manner wherever possible, using the same sampling network as for the other parameters" [*15*].

The third group then decided that the following bacteriological data should be collected through mandatory measurements on individual samples [*15*]. They are (*a*) total coliforms as number of colonies/100 ml, (*b*) fecal coliforms (*E. coli*) as number of colonies/100 ml, and (*c*) fecal streptococci (enterococci) as number of colonies/100 ml.

Shellfish Bed Waters

The last mentioned group of experts was the only one within the framework of WHO that made a direct and detailed reference to the need for surveying the bacterial quality of waters in culture areas of shellfish. This subject is also discussed in the "Guide to Shellfish Hygiene" [*16*]. This part of the program calls for the routine surveillance of the various related components, including water. The bacterial determinants always to be measured include total coliform, *E. coli*, fecal streptococci, and total heterotrophic bacteria. In relation to the epidemiological conditions, it is suggested that *Salmonella* and *Vibrio* (*cholerae*, NAG, *parahaemolyticus*) should also be looked for [*15*]. The detailed laboratory techniques have been worked out by another group and published already by WHO in 1974 [*17*].

Conclusion

In summary, one may conclude that an international consensus on bacterial water quality has developed over the years through work done at a number of national institutions and through review and consultation among experts, much of it under WHO auspices. It is clear that drinking water quality criteria should emphasize freedom from biological pollution.

As for surface and, particularly, recreational waters, the bacterial load attributable to outside sources should be kept to a minimum. However, whereas some guidelines have been proposed for recreational coastal waters, the prevailing scientific controversy, the present lack of information, and inconclusive epidemiological evidence call for further study of the subject before definite views can be put forward.

The activities which have been carried out under the auspices of WHO have, no doubt, stimulated a greater awareness of the need for effective action to confront the problems in question and emphasize their urgency to policy makers. As stated in a recent progress report on WHO's human health and environment program, submitted by the Director-General to the fifty-seventh session of the Executive Board, "today, millions of people of the developing countries have water supply and waste disposal facilities they did not enjoy ten years ago" [*18*]. This points clearly to the importance of WHO's continuing role in not only advocating safe drinking water quality criteria but also in stimulating efforts in achieving them through collaborative actions with Member States.

Country projects, where WHO has collaborated with the governments concerned, at first concentrated only on water supply and waste disposal, but they have increasingly extended their efforts in the last ten years into water pollution control and are now beginning to deal also with recreational coastal waters and beaches. WHO has repeatedly convened meetings of various kinds (study courses, seminars, working and study groups, or symposia) during which the question of bacterial water quality has been reviewed, and, moreover, the conclusions reached and the recommendations made were communicated to the Member States for their consideration and appropriate follow-up.

APPENDIX I—INTERNATIONAL STANDARDS FOR DRINKING-WATER[5]

2. Bacteriological Examination

2.3 Standards of Bacterial Quality Applicable to Supplies of Drinking-Water

Some supplies of drinking-water are chlorinated or otherwise disinfected before being distributed; others are not. There does not, however, appear to be any logical reason for setting different bacteriological standards for supplies that are disinfected and for those that are not so treated. Efficient chlorination yields a water that is virtually free from coliform organisms, and, if supplies that are distributed without chlorination or other form of disinfection cannot be kept up to the bacteriological standard that can reasonably be expected of disinfected water, steps should be taken to chlorinate this water or disinfect it in some other way.

It would seem to be reasonable, however, to make a distinction between water from supplies distributed by means of a piped distribution network and water from supplies not so distributed, since it may not be practicable to keep the latter up to the standards proposed for supplies distributed through a piped network.

In the consideration of bacterial standards for supplies of drinking-water distributed through a piped network, it must be remembered that the quality of the water in the distribution system itself may not be the same as that of the water entering the system, since a water that is perfectly satisfactory when it enters the system may undergo some deterioration before it reaches the consumer's tap. Two points should be stressed: (1) the necessity of maintaining a sufficiently high pressure throughout the whole distribution system to prevent contamination from entering the system along the length of the mains by back-syphonage; and (2) the necessity for every distribution system to have available a means of chlorination to deal with accidental pollution, which is always a possibility.

2.3.1 Piped Supplies

2.3.1.1. Water entering the distribution system

(*a*) *Chlorinated or otherwise disinfected supplies.* Efficient treatment, culminating in chlorination or some other form of disinfection, should yield a water free from any coliform organisms, however polluted the original raw water may have been. In practice this means that it should not be possible to demonstrate the presence of coliform organisms in any sample of 100 ml. A sample of the water entering the distribution system that does not conform to this standard calls for an immediate investigation into both the efficacy of the purification process and the method of sampling. It is important, however, in testing chlorinated waters, that presumptive positive tubes should always be subjected to appropriate confirmatory tests.

(*b*) *Non-disinfected supplies.* Where supplies of this sort exist, no water entering the distribution system should be considered satisfactory if it yields *E. coli* in 100 ml. If *E. coli* is absent, the presence of not more than 3 coliform organisms per 100 ml may be tolerated in occasional samples from established non-disinfected piped supplies, provided that they have been regularly and frequently tested and that the catchment

[5]Reprinted with permission of the World Health Organization, Geneva, Switzerland, 1971.

area and storage conditions are found to be satisfactory. If repeated samples show the presence of coliform organisms, steps should then be taken to discover and, if possible, remove the source of the pollution. If the number of coliform organisms increases to more than 3 per 100 ml, the supply should be considered unsuitable for use without disinfection.

2.3.1.2 Water in the distribution system

Water that is of excellent quality when it enters the distribution system may undergo some deterioration before it reaches the consumer's tap. Just as much deterioration may occur in the distribution system of a chlorinated supply in which there is little or no residual chlorine in the water reaching the consumer as in that of a non-disinfected supply, so that in this respect the two are on the same footing. Coliform organisms may gain access to the water in the distribution system from booster pumps, from the packing used in the jointing of mains, or from washers on service taps. In addition, the water in the distribution system may become contaminated from outside, for example, through cross-connexions, back-syphonage, defective service reservoirs and water tanks, damaged or defective hydrants or washouts, or through inexpert repairs to domestic plumbing systems. Although coliform organisms derived from tap washers or the jointing material of mains may be of little or no sanitary significance, the entry of contamination into the water in the distribution system from outside is at least as potentially dangerous as the distribution of originally polluted and insufficiently treated water.

Ideally, all samples taken from the distribution system, including consumers' premises, should be free from coliform organisms. In practice, this standard is not always attainable, and the following standard for water collected in the distribution system is therefore recommended:

(1) Throughout any year, 95% of samples should not contain any coliform organisms in 100 ml.

(2) No sample should contain *E. coli* in 100 ml.

(3) No sample should contain more than 10 coliform organisms per 100 ml.

(4) Coliform organisms should not be detectable in 100 ml of any two consecutive samples.

If any coliform organisms are found the minimum action required is immediate re-sampling. The repeated finding of 1 to 10 coliform organisms in 100 ml, or the appearance of higher numbers in individual samples suggests that undesirable material is gaining access to the water and measures should at once be taken to discover and remove the source of the pollution.

The presence of any coliform organisms in a piped supply should always give rise to concern, but the measures—apart from the taking of further samples—that may be considered advisable in order to safeguard the purity of the water supplied to consumers will depend on local conditions.

The degree of contamination may be so great that action should be taken without delay, even before the result of the examination of a repeat sample is known. This is a matter for decision by those who know the local circumstances and who are responsible for safeguarding the health of the community.

2.3.2 Individual or Small Community Supplies

Where it is economically impracticable to supply water to the consumers through a piped distribution network and where reliance has to be placed on individual wells, bores, and springs, the standard outlined above may not be attainable. Such a standard should, however, be aimed at and everything possible should be done to prevent pollution of the water. By relatively simple measures, such as the removal of obvious sources of contamination from the catchment area and by attention to the coping, lining, and covering, it should be possible to reduce the coliform count of water from even a shallow well to less than 10 per 100 ml. Persistent failure to achieve this, particularly if *E. coli* is repeatedly found, should, as a general rule, lead to condemnation of the supply.

APPENDIX II—EUROPEAN STANDARDS FOR DRINKING WATER[6]

2. Bacteriological Examination

2.3 Standards of Bacterial Quality Applicable to Piped Supplies of Drinking-Water

Some piped supplies of drinking-water are chlorinated or otherwise disinfected before being distributed; others are not. There does not, however, appear to be any logical reason for setting different bacteriological standards for piped supplies which are chlorinated or otherwise disinfected and for those which are not so treated. Efficient chlorination yields a water which is virtually free from coliform organisms, and if piped supplies which are distributed without chlorination or other form of disinfection cannot be kept up to the bacteriological standard which can reasonably be expected of disinfected water, steps should be taken to chlorinate this water or disinfect it in some other way.

In considering bacterial standards for piped supplies of drinking-water, it is necessary to have regard to the quality both of the water entering the distribution system and of that in the distribution system itself. A water which is perfectly satisfactory when it enters the distribution system may undergo some deterioration before it reaches the consumer's tap. Coliform organisms may gain access to the water in the distribution system from booster pumps, from packing used in the jointing of mains, or from washers on service taps. In addition, contamination from outside the distribution system may gain access to the water in the distribution system, for example, through cross-connexions, back-siphonage, defective service reservoirs and water tanks, damaged or defective mains, or defective hydrants, or through inefficient repairs to domestic plumbing systems. Although coliform organisms derived from tap washers or the jointing material in mains may be of little or no sanitary significance, contamination which gains access to the water in the distribution system from outside is at least as potentially dangerous as contamination which enters the distribution system in polluted or insufficiently treated water. It is advisable to draw attention to two points: the necessity of maintaining a sufficiently high pressure throughout the whole distribution system to prevent contamination getting into the system along the length of the mains by back-siphonage, and the necessity for every distribution system to have available a means of chlorination to deal with accidental pollution, which is always a possibility.

The precise action to be taken when coliform organisms are found in a sample taken from the distribution system will depend on local circumstances, but it should be borne in mind that just as much deterioration is liable to occur in the distribution system of a chlorinated supply as in that of a non-chlorinated supply, and that, in this respect, the two should be considered on the same footing.

It cannot be stressed too strongly that bacteriological examination has its greatest value when it is frequently repeated. The examination of a single sample can indicate no more than the conditions prevailing at the moment of sampling at that particular point in the supply system. For adequate control of the hygienic quality of the water

[6]Reprinted with permission of the World Health Organization, Geneva, Switzerland, 1970.

supply it is necessary to have frequent bacteriological examinations of samples collected from carefully selected points throughout the entire supply system, including dead ends.

2.3.1 Recommendations

It is of the utmost importance for the control of the hygienic quality of the water supply that bacteriological examination of both the water entering the distribution system and the water in the distribution system itself be carried out frequently and regularly.

When one 100-ml sample shows the presence of coliform organisms, a further sample from the same sampling point should be examined immediately. This is the least that should be done; it may be considered wise to examine samples also from other points in the distribution system and to supplement these with samples from pumping stations, reservoirs, or treatment plants. The presence of any coliform organisms in a piped supply should always give rise to concern, but what steps—apart from the taking of further samples—it may be considered advisable to take to safeguard the purity of the water supplied to consumers will depend on local conditions.

The degree of contamination may be so great that action should be taken without waiting for the result of the examination of a repeat sample. This is a matter for decision by those who have a knowledge of the local circumstances and the duty to safeguard the health of the community.

The following bacteriological standards are recommended for piped supplies of drinking-water:

Coliform organisms must be absent from any water entering the distribution system, whether the water be disinfected or naturally pure. In a disinfected water the presence of coliform organisms must always lead to suspicion about the efficiency of the disinfection process. The appearance of coliform organisms in a water which is normally naturally pure calls for immediate investigation. Ideally, the same standard should be applied to any water in the distribution system, but in the aggregate results a limit of tolerance of the presence of one or more coliform organisms in a 100-ml sample of water can be permitted in 5% of the samples examined, providing that a positive result is not obtained in two or more consecutive samples and that at least 100 samples of 100 ml each, regularly distributed over the year, are examined. The standard suggested, i.e., that 95% of the 100-ml samples taken throughout the year should not show the presence of any coliform organisms, corresponds to an average density of about one coliform organism in 2 litres of water. This is a statistical concept indicating the generally satisfactory bacterial quality of the supply, but, clearly, whatever action is appropriate should be taken when one bad sample is obtained without waiting to see whether more than 5% of the samples examined during the year are unsatisfactory.

APPENDIX III—AN EXCERPT FROM THE SUMMARY REPORT OF A WORKING GROUP ON BACTERIOLOGICAL EXAMINATION OF WATER (MAINZ, APRIL 1975)[7]

It was agreed that the Chapter should not include methods for total coliform bacteria, since the total coliform group has little significance in the context of water pollution con-

[7]Reprinted with permission of the World Health Organization, Regional Office for Europe, Copenhagen, Denmark, 1975.

trol, although it is important in drinking water analysis. Methods for faecal coliforms, faecal streptococci, and anaerobic sulfite-reducing spore formers should be included in the Chapter.

Because equivalent results are obtained by both the membrane filter and multiple tube dilution techniques, neither was designated as Reference Method but both were designated as alternative methods for faecal coliform bacteria, faecal streptococci and anaerobic sulfite-reducing spore formers. Using the membrane filter procedures, M-FC medium for faecal coliform bacteria, M-enterococcus medium for faecal streptococci and iron sulfite agar for the anaerobes, were recommended. It was recommended that when applying the tube technique, lauryl tryptose broth with confirmation in EC medium be used for faecal coliform bacteria, glucose azide broth with confirmation in ethyl violet azide broth be used for faecal streptococci, and iron sulfite agar be used for the anaerobic sulfite-reducing spore formers. Where alternate media are found to give corresponding agreement through adequate evaluation, they may be classified as being equivalent.

For detection of a broad spectrum of faecal streptococci, a pour-plate method using Pfizer Selective Enterococcus (PSE) agar was recommended. Similarly, a pour-plate method with iron sulfite agar may be used for the anaerobic sulfite-reducing spore formers.

The recommended diluent in enumerating bacteria is the phosphate diluent (phosphate buffer with added magnesium sulfate), as described in the 14th Edition of US Standard Methods (in press).

Pathogenic Agents

It was agreed that this section of the Chapter should not include methods for Shigella, since in most cases recovery attempts are unsuccessful. Methods for Salmonella, Vibrio cholera, NAG-Vibrios, Vibrio parahaemolyticus and pathogenic Leptospirae should be included in the Section.

The Reference Method for Salmonella includes inoculation of an appropriate sample into enrichment broths (potassium tetrathionate and strontium selenite broths), streaking onto primary selective agars (brilliant green lactose saccharose and bismuth sulfite agar), streaking onto secondary selective agars (phenol red and desoxycholate citrate agars), and screening by biochemical and serological examinations.

The Reference Method for Vibrio cholera and NAG-Vibrios includes inoculation of an appropriate sample into selective fluid media (alkaline peptone water and taurocholate-tellurite peptone water), streaking onto primary selective agars (thiosulfate-citrate-bile-salt-sucrose and taurocholate-tellurite-gelatine agars), streaking onto secondary non-selective agar (blood agar), and screening by biochemical and serological examinations.

The Reference Method for Vibrio parahaemolyticus includes inoculation of an appropriate sample into selective fluid media (supplemented meat broth and salt-colistin broth), streaking onto primary selective agars (thiosulfate-citrate-bile-salt-sucrose and bromothymol blue-teepol-salt agar), streaking onto secondary non-selective agar (blood agar) and screening by biochemical and serological examinations.

The Reference Method for pathogenic Leptospirae includes inoculation of an appropriate sample into young guinea pigs, serological testing of the guinea-pig blood, and cultivation of the guinea-pig kidney (on Stuart's medium, Korthoff's medium, and Tween 80 albumin medium).

The Working Group emphasized that pollution from a variety of animals and the direct or indirect contamination from such sources affected the quality of water. Although there must always be a high priority on health effects on man, the health of other living organisms must also be considered, since pathogens infective to these organisms might be present in the water. Moreover, it was recognized that the extent of international trade, tourism, exchange of workers, and the environmental circumstances of certain countries, now made it essential to recognize and consider the significance and detection of possible infections and other diseases previously thought to be of little importance in Europe. These include infestations with helminths, amoebiasis (including Naegleria and Hartmanella), giardia, etc., and possibly certain fungi, which may be spread by water, among other media.

Comparative Bacteriological Tests

To resolve unsettled questions of method reliability and media selection, the Working Group has proposed a comparative study of bacteriological tests. It was proposed that the running of comparative bacteriological tests for water analysis through the collaboration of several laboratories represented at the Working Group (Dr Barrow, Mrs Ormerod, Dr Müller and Dr Vial) would be conducted.

The possibility of using the Quality Control Programme under development by the Public Health Laboratory Service in the United Kingdom as a basis for such comparative tests, particularly in relation to the water samples preserved for mailing, will be investigated by Dr Wahba and Dr Barrow.

The results of the comparative tests organized by Dr Vial for the Commission of the European Communities will be made available to WHO and may provide further material for the Chapter on Bacteriological Examination.

The deliberations of the Working Group revealed that additional topics should be covered by the Manual:

1) A glossary of microbiological terms used in these Chapters should be compiled.

2) A section on microbiological statistics should be eveloped to cover the rationale for data analysis and be added to the Chapter on data analysis, processing and retrieval.

3) A section on microbiological laboratory quality control should be developed in addition to material already prepared for physical and chemical laboratory quality control, and be incorporated into the Chapter on analytical errors.

4) It should be recognized that many pathogens and potential pathogens may be transmitted by the water route. Therefore, a special chapter on certain agents (such as fungi, antinomycetes, pseudomonas, protosae (Naegleria, Hartmanella and giardia) and helminths), the public health significance of which was increasing, could usefully be developed in collaboration with the units concerned at WHO Headquarters and other Regional Offices.

References

[*1*] "The WHO Environmental Health Criteria Programme," Document EP/73.1, World Health Organization, Geneva, Switzerland, 1973.

[*2*] "Environmental Health Criteria," Document EP/73.2, World Health Organization, Geneva, Switzerland, 1973.

[*3*] "Standards of Drinking Water Quality and Methods of Examination Applicable to European Countries," Document MH/EUR/46.56, World Health Organization, Regional Office for Europe, Geneva, Switzerland, 1956.

[*4*] "International Standards for Drinking-Water," World Health Organization, Geneva, Switzerland, 1958.

[*5*] "Second Seminar for European Sanitary Engineers," Document EURO 9.2, World Health Organization, Regional Office for Europe, Geneva, Switzerland, 1951.

[*6*] "International Standards for Drinking-Water," 3rd ed., World Health Organization, Geneva, Switzerland, 1971.

[*7*] "European Standards for Drinking-Water," 2nd ed., World Health Organization, Geneva, Switzerland, 1970.

[*8*] "Guide to Hygiene and Sanitation in Aviation," World Health Organization, Geneva, Switzerland, 1960.

[9] Lamoureux, V. B., "Guide to Ship Sanitation," World Health Organization, Geneva, Switzerland, 1967.

[10] "Water Pollution in Europe—Fourth European Seminar for Sanitary Engineers," Document EURO 9.4, World Health Organization, Regional Office for Europe, Geneva, Switzerland, 1956.

[11] "Trends and Developments in Water Pollution Control in Europe," Document EURO 0415, World Health Organization, Regional Office for Europe, Copenhagen, Denmark, 1970.

[12] "Bacteriological and Virological Examination of Water," Document ICP/CEP 206(7), World Health Organization, Regional Office for Europe, Copenhagen, Denmark, 1975.

[13] "Health Criteria for the Quality of Recreational Waters with Special Reference to Coastal Waters and Beaches," Document W.POLL/72.10, World Health Organization, Geneva, Switzerland, 1972.

[14] "Guides and Criteria for Recreational Quality of Beaches and Coastal Waters," Document EURO 3125(1), World Health Organization, Regional Office for Europe, Copenhagen, Denmark, 1975.

[15] "Coastal Water Quality Control Programme in the Mediterranean," Document EHE/76.1, World Health Organization, Geneva, Switzerland, 1976.

[16] Wood, P. C., "Guide to Shellfish Hygiene," Offset Publication No. 31, World Health Organization, Geneva, Switzerland, 1976.

[17] "Fish and Shellfish Hygiene," Technical Report, Series No. 550, World Health Organization, Geneva, Switzerland, 1974.

[18] "WHO's Human Health and Environment Programme Progress and Future Development," Document EB57/23, World Health Organization, Geneva, Switzerland, 1976.

M. P. Kraus[1]

Bacterial Indicators and Potential Health Hazard of Aquatic Viruses

REFERENCE: Kraus, M. P., **"Bacterial Indicators and Potential Health Hazard of Aquatic Viruses,"** *Bacterial Indicators/Health Hazards Associated With Water, ASTM STP 635*, A. W. Hoadley and B. J. Dutka, Eds., American Society for Testing and Materials, 1977, pp. 196–217.

ABSTRACT: A fundamental difference exists between bacterial (living) and viral (nonliving) organisms in that a virus is a parasite depending on living host cells for material and energy to replicate viral entities. Therefore, an "indicator organism" for virus hazards must take into account the fact that a virus must be able to travel among hosts in order to remain virulent. In this paper, the author modifies the usual connotation of indicator organism to include an "indicator system" which consists of virus and the host on which it is assayed. The state of the art on virus hazards is examined in the light of the advanced technology useful to engineers in the solution of virus problems in water reuse. It is generally accepted that a virus is the most suitable "indicator organism" for a virus. The research use of various virus indicators is here briefly reviewed and requirements for a timely and realistic indicator system is set forth. A system containing multiple hosts is suggested, and data are given as to its effectiveness in monitoring virus quality in reuse water, sludges, and soil.

KEY WORDS: bacteria, water, coliform bacteria, indicator organism, virus hazard, sludge, spray irrigation, cyanophage, enteric virus, ocean dumping, oxidation ponds, DNA.

A task force to write standard methods or practices is expected to employ "the best available technology." The purpose of this paper is to explore and evaluate available technology by which potential virus hazards in the repeated reuse of water may be assayed. Potential is the important word in this consideration. It tends to imply, in addition to conventional enteric pathogens, viruses and virally maneuvered events whose toxicities are manifested in a less direct manner. Apparent increased incidence of cancer correlating with poor water quality, indications of numerous "slow" (or latent) virus diseases, and the findings of more and more evidence of DNA recombination across species lines give cause for concern that reused water may have virus quality

[1]Director, Algal Research Center, Landenberg, Pa. 19350.

parameters whose comprehension and control demand our best efforts.

In 1970, a committee, set up by the Environmental Quality Division of the American Society of Civil Engineers was charged with the task of evaluating virus hazards in water. The committee was presented with the following questions for study [*1*].[2] Let us reconsider these questions along with the committee's conclusions to find what recent advances in technology may be used to extend the conventional wisdom and to solve current new questions.

What Viral Infections Are Waterborne?

Conclusions: 1970

Infectious hepatitis is waterborne; gastroenteritis and more severe syndromes (polio, Coxsackie, and ECHO viruses) and respiratory infections and conjunctivitis (adenoviruses and reoviruses) may be waterborne.

Appraisal: 1976

Enteric Viruses—Hepatitis A is still considered the only true waterborne virus disease in the sense that a virus shed into the water is transmitted via shellfish. Other enteric virus diseases may originate in water but appear to be easily transmitted by contact once the disease is established. The symptoms (headache, skin rash, gastrointestinal upset, muscular pain, and other disorders) associated with enteric virus diseases overlap one another and cover a range of organs. The relation of specific virus to specific syndrome is inexact.

Waterborne Infection by Organisms Other than Escherichia coli—Many water treatment operators interviewed by the author feel that (soil) bacteria in distribution systems are responsible for outbreaks of subclinical disease among the populace. Klebsiella-like organisms are frequently isolated from water treatment plants. Vlassoff[3] gives details on the taxonomy and a clinical picture of klebsiella behavior. Its variable pathogenicity and other features suggest that further study of its host/virus relationships might be profitable.

Hoadley,[4] in a thorough treatment of the literature on pseudomonas, has touched on their relation to such waterborne hazards as (*a*) outbreaks of disease in calves, mink, rabbits, and chinchillas, (*b*) ear infections in swimmers and divers, and (*c*) infections in swimming pools and whirlpool baths. The implication of viruses and virally-maneuvered events (for example, virus transduction, transformation, or transfection) in such cases has had some study but not in great depth. Pseudomonaphages were present in oysters and mussels. Seasonal variation of the phage did not correspond with the seasonal variation of the bacteria, which is not unlike viral behavior in nature. In whirlpool baths, serotypes varied—bacteria from the floor differing from those segregated from the baths. One should note, in all these situa-

[2]The italic numbers in brackets refer to the list of references appended to this paper.
[3]This publication, pp. 275–288.
[4]This publication, pp. 80–114.

tions, that crowding and lowered sanitary conditions may be conducive to viral recombinations. Improved sanitation, by removal of host bacteria, lowers chances for virus multiplication.

How Important Is Virus Transmission by Water?

Conclusions: 1970

Enteric diseases can be directly transmitted by drinking water. Transfer also may be possible via excretions from human carriers. Contact recreation poses a more serious hazard than drinking water.

Appraisal: 1976

The direct transfer of enteric disease by "properly treated" drinking water is not well documented. However, the presence of viruses (including bacteriophage) in disinfected reuse water is well established. As well as viruses shed into water by human carriers, animal (and plant) viruses may have a potential virus hazard. Storm runoff contains excreta from mammals, birds, insects, etc. and leachate from vegetation and woodland composts, feed lots, and agricultural industries—all of which contain a large assortment of viruses. Also, streams receive the backwash from filters used in water treatment processes, allowing reactivation of adsorbed or absorbed ("inactivated," but not killed) viruses which have been removed from the treated water by physical or biological processes. Evidence for interaction among the components of various types of viruses is increasing rapidly. It cannot be assumed that humans are the sole source of pathogenic viruses in water.

Hazards from Latent Virus Infection and Defectively Lysogenized Hosts—In evaluating waterborne virus hazard, some concern must be paid to mechanisms similar to those of the "slow" viruses mentioned by Cookson [*2*]. Some of these are akin to lysogenized defective viruses which lack a complete set of virus genes for their reassembly. Hence, they may be able, for example, to enter a cell, recombine, and multiply but remain latent. Such viruses can be transmitted among cells with little evidence of their presence. Often superinfecting viruses may recombine with viral genes carried by such cells, replicating virus progeny with new properties. (M. P. Kraus, in preparation). The prevalence of lysogeny in nature [*3*] allows much interplay among coliphages, bacteriophages, cyanophages, and their hosts in polluted water. The process is not random, however, but proceeds by specific stages which can be identified. [*4*]. The author's observations show that development of cyanophage in polluted streams and oxidation ponds often proceeds by means of a bacterial helper. It appears that some complementation of gene function occurs between defective bacteriophage and cyanophage, resulting in a virulent lysis of algal hosts in situations where neither bacterium nor alga could produce a plaque alone. In repeatedly reused water, the possibility exists that a latent virus infection could pose a greater threat than direct infection from human carriers. While new genetic interactions provide the mechanisms for environmental adaptation, there is a certain risk of interactions unfavorable to man.

Viral Hazards in Polluted Water May Be Induced—The lysogenic state (which provides an immunity against virus infection) is amenable to induction of gene function (release of immunity) by various stresses such as insufficient nutrient, heat, radiation, antibiotics released by other organisms, and factors accompanying overcrowded conditions. The buildup of metals, organics (including nucleotides), and other industrial by-products in repeatedly reused water may give rise to metabolic irregularities which upset cellular controls and may induce a cancerous condition [*5-6*]. The increased production of new industrial products introduces into water-treatment plants a variety of new factors which may break down established immunities and lead to new viral variables to which the system must readjust. The general instability of enteric organisms [*7*] does not ensure that new viruses will not appear in reuse water nor that current assays for enteric virus are sufficient to detect new pathogens.

Is the Coliform Test an Adequate Indicator of Virus?

Conclusions: 1970

Since the ratio of coliforms:coliphage:enteric virus was found to be roughly 50 000:10 to 100:1, removal of coliforms was initially considered a suitable indicator of virus removal. It was then shown in treated water that when coliforms were absent, viruses were still present. Thus, a virus indicator was deemed desirable.

Appraisal: 1976

The concensus, from numerous studies on the resistance of viruses to chlorination, is that *E. coli* counts have little relation to presence of enteric virus.

Studies to find a more suitable bacterial indicator than *E. coli* were carried out by Englebrecht et al [*8*]. After examining many types of organisms isolated from wastewater, an unidentified yeast and two acid-fast bacteria were found to resist chlorination to a greater degree than enteric virus. However, differences in behavior between bacteria and viruses and their mode of replication and stability, dictate that a virus indicator is the only suitable model for virus behavior. The bacterial indicator then becomes the host organism by which the viral activity is recognized. The most obvious virus selection is coliphage, for which the various strains of *E. coli* are hosts.

Coliphage Indicators—There are many types of coliphage. The pathogenic viruses in the gut are termed enteric viruses, and they differ from the bacteriophages and coliphages in the gut by being assayed on animals or mammalian tissue culture cell lines. To be used as an indicator, some correspondence between coliphage and enteric pathogens must be established. Seasonal variation in the coliform:coliphage:enteric virus ratio was measured by Kott et al. Using an experimental oxidation pond, decreases in virus concentration due to chlorination showed that coliphage MS2 and f2 were more resistant than polio and other human enteric viruses [*9*] and, therefore, were suitable indexes of virus quality. By counting enteric viruses on three lines of monkey kidney cells, Buras [*10*] showed seasonal variations with counts highest in the summer. She believed more virus types were present than could be counted.

Bacteriophage as Indicators of Enteric Bacteria—Conversely, bacteriophage has been proposed as an indicator for enteric bacteria on grounds of giving rapid results. Good correlation with enteric virus was not ensured, and the choice of an indicator host presented problems.

Cyanophage has demonstrated good correlation with sewage effluent into streams and the ability to pinpoint sources of pollution [*4*].

What Is the Status of Control of Hepatitis Carried by Shellfish?

Conclusions: 1970

The coliform index is adequate. Virus transmission is mechanical. A virus index rather than a *bacterial* index would be desirable.

Appraisal: 1976

Coliform counts are still the index used in the determination of shellfish closures. Its adequacy has been questioned by those who have consumed without apparent ill effect shellfish from restricted areas and desire a test more specific for pathogenicity. The hypothesis of purely mechanical transmission of virus in oysters and security in the effectiveness of depuration operations was shaken by the work of Fries and Tripp who demonstrated by electron microscopy the uptake of cyanophage by oyster leucocytes [*13*]. Progress in the study of hepatitis as a waterborne disease proceeds slowly. Bryan [*14*] has described a recent outbreak of hepatitis A. Others [*15-16*] conducted epidemiological studies of an outbreak and obtained electron micrographs of a virus taken from stools of infected persons.

Coliphage has been studied as an indicator of enteric virus in shellfish but was considered not to give significant results [*17*] because of poor correlation with enteric virus and shifts in host sensitivity, as measured on three different hosts.

What Is the Status of Virus Sampling and Detection?

Conclusions: 1970

Methods of detecting, enumerating, and identifying a virus in the aquatic environment are inadequate. New methodologies are needed.

Appraisal: 1976

The Committee's conclusion that technology was needed to recover small quantities of enteric viruses from large quantities of water prompted a large amount of research testing physical and chemical systems of virus concentration. A large variety of filter types, flocculation systems, elution procedures, etc., have been examined [*17-27*]. Several papers [*28-130*] suggest the use of filtration methodology for the standard examination of drinking water. The filtration of large quantities of water for drinking or specialized uses certainly has merit, but the standard detection of virus by concentration from large volumes of reuse water is a mathematical game. The mathematics says that there is "no confidence" in the finding of a single "tissue culture infectious unit" in 1000 gal of water. Added to this, is the uncertainty that serological identification from tissue culture cell lines or laboratory animals is sufficient to indicate the pathology of the isolate for humans. No epidemiology is available to equate the single tissue culture plaque with the ability to produce an epidemic disease. No economically-based calculations have been done to justify the costs of isolation with respect to the risk involved.

Lack of mathematical and serological security, however, does not negate the effectiveness of filtration techniques to obtain excellent water for high quality uses. In examinations of water used for aquaculture, the author's laboratory has found no natural cyanophage in any water which has been properly filtered by these methods.

The interest in virus concentration also has brought about a general use of tissue culture techniques for the enumeration of enteric viruses. A large number of cell lines are available for a variety of virus identifications [*31*].

Cyanophage as Indicator of Virus Quality—The high cost of filtration and tissue culture techniques makes it advisable to search for a general monitor which is rapid and inexpensive. For the last several years, cyanophage has been continually upgraded as an indicator of virus quality in water. Originally, a single blue-green algal host was used to detect cyanophage by a procedure [*32*] similar in principle to the use of bacteriophage by Buris [*33*] and Kenard and Valentine [*10*]. With the isolation and axenic culture [*4*] of a variety of genetically differing strains of a single species of algal host, a panel of hosts was developed whereby differences among viruses could be detected. As information is gained, the panel of hosts becomes continually more refined. Hosts of different species may be used in the panel, and bacterial and tissue culture can be compared. Such a methodology has the advantage of being able to follow gene transfer and detect the induction of gene function.

How Effective Are Virus Removal Systems?

Conclusions: 1970

Viruses are not completely removed by any process.

Appraisal: 1976

Safferman and Moriss [*34*] measured coliphage and enteric virus removal at a multistage activated sludge process. They found about 15 percent virus removal at the primary stage, 58 to 78 percent during nitrification, 96.9 percent during denitrification, and 99.98 percent for the complete process. Incidentally, they observed that coliphages and enteric viruses have no direct relation to each other.

Problems of Virus Removal—Much experimentation has been carried out to find a more acceptable method than chlorination for destroying virus. Table 1 summarizes research on parameters of various virus removal treatments, measuring either natural or seeded coliphages or enteric virus.

It is well documented that a sewage treatment facility, no matter how well constructed and efficiently run, technologically will not exceed a 10 000 fold

TABLE 1—*Parameters studied in virus removal treatments.*

Treatment	Virus	Remarks	Reference
Biological treatment: bacterial-algal mixtures	polio	adsorption and microbial activity reduce virus counts	*35*
Chlorination	polio	pH studies on effectiveness of OCl– versus HOCl; HOCl more effective	*36,37*
Chlorination	polio; coliphage f2, MS2	kinetic studies: time, pH, aggregation	*38*
Chlorination	trachoma	short-lived effectiveness in clean water	*39*
Chlorination	enterovirus	unable to destroy virus with reasonable chlorination	*40*
Halogenation	coliphage f2; polio; ameba	series of basic papers; effective contact made by acid feed and flash mixing; iodine is virucidal at high pH; the active species may be a hydrated ion. Globaline is a questionable virucide; some effect on amebic cysts	*41–44*
Chlorination	natural entervirus	advanced waste treatment and turbidity control	*45*
Coagulation	polio; coliphage f2	high pH alum and lime treatments compared; solids contact up-flow clarifier gives 99.98% virus removal	*46*
Coagulation	polio	fewer viruses removed from sludge than from raw river water; coliforms more readily removed than viruses	*47*
Coagulation	polio; coliphage f2	turbidity removal parallels virus removal; coagulants evaluated; f2 a good model for enterovirus	*48,49*
Ozonization	polio; f2	rate equations for inactivation of virus by ozone	*50*
Ozonization	chemical study	discussed oxidation potential of OH radical and its effect on viruses in clean water and in water with high organics	*51*
Ozonization	bacteriophage T2; polio	aggregation a factor in interpreting survival curves	*52*
UV irradiation	no viruses; measured	effects in shellfish; turbidity problems	*53,54*

reduction in virus. An important facet of the problem may be explained by experimentation from two quite different approaches—the examination of the state of virus aggregation as it relates to the effectiveness of halogenation and the effectiveness of radiation on virus destruction as reflected by measurements on genetically differing hosts.

A series of excellent papers by Sharp and colleagues [*55-57*] attributes the failure to remove viruses completely to the state of aggregation of the virus particles. Using various enteric viruses and various conditions of disinfection, properties of the state of aggregation were compared by electron microscopy, high speed centrifugation, and plaque formation in tissue culture monolayers. They noted that aggregation, which cannot be entirely prevented by the best techniques of dispersion, is responsible for a change in the virus inactivation efficiency. Since the natural release of virus is in the form of tightly packed cytoplasmic crystals, and since aggregates also may be protected on particulate matter, unaggregated particles will seldom be available in the complex environment of polluted water. Interesting data with regard to special behavior of aggregates was obtained in an early paper by Kim and Sharp [*58*] using rabbit pox virus. For highly dispersed single particles, no plaques occurred on the chosen host; with increased aggregation, plaques were formed. Since these virus particles were prepared as unpurified mechanical lysates, more than one virus type may have been present, and complementation and recombination among virus particles may have resulted in plaque formation.

Extended Survival of Cyanophage to Ionizing Radiation [59,60]—Failure to destroy cyanophage in sewage and oxidation pond water by irradiation with large doses of gamma radiation brought about studies of radiation effects on cyanophage as determined by plaque formation on genetically differing hosts. Figure (A,B,C, and D) shows survival curves of an irradiated cyanophage as plated out on a sensitive host after incubation in four different hosts. A is a virus-resistant host which does not absorb virus; B is a partially resistant host which absorbs virus and produces turbid plaques; C is a lysogenized host which carries the irradiated virus as a prophage; and host D contains multilysogeny of unknown character. It is clearly shown in curve A that the free virus (unable to interact with the incubating host) is destroyed by irradiation along a curve with a steep slope. Curve B, on a host which absorbs virus and allows the virus to integrate, shows a rise after a steep slope with accompanying change in plaque morphology. Curve C, on a host already lysogenized by a virus of the same serotype as the superinfecting virus, demonstrates the ability of the superinfecting irradiated virus to interact with the carried virus to produce progeny which extends the survival curve. Curve D shows two regions characterized by new plaque morphology. These curves serve to demonstrate, in a comparatively simple bacteria-free system, only a small part of the intricate interaction which can proceed with greater variation in a polluted environment. It is, therefore, not reasonable,

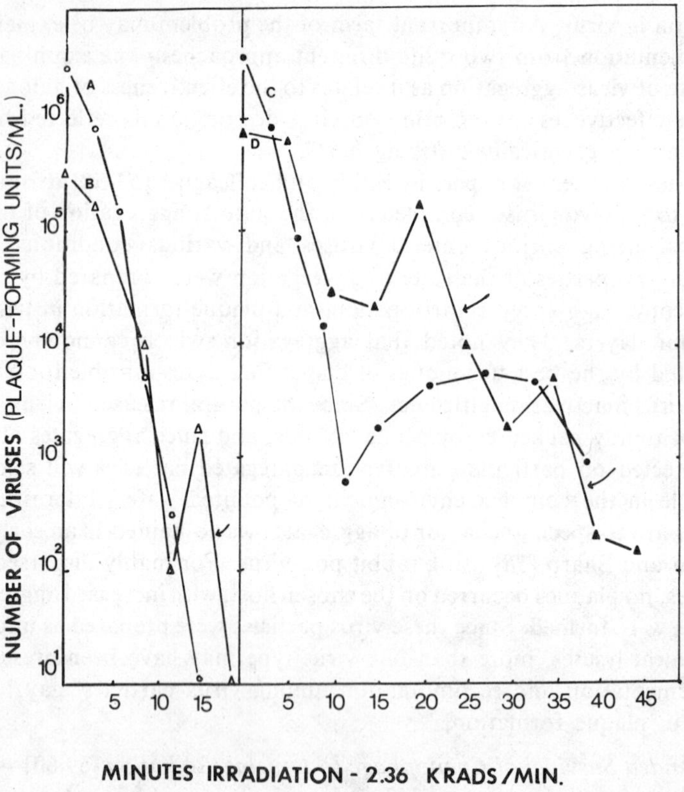

FIG. 1—*Survival curves of S3 cyanophage as plated-out after incubation on various hosts; A: o—incubation on cyanophage-resistant host; B:△—incubation on turbid-plaquing host; C: • incubation on host lysogenized by S3 cyanophage; D:△—incubation on host with unknown lysogeny; arrow points to regions of curves where new plaque morphology is noted.*

to expect a single, laboratory-adapted model virus system in clean water to give a realistic answer to the question of effective sterilization of sewage or reuse water. The radiation model illustrates some of the difficulties which exist in translating results of experiments on clean water systems, using known added virus, to field conditions containing unknown viruses and a plurality of hosts.

What Risks Are Associated With Spray Irrigation of Crops?

Conclusions: 1970

There are no data to date incriminating virus; the situation is unknown.

Appraisal: 1976

The effluents from waste treatment contain nutrients which energy-mindful conservationists maintain should be returned to local areas or other areas within economical limits. Spray irrigation is one mechanism for the conservation of a nutrient from organic wastes. From a practical viewpoint, spray irrigation has had no evident degrading influence on health either to the crop or the consumer. Problems such as nitrogen buildup in ground water must be watched; the effect of aerosols is under study.

Dugan et al have measured the nitrogen inventory (increase) in lysimeters testing land disposal of wastewater in Hawaii. Examination of enteric viruses (reoveruses, Coxsackie, polio, and ECHO) shows from 27 to 19 000/litre in raw sewage, 2 to 750 in the chlorinated effluent and a reduction to 1 per litre after passage through 5 in. of soil [*61*].

Chlorination before land application, as favored by Bernarde [*62*], is a controversial topic. As shown previously, even high doses of disinfectant do not guarantee virus destruction. It is also likely that inactivation in soil retains viable viruses that can be eluted.

The principles of virology do not give any assurance that no potential virus hazards exist in the land application of sewage wastes, be it by spraying or other methods. The capacity of the soil and soil organisms to absorb, alter, or destroy viruses must be assumed to have finite limits. Kardos and Sopper [*63*] have studied the effects of spray irrigation on cropland and woodlands and have measured limiting rates for good crop management. Foster and Englebrecht [*64*] have outlined possible health hazards which should be considered in the land application of treated sewage. While there is no evidence that these hazards are greater than those encountered in the daily routine of most persons, a philosophical answer is insufficient. Some of the physical parameters dealing with viruses in soil have been referred to previously, and the current literature is receiving many papers on these topics. Little attention is being given, however, to biological activity which is equally as important, if not more so, since the physical parameters are largely reversible. Cyanophage promises to be an effective biological model of mechanisms of host/virus interaction among organisms in the passage of reuse water through soils and over croplands [*65,66*].

What Are the Contributions From Animals and Birds?

Conclusions: 1970

The contributions from animals and birds are relatively unimportant in human disease.

Appraisal: 1976

The discharge from animal feed lots and the disposal of wastes from agricultural industries are becoming increasingly problematical. Evidence is also increasing that some animal viruses can also infect humans. New Castle disease of fowl often produces conjunctivitis in man, for example. Duck ponds, pig pens, chicken slaughter house wastes, mushroom compost leachates, and tomato cannery wastes, all have high-cyanophage titers [*32*].

Blue-green algal blooms often have high cyanophage counts and are associated with toxicities in water which produce fish and animal kills. Evidence for increased tumorous fish in polluted water as compared with clean water has been documented [*67*].

Studies [*68*] show that recombination can occur between influenza viruses from man and lower animals and may offer an explanation for new strains of pandemic influenza virus. Some workers, studying transmission of disease across species lines, show results [*69*] using DNA-DNA and DNA-RNA hybridization techniques consistent with the acquisition of bovine leukemia virus genome as the result of horizontal transfer from other species. Such transmission could be accomplished by animal discharges into polluted water.

Kraus [*66*] shows cyanophage counts up to 2×10^4 in various animal and agricultural lagoons. Amebas and other protozoans are frequent in such water, and protozoan cysts may be virus carriers. There is also considerable reason to believe that "killing" plaques (showing lysis but no multiplication of cyanophage) obtained frequently from concentrated agricultural wastes are due to bacteriocins (cyanocins) which are products of Col factors or plasmids [*4*]. In the author's laboratory, colicin tranfer from *Klebsiella pneumoniae* to algal hosts has been demonstrated. Similar plasmid tranfer has been reported by others [*70*].

The previous examples, taken at random from a host of similar reports, typify the complicated and yet specific nature of virus transmission and illustrate the need for an economical and rapid monitoring system that can act as a neutral recipient in modeling mechanisms of gene transfer capable of initiating new diseases.

Do Enteric Viruses Multiply in Water?

Conclusions: 1970

No.

Appraisal: 1976

The answer is still no. In general, however, enteric viruses in tissue culture demonstrate the same virological principles which apply to bacteriophage and other viruses: they may become defective, enter new hosts, and transduce new characteristics. As with any other plaque-type analysis, a zero plaque does not constitute absence of virus. The host may be immune or nonreceptive, or the virus may exist in a latent form.

The introduction to the recent conference on Recombinant Molecules at Massachusetts Institute of Technology (MIT) stated, "the ability of the genetic apparatus of living cells to incorporate or substitute foreign genetic material from other plant or animal species has been demonstrated both in nature and in the laboratory." The author's experience with cyanophage from sewage, leachates of organic composts, and various polluted waters gives plentiful evidence that the previous quotation is true. At the conference, Falkow [*71*] presented evidence that plasmid DNA, transferable by molecular mechanisms of transduction, transfection, or transformation, is the unit responsible for pathogenicity in certain cases of *E. coli* diarrhea.

In the same conference, Natalie Teich [*72*] presented graphic displays of the ways RNA C type tumor viruses in nature could become infective. Diseases of this type (for example, Avian leukosis and Rous Sarcoma) are often nonlytic (chronic) in one animal and carry genes for infectivity in another. Such viruses are ubiquitous in nature. They are membrane viruses. Initiated as a core against a membrane, they can become integrated into a host, yield messenger RNA, and transcribe protein. They can recombine, complement defective genes and produce mixtures which yield new host ranges, new serological types, etc. The protection against infection of a host, due to an established immunity, can be suppressed by chemicals, X-rays, ultraviolet light, heat, aging, and other environmental stresses. Therefore, in spite of the fact that enteric viruses may not multiply as such in water, the factors for disease that they may carry may find a means for propagation, particularly in the polluted aquatic environment. This is the essence of what is currently being termed a potential aquatic virus hazard.

What Is the Extent of Virus Concentration in Water?

Conclusions: 1970

Viruses are present in low numbers.

Appraisal: 1976

A 0.02 to 0.05-ml specimen is the volume used in a cyanophage assay of drinking water. Data [*66*] indicate that from less than 1 up to many thousand cyanophage may be taken in in a glass of treated drinking water. Aquatic virus counts may be much larger than we imagine. Virus replication, which can sometimes multiply a virus 10 000 fold per cell, can easily produce a count of 1 plaque forming unit/ml in a large body of water in a short time. Both host and virus can exert cellular controls that can put an end to continued high multiplication which would destroy the host completely. Water which is receiving biological treatment is expected to have a rapidly fluctuating virus character during the course of successive treatments. The data of Vaughn and Metcalf [*17*] show a host shift. While this is regarded as an annoying inconsistency, there is a fundamental reason. In the cyanophage/host system, the fluctuation is reflected in the plaque morphology [*4*] which varies from host to host according to genetic differences but is uniform on a single host.

How Stable Are Viruses in Water?

Conclusions: 1970

Enteroviruses can survive for a long time; detention cannot be used as a safety factor.

Appraisal: 1976

Viruses are very stable in clean water; in polluted water where bacteria and other organisms are active, the situation is much more complex.

Herrmann et al [*73*], by comparing data on the "die-off" of polio and Coxsackie virus in natural lake water and in membrane-filtered lake water, found more rapid die-off in natural lake water. Increased rates of die-off of enteric bacteria were observed in the presence of mixed algal cultures, and a higher die-off rate was observed when the wastewater was anaerobically pretreated before entrance into facultative maturation ponds [*74*]. Berry and Noton [*75*], measuring the survival of coliphage T2 in seawater, found that activation was temperature dependant and enhanced by sewage pollution and sunlight. It had the characteristics of a biological inactivation dependent on microorganisms. The sunlight implicates photosynthetic organisms.

Britton and Mitchell noted protective effects of colloidal montmorillonite clay and of *E. coli*, type K (which is not lysed by T7) on the survival of coliphage T7 in seawater [*76*]. One might consider the fact that algal hosts, in general, are very resistant to coliphages T7, and cyanophage is stable in the presence of the hosts both in seawater and fresh water. Any similarity between coliphage and cyanophage is not known. However, in the case of protection of T7 by *E. coli*, type K, the protective effect might be more biochemical than chemical (organic).

By means of host-range and plaque-morphology studies, Kraus [*65*] described increased virulence to varied hosts under overcrowded conditions in an experimental polishing pond fed by the effluent from a trickling filter. The algal growth on the trickling filter is resistant to the viruses in the wastewater by virtue of lysogenization. Cyanophage released under stress is able to initiate virulence among the varied population in the polishing pond. This result is demonstrated as virulence towards new species of algae. In spite of the increased virulence, the cyanophage carries a strong trait for lysogenization, which under favorable nutrient conditions rapidly "immunizes" new hosts in downstream recovery. The importance of virally manipulated biochemical and genetic warfare among species cannot be underestimated in understanding the succession of microorganisms in oxidation ponds, stabilization lagoons, and polluted lakes and streams. There is virally manipulated, but very specific, interplay among blue-green algae and certain bacteria. The capability of enteric viruses to infect foreign hosts and transfer pathogenicity remains an open question which should receive further investigation.

What Health Hazards Exist in the Recreational Use of Impounded Water Supply?

Conclusions: 1970

Prudence justifies caution; chlorination treatments are judged adequate.

Appraisal: 1976

Other papers in this publication have taken up the problem of potential health hazards related to swimming. Swimming and boating activities cannot put into reservoirs many more problems than are already considered in the treatment of polluted and toxic river water, problems of virus hazard included. Though present chlorination procedures are questioned as to efficiency of virus removal, the conclusions stand at the 1970 level; whether to use a reservoir for recreation is a socioeconomic problem, with caution as to potential virus hazards.

How Effective Are Water Purification Processes in Removal or Destruction of Viruses?

Conclusions: 1970

The record gives assurance that recommended sanitation and available technology can provide protection. Vigilance is urged in chlorination and turbidity control.

Appraisal: 1976

Results from the concentration of large samples of drinking water show essentially no enteric viruses in properly treated drinking water [*77*].

In the examination by Kraus [*66*] for cyanophage in very small samples of drinking water from various private and public sources, it should be pointed up that no previous treatment was given the sample (such as treatment with ether or chloroform to destroy interfering bacteria) and that the sample was kept at room temperature overnight. One-millilitre samples were then sucked up from the bottom of the vial and used to spot test a series of host plates. Bacteria in the sample often influenced the development of virulence on the algal hosts. Comparisons of specimens from the top of the vial with those from the bottom of the vial indicated that most of the virus activity is in the sediment. Since it would be impossible to maintain bacteria-free conditions throughout a water treatment plant-distribution system, bacteria may influence bacteriophage and cyanophage concentrations. Unpublished studies (Lisa Lennihan and M. P. Kraus) on a single treatment plant showed that the blue-green algae on the walls of the treatment tank were resistant to the cyanophage in the influent water and that the host-range and plaque-morphology of the cyanophage in the treated water differed from those in the influent water.

New Questions

What About Viruses in Sludge?

Current research has been concerned more with the effect of waste treatment on seeded viruses than with biodegradation of natural viruses. Gaby [*78*] measured virus removal by aerobic biodegradation in windrows of a composting mixture of refuse and sewage sludge reaching a temperature of 49 to 74° C. Enteroviruses were not isolated, but polio was added and found to be inactivated after three to seven days. Parasitic ova remained intact. Good quality land conditioner was obtained if the windrows were turned and the temperature was sufficiently high.

Malina et al [*79*], using added tritium-labeled poliovirus, compared viruses attached to activated sludge with those in the supernatant as a function of time. Most of the viruses were associated with sludge particles and could be recovered as infectious units in spite of a 10^4 reduction of viruses in the supernatant.

Careful work by Wellings et al demonstrated the importance of solids-associated virus in the viral examination of wastewater, sludge, and soils. Their data emphasized the inadequacy of virus concentration techniques that do not include the comminution of solids in evaluating viruses [*80*].

Ward and Ashley [*81*] used "advanced technology" in the study of mechanisms of RNA inactivation in anaerobically digested sludge seeded with tritium-labeled poliovirus. By sedimentation analysis of the RNA and SDS gel electrophoresis of the protein of the sludge-associated particle, they determined that viral inactivation was possibly due to a nicking of the RNA by penetration, through the capsid, of a sludge component present only in the anaerobically digested sludge.

The ability of enteric virus to survive anaerobic digestion was tested by Meyer et al [*82*] using germ-free piglets and porcine enterovirus (ECPO). Virus was added to the sludge at known density and measured after digestion. Zero virus counts were obtained after five days digestion. The epidemiology of virus effect on the piglets could not be determined, however, because the piglets succumbed to *Salmonella*. (Since the piglets were germ-free, they did not carry the necessary immunities!)

When large amounts of water were filtered [*83*], nematods and amebas were detected in all finished water samples, but no public health significance was attached to them. Protozoans as vectors of virus in polluted waters should not go unexamined. Large numbers of protozoans are present in sludge and are effective in removing *E. coli* [*84*]. When protozoans feed on bacteria, what happens to the bacteriophage? I am familiar only with cyanophages, but perhaps they are typical. In my laboratory, bacteria-free amebas were fed bacteria-free, cyanophage-infected algae and allowed to encyst.

The host range and plaque morphology of the original cyanophage was known. The cysts were well washed, collected by centrifugation, put into fresh growth medium, and allowed to excyst. The plaque morphology and host range of the cyanophage recovered after excystment was different from the original. Viral recombination within the ameba is implied but further experimentation is necessary. The evidence is firm, however, that the virus is viable and not destroyed by digestive or other metabolism.

In sewage or water treatment plants, particles from flocculation treatment, sand or trickling filters, and tanks and equipment used become coated with bacterial slimes, fungi, algae, and protozoans, etc. The biological activity which occurs is responsible for some inactivation and loss of virus count. More often than not, as we have seen in the foregoing discussions, the inactivation is not destruction of virus. Viruses absorbed on sedimenting floc or sand are readily eluted. On the backwash of filters or disposal of sludge, viruses deposited therein are again available to the environment. As we have already seen, viruses are protected as physical aggregates, within their own hosts or within vector organisms. A single virus is readily disinfected, but an aggregate is not. A free virus with no means for recombination is readily destroyed by radiation, but, within a cell, even a defective virus can carry out maneuvers to stay alive.

Albeit, in spite of an unavoidable persistance of virus, it appears that with suitable sanitation a tolerable situation exists. What defines this condition, what causes its breakdown, and how do we measure its limits? These are questions we must answer if we wish to utilize oxidation ponds for sewage treatment and employ conservative sludge disposal on land and at sea.

What Will Happen When Sludge Is Dumped into the Ocean?

It is desirable, conservationwise, to utilize sewage eutrophication by photosynthetic solar energy conversion and the ensuing production of marine food. We have information on die-off of enteric viruses and coliphages in the sea, and we know that bacterial activity are factors in their demise. It is at this point where the increase in cyanophage virulence is found. It is important to learn what relation may exist between cyanophage and coliphage and enteric virus. In host range, plaque-morphology studies of cyanophage in the St. Jones River, Delaware, increased cyanophage virulence occurs at sewage outfalls (Beverly Slack and M. P. Kraus). As the eutrophic stream flows through marshland toward the estuary, recovery of algal hosts occurs, immune to the viruses in the water. As protozoans consume the algae and increased diversity develops along a food chain, a number of events can betake the virus, such as protein and DNA-degrading enzymes, host alterations, etc.

In polluted water, there are myriad bacteria multiplying at a high rate converting nutrients and converting themselves. How do we tell one from another? Farmer and Brenner[5] have demonstrated one way, utilizing "best

[5]This publication, pp. 37–47.

available technology." This technology is at the genetic level and works suitably for any DNA (or RNA). Actually, the DNA does not care whether it is in an alga, a bacterium, or a virus. What we need to know about it is what it codes for and how it functions. Bacterial indicators, including tissue culture, of human enteric virus pathogens are specifically useful only when the disease can be shown to be carried by a specific virus or virus component. The relation of enteric virus to enteric bacteria is a hazy area.

It has been shown, for example, that an *E. coli* diarrhea is under the control of a plasmid [*71,85*]. Such plasmid DNA can be transferred from one organism to another and may not be too fastidious about the serotype of the host, once it has found a means of entrance. Thus, serotype and other markers may not necessarily denote the plasmid presence. What we need to know is how a plasmid is transferred in the aquatic environment. We are also interested in disease effects along the food chain: fish toxicities, hepatitis, tumors, and cancers. It has been documented that tumors in fish are more numerous in polluted water [*67*]. How are these tumors initiated? Are they initiated in pollution effects which pass along a food chain, or do specific paths develop at random?

Interactions of Viruses on Soil

In placing sludge and wastewaters on soil, we are playing the same games of recombinant DNA as those so hotly in current debate. The full potential is unknown, but it is doubtful that we create in the laboratory anything that is not already being done, or at least attempted, in nature.

Gilbert et al [*86*] have studied the reduction of seven or more types of enteric viruses after wastewater discharge through soils. No viruses were found in any examinations of large quantities of filtered water from lysimeter wells. However, delayed plaques and cytopathic effects were mentioned. It is possible that these cytopathic effects (even though they were reduced to zero after several blind passages) are of potential viral consequence. Delayed plaques could result after the accomplishment of several undetected recombinations and may even be worthy of investigation as a slow-type virus phenomenon.

Fetter and Holtzmacher [*87*], Wellings et al [*88*] and Schaub and Sagik [*89*] deal with virus movement after wastewater discharge into soil and the infectivity of colloidally-bound virus particles. Lund and Ronne [*90*] and Lund [*91*] find no inactivation of virus on chemical sludge and warn against the removal or pretreatment of any part of the sample before assay.

Cooper el al [*92*] and Peterson [*93*] point up an increase of enteric viruses in landfill leachates and recreation areas due to soiled disposable diapers.

The papers mentioned previously illustrate parameters which must be considered in the assay of virus hazards in reuse water and sludge disposal. It is obvious that the physical parameters have been emphasized to the almost complete neglect of the biological parameters. In giving more attention to this aspect cyanophage may be a suitable monitor for studying mechanisms of biological activity in the passage of reuse water through soils. Table 2 [*66*] presents data on the numbers, host range, and plaque morphology of cyanophages recovered from suction lysimeter wells set at various depths in spray-irrigated reed canary grass. Five millilitres of the lysimeter sample were incubated with an equal amount of growth medium to allow natural multiplication of cells and cyanophages present; 0.02 ml was then spotted on a range of algal hosts and cyanophages recovered. While no virus hazard can be assigned to the presence of cyanophages under these conditions, it is interesting to note that at 24-in. depth in each plot there is increased virus activity as compared to the control plot D, which received no sewage. In Table 2, a significant increase is found in the number of hosts which are sensitive. This illustrates the probable elution of viruses which may have been attached to soil particles in the upper layers of soil and were subsequently eluted and carried further as successive liquid applications were applied. The host range, plaque morphology of the viruses in the original sewage differed from that found in the lysimeter specimens. Controls carried out on tissue culture monolayers and bacteria-free conditions on algal lawns showed no alteration of host range and plaque morphology.

It is useful to compare the work of Casida et al [*94*] who also found that incubation was necessary to detect bacteriophage activity in soil. These authors, using a limited range of hosts, do not defend the phenomenon of lysogeny as a possible explanation of their results. However, the present author's observations, using a wide range of cyanophage hosts, support

TABLE 2—*Lysimeter studies of spray irrigation.*

Sample		Number Virus / Unit Volume	Number Hosts Affected	Virulence
Applied spray 1		25	16	s,+
Applied spray 2		25	17	s,+
Plot AI	depth 6 in.	2 to 10	3	s
Plot AII	depth 12 in.	7 to 100	8	+
Plot AIII	depth 24 in.	50 to 200	10	+
Plot AIV	depth 48 in.	7 to 50	7	+
Plot BI	depth 6 in.	15	1	+
Plot BII	depth 12 in.	7 to 25	4	s
Plot BIII	depth 24 in.	1 to 50	11	s
Plot BIV	depth 48 in.	1 to 50	9	+
Plot CI	depth 6 in.	10 to 50	3	s
Plot CII	depth 12 in.	10 to 25	3	s
Plot CIII	depth 24 in.	7 to 50	10	+
Plot CIV	depth 48 in.	1 to 25	4	s
Plot DI[a]	depth 6 in.	15 to 100	9	+
Plot DII	depth 12 in.	15 to 100	7	+
Plot DIII	depth 24 in.	7 to 25	2	+
Plot DIV	depth 48 in.	1 to 25	6	s

NOTE—Symbols: +, high virulence; s, restricted virulence; o, turbid plaques, low virulence; spray 1, chlorinated secondary sewage effluent; spray 2, chlorinated secondary sewage plus anaerobically digested liquid sewage sludge.

[a] Plot D is a control plot receiving no spray irrigation.

lysogeny as a mechanism. Lysogenized hosts, commonly derived from domestic sewage, demonstrate a release of high titers of cyanophage which lyse and destroy sensitive hosts within a 7-h period.

Conclusion

It appears that best available technology for determining potential aquatic virus hazard may properly include a genetically based test which is applicable regardless of species differences. Each potential disease cannot be studied individually. What the sanitary engineer and the public health official must have is an easily maneuvered, inexpensive, rapid, and universal assay system capable of responding to environmental viral phenomena. By means of such a system, the engineer becomes familiar with general principles and mechanisms of aquatic virus transfer, and, since the genetic code is universal, what he learns in one system can be translated to special systems. Thus, the engineer becomes equipped to troubleshoot special problems of public safety, environmental protection, and economic concern.

In indicator organism must be ubiquitous in the aquatic environment. It must be able to sit at a pivital point and reflect viral activity as it enters and leaves a given situation or a given organism. It must be free from ambiguity by contaminating organisms.

In this paper, the author has cited cyanophage data on many different aspects of aquatic virus assay and compared it with other data from other assay techniques. A cyanophycean host/virus system meets the foregoing requirements. The algal host/virus system works at room temperature on simple medium using sunlight as its energy source. Being photosynthetic, it can easily distinquish virus transfer which might be difficult in bacterial systems. It is less rapid in its replication than bacterial systems; hence, individual stages can be more easily separated for study. By means of a long host range it can identify differences among viruses encountered in the polluted aquatic environment. Biochemically, it can respond to metal toxicities and carcinogens. It also may be a "safe" system for engineers, since the virus attenuation afforded by a photosynthetic system may be greater than that of a coliphage or other phage. Most of all, it is inexpensive and easily manipulated.

In the field of aquatic virology, a much better background is needed in basic and environmental virology before we can responsibly act to set standards.

Acknowledgment

The author was supported by a Gerry fellowship and thanks referees and many others who gave assistance in the preparation of this paper.

References

[*1*] Berger, B. B., *Journal of the Sanitary Engineering Division,* American Society of Civil Engineers, Vol. 96 SAI, 1970, p. 111.

[2] Cookson, J. T., Jr., *Journal of the American Water Works Association*, Vol. 66, 1974, p. 707.
[3] Hayes, W., *The Genetics of Bacteria and Their Viruses*, John Wiley and Sons, New York, 1964.
[4] Kraus, M. P., "Host-range and Plaque-morphology of Blue-green Algal Viruses," Technical Report DEL-SG-1.74, College of Marine Studies, University of Delaware, Newark, Del.
[5] Ember, L., *Environmental Science and Technology*, Vol. 9, 1975, p. 1116.
[6] Page, T., Harris R. H., and Epstein, S., *Science*, Vol. 193, 1976, p. 55.
[7] Davis, B. D., Dulbecco, R., Eisen, H. N., Ginsberg, H. S., and Wood, W. B., Jr., *Microbiology*, 2nd Ed., Harper and Row, Hagerstown, Md.
[8] Englebrecht, R. S., Foster, D. H., Greeing, E. O., and Lee, S. H., "New Microbial Incidators of Wastewater Chlorination Efficiency," *Project 17060 EYZ*, Office of Research and Development, U.S. Environmental Protection Agency, 1974.
[9] Kott, Y., Roze, N., Sperber, S., and Betzer, N., *Water Research*, Vol. 8, 1974, p. 165.
[10] Buras, N., *Water Research*, Vol. 10, 1976, p. 295.
[11] Kenard, R. P. and Valentine, R. S., *Applied Microbiology*, Vol. 27, 1974, p. 484.
[12] Kraus, M. P., *Journal of Phycology*, Vol. 6, 1970, p. 5.
[13] Tripp, M. M. and Fries, C., *Journal of Internal Pathology*, Vol. 15, 1970, p. 136.
[14] Bryan, J. A., *American Journal of Epidemiology*, Vol. 99, 1974, p. 320.
[15] Leger, R. T., Boyer, K. M., Pattison, C. P., and Maynard, J. E., *Journal of Infectious Diseases*, Vol. 131, 1975, p. 167.
[16] Gravelle, C. R., Hornbeck, C. L., Maynard, J. E., Schable, G. A., Cook, E. N., and Bradley, D. N., *Journal of Infectious Diseases*, Vol. 131, 1975, p. 163.
[17] Vaughn, J. M. and Metcalf, T. G., *Water Research*, Vol. 9, 1974, p. 613.
[18] Berg, G., Dahling, D. R., and Berman, D., *Applied Microbiology*, Vol. 22, 1971, p. 608.
[19] Sweet, G. H. and Ellender, R. D., *Water Research*, Vol. 6, 1972, p. 775.
[20] Rao, V. C., Chandorkar, U., Rao, N. U., Kumaran, P. and Lakhe, S. B., *Water Research*, Vol. 6, 1972, p. 1565.
[21] Sorber, C. A., Malina, J. F., Jr., and Sagik, B. P., *Environmental Science and Technology*, Vol. 6, 1972, p. 438.
[22] Sobsey, M. D., Wallis, C., Henderson, M. and Melnick, J. L., *Applied Microbiology*, Vol. 26, 1973, p. 529.
[23] Hill, W. F., Jr., Akin, E. W., Benton, W. H. and Metcalf, T. G., *Applied Microbiology*, Vol. 23, 1972, p. 880.
[24] Jakubowski, W., Hoff, J. C., Anthony, N. C. and Hill, W. F., Jr., *Applied Microbiology*, Vol. 28, 1974, p. 501.
[25] Farrar, L. and Hedrick, H. G., *Developments in Industrial Microbiology*, Vol. 14, American Institute of Biological Sciences, Washington, D.C., 1973, p. 376.
[26] Konowalchuk, J., Speirs, J. I., Pontefract, R. D. and Bergeron, G., *Applied Microbiology*, Vol. 28, 1974, p. 717.
[27] Oza, P. D. and Chaudhuri, M., *Water Research*, Vol. 9, 1974, p. 707.
[28] Hill, W. F., Jr., Akin, E. W., Benton, W. H., Mayhew, C. J., and Jakubowski, W., *Applied Microbiology*, Vol. 27, 1974, p. 1177.
[29] Clarke, N. A., Hill, W. R., Jr., and Jakubowski, W., *Proceedings, AWWA Technology Conference*, 1–4 Dec. 1974, Dallas Tex., American Water Works Association.
[30] Gerba, C. P., Wallis, C. and Melnick, J. L., *Environment, Science and Technology*, Vol. 9, 1973, p. 1122.
[31] W. Alton Jones Cell Science Center, P.O. Box 631, Lake Placid N.Y. 12946.
[32] Kraus, M. P., *Proceedings*, 4th Mid-Atlantic Industrial Waste Conference, University of Delaware, Dept. of Civil Engineering. Newark, Del., p. 351.
[33] Buras, N., *Water Research*, Vol. 8, 1974, p. 19.
[34] Safferman, R. S. and Morris, M. E., *Water Research*, Vol. 10, 1976, p. 413.
[35] Sobsey, M. D. and Cooper, R. C., *Water Research*, Vol. 8, 1974, p. 869.
[36] Kott, Y., Nupen, E. M. and Ross, W. R., *Water Research*, Vol. 8, 1974, p. 869.
[37] Scarpino, P. V., Berg, G., Chang, S. L., Dahling, D., and Lucas, M., *Water Research*, Vol. 6, 1972, p. 959.
[38] Podoska, R. A. and Hershey, D., *Journal of Water Pollution Control Federation*, Vol. 44, 1972, p. 738.
[39] Nabli, B. *Archives of Institute Pasteur Tunis*, Vol. 49, No. 1–2, 1973, p. 15.

[40] Kott, Y., *Water Research*, Vol. 7, 1973, p. 853.
[41] Kruse, C. W., Olivieri, V. P. and Katawa, K., *Water and Sewage Works*, Vol. 118, 1972, p. 187.
[42] Kruse, C. W., Kawata, K., Olivieri, V. P. and Langley, E. K., *Water and Sewage Works*, Vol. 120, 1973, p. 57.
[43] Cramer, W. N., Kawata, K. and Kruse, C. W., "Clorination and Iodination of Poliovirus and f2 coliphage. *Journal of Water Pollution Control Federation*, Vol. 48, 1976, p. 61.
[44] Stringer, R. and Kruse, C. W., *Journal*, Environment, Engineering Division, *Proceedings*, American Society of Civil Engineers, Vol. 99, No. EE2, 1973, p. 156.
[45] Culp, R., *Journal of American Water Works Association*, Vol. 66, 1974, p. 698.
[46] Wolf, H. W., Safferman, R. S., Mixson, A. R. and Stringer, C. E., *Journal of American Water Works Association*, Vol. 66, 1974, p. 526.
[47] Sobsey, M. D., Wallis, C., Hobbs, M. F., Green, A. C. and Melnick, J. L., *Journal*, Environment, Engineering Division, *Proceedings*, American Society Civil Engineering, Vol. 99, No. EE3, 1973, p. 245.
[48] York, D. W. and Drewry, W. C., *Journal of American Water Works Association*, Vol. 66, 1974, p. 711.
[49] Sheldon, S. P. and Drewry, W. A., *Journal of American Water Works Association*, Vol. 65, 1973, p. 627.
[50] Majundar, S. B., Cecker, W. H. and Sproul, O. J., *Journal of Water Pollution Control Federation*, Vol. 45, 1973, p. 2433.
[51] Peleg, M., *Water Research*, Vol. 10, 1976, p. 361.
[52] Katzenelson, E., Klatter, B. and Schuval, H. I. *Journal of American Water Works Association*, Vol. 66, 1974, p. 725.
[53] Presnell M. W. and Cummins, J. M., *Water Research*, Vol. 6, 1972, p. 1203.
[54] Mone, J. C., *Pollution Engineering*, Vol. 5, 1973, p. 33.
[55] Sharp, D. G., Floyd, R. and Johnson, J. D., *Applied and Environmental Microbiology*, Vol. 31, 1976, p. 127.
[56] Sharp, D. G., Floyd, R. and Johnson, J. D., *Applied Microbiology*, Vol. 29, 1976, p. 94.
[57] Floyd, R., Johnson, J. D. and Sharp, D. G., *Applied and Environmental Microbiology*, Vol. 31, 1976, p. 298.
[58] Kim, K. S. and Sharp, D. G., *Virology*, Vol. 30, 1966, p. 724.
[59] Kraus, M. P., "Interaction of a Bluegreen Algal Host / Virus System with Ionizing Radiation," AEC Document COO-3007-23, Atomic Energy Commission, 1972.
[60] Kraus, M. P., "Blue-green Algal Genetics: A New Methodology," AEC Document COO-3007-24, Atomic Energy Commission, 1973.
[61] Dugan, G. L., Young, R. H. F., Lau, L. S., and Loh, R. C. S., *Journal of Water Pollution Control Federation*, Vol. 17, 1975, p. 2067.
[62] Bernarde, M. A., *Journal of American Water Works Association*, Vol. 65, 1973, p. 432.
[63] Kardos, L. T. and Sopper, W. E., in *Recycling Treated Municipal Wastewater and Sludge through Forest and Cropland*, W. E. Sopper and L. T. Kardos, Eds., Penn State University Press, State College, Pa., 1973, pp. 148–163.
[64] Foster, D. H. and Engelbrecht, R. S., *Recycling Treated Municipal Wastewater and Sludge through Forest and Cropland*, W. E. Sopper and L. T. Kardos, Eds., Penn State University Press, College Park, Pa., 1973, pp. 247–270.
[65] Kraus, M. P., *Proceedings*, 2nd National Conference on Complete Water Reuse, American Institute of Chemical Engineers and EPA Technology Tranfer, Water's Interface with Energy, Air and Solids, Chicago, 4–8 May 1975. pp. 401–
[66] Kraus, M. P., *Proceedings*, Third National Conference on Complete Water Reuse, Symbiosis as a Means of Abatement for Multi-Media Pollution, American Institute Chemical Engineers and EPA Technology Tranfer, L. K. Cecil and P. C. Welch, Eds., 27–30 June 1976, Cincinnati, Ohio, pp. 476–482.
[67] Brown, E. R., Hazdra, J. J., Keith, L., Greenspan, I., Kwapinski, J. B. G. and Beamer, P., *Cancer Research*, Vol. 33, 1973, p. 189.
[68] Webster, R. G., Campbell, C. G., and Granoff, A., *Virology*, Vol. 51, 1973, p. 149.
[69] Callahan, R., Lieber, M. M., Todaro, G. J., Graves, D. C. and Ferrer, J. F., *Science*, Vol. 192, 1976, p. 1005.
[70] Shanmugam, K. T. and Valentine, R. C., *Science*, Vol. 187, 1975, p. 919.
[71] Falkow, S., *Proceedings*, 10th Miles International Symposium: Recombinant DNA, Massachusetts Institute of Technology, Cambridge, Mass., 9–11 June 1976.

[72] Teich, N. M., *Proceedings*, 10th Miles International Symposium: Recombinant DNA, Massachusetts Institute of Technology, Cambridge, Mass., 9–11 June 1976.
[73] Herrmann, J. E., Kostenbaden, K. D., Jr. and Cliver, D. O., *Applied Microbiology*, Vol. 28, 1974, p. 895.
[74] Davis, E. M. and Gloyna, E. F., *Journal*, Environment, Engineering Division, *Proceedings*, American Society Civil Engineers, Vol. 99, No. EE3, 1973, p. 379.
[75] Berry, S. A. and Noton, B. G., *Water Research*, Vol. 10, 1976, p. 323.
[76] Bitton, G. and Mitchell, R., *Water Research*, Vol. 8, 1974, p. 227.
[77] Jakubowski, W. and Clarke, N. A., *U.S. EPA Water Supply Research Bulletin*, 10 Feb. 1976.
[78] Gaby, W. L., "Evaluation of Health Hazards associated with Solid Waste/Sewage Sludge Mixtures," Techical Report 1DB064, U.S. EPA, Cincinnati, Ohio, U.S. Government Printing Office, Washington D.C.
[79] Malina, J. F., Jr., Ranganathan, K. R., Sagik, B. P. and Moore, B. E., *Journal of Water Pollution Control Federation*, Vol. 17, 1975, p. 2178.
[80] Wellings, F. M., Lewis, A. L. and Mountain, C. W., *Applied Environmental Microbiology*, Vol. 31, 1976, p. 334.
[81] Ward, R. L. and Ashley, C. S., *Applied Environmental Microbiology*, Vol. 31, 1976, p. 921.
[82] Meyer, R. C., Hinds, F. C., Isaacson, H. R. and Hinesly, T. D. *Proceedings*, International Symposium on Livestock Wastes, American Society of Agricultural Engineers, St. Joseph, Mo., p. 183.
[83] Clarke, N. A., Akin, E. W., Liu, O. C., Hoff, J. C., Hill, W. F., Jr., Brashear, D. S., and Jakubowski, W., *Journal of American Water Works Association*, Vol. 67, 1975, p. 192.
[84] Enzinger, R. M. and Cooper, R. C., *Applied Environmental Microbiology*, Vol. 31, 1976, p. 758.
[85] Smith, H. W. and Huggins, M. B., *Journal of General Microbiology*, Vol. 92, 1976, p. 335.
[86] Gilbert, R. G., Rice, R. C., Bouwer, H., Gerba, C. P., Wallis, C. and Melnick, J. L., *Science*, Vol. 192, 1975, p. 1004.
[87] Fetter, C. M., Jr., and Holtzmacher, R. D., *Journal Water Pollution Control Federation*, Vol. 46, 1974, p. 264.
[88] Wellings, F. N., Lewis, A. L., Mountain, C. W., and Pierce, L. V., *Applied Microbiology*, Vol. 29, 1973, p. 751.
[89] Schaub, S. A. and Sagik, B. P., *Applied Microbiology*, Vol. 30, 1975, p. 212.
[90] Lund, E. and Ronne, V., *Water Research*, Vol. 7, 1973, p. 863.
[91] Lund, E., *Water Research*, Vol. 7, 1973, p. 873.
[92] Cooper, R. C., Potter, J. L., and Leong, C., *Water Research*, Vol. 9, 1974, p. 733.
[93] Peterson, M. L., *American Journal of Public Health*, Vol. 64, 1974, p. 912.
[94] Casida, L. E., Jr., and Liu, K., *Applied Microbiology*, Vol. 28, 1975, p.28.

D. J. Ptak[1] *and W. Ginsburg*[1]

Bacterial Indicators of Drinking Water Quality

REFERENCE: Ptak, D. J. and Ginsburg, W., "**Bacterial Indicators of Drinking Water Quality**, *Bacterial Indicators/Health Hazards Associated With Water, ASTM STP 635*, A. W. Hoadley and B. J. Dutka, Eds., American Society for Testing and Materials, 1977, pp. 218-221.

ABSTRACT: The papers presented at the ASTM International Symposium on Bacterial Indicators of Potential Health Hazards Associated with Water in June of 1976 were reviewed and are summarized in this paper as they pertain to bacterial indicators of drinking water quality.

KEY WORDS: bacteria, water, fecal coliforms

Drinking water of good bacteriological quality in the United States is indicated by the absence of coliforms when a minimum of 100 ml is tested by the membrane filter (MF) technique or 50 ml by the most probable number (MPN) method. Due to limitations such as media selection, temperature, and incubation time, some ambiguity exists regarding the coliform group. Consequently, pathogens and secondary invaders are not recognized even though their numbers sometimes appear to exceed those of the coliform group.

Differentiation of coliforms, although valuable in certain situations, is of limited value in determining the suitability of water for human consumption, since contamination with any type of coliform renders the water potentially unsafe. This view is also supported by Farmer [*1*][2] who cautions writers of microbiological standards against adopting bacterial species as indicators of pollution. Because of biochemical inconsistency and the time required to process each colony to fit a species definition, Farmer advocates a simple operational definition such as coliform. Since most countries have developed adequate means for production of coliform-free water, let us focus our attention on the remaining aquatic bacterial population.

[1]Water bacteriologist III and chief water bacteriologist, respectively, Microbiology Unit, Water Purification Laboratories, Chicago, Ill. 60611.

[2]The italic numbers in brackets refer to the list of references appended to this paper.

Data have been submitted at the International Symposium on Bacterial Indicators of Potential Health Hazards Associated with Water by numerous investigators showing the presence of *Aeromonas, Streptococcus, Clostridium, Pseudomonas, Vibrio, Staphylococcus, Candidus, Mycobacterium, Salmonella*, and a wide spectrum of other microorganisms including viruses in drinking water. Although waterborne outbreaks of these microorganisms are a rare occurrence, they do occur, and the coliform bacteria are attacked as being inadequate indicators of pollution. Mack [*2*] agrees that a better indicator system is required, but what alternatives are available? Dufour [*3*] has provided convincing evidence that *Eschericia coli* is the only coliform of undoubted fecal origin and feels it should be used as the indicator. This takes us back to Farmers' [*1*] caution against using a species as an indicator. The argument can be avoided by not permitting coliforms of any type to be present in drinking water.

The enterococci or fecal streptococci have been proposed as an indicator of the quality of drinking water, but the data presented at the International Symposium show coliforms better suited since the fecal streptococci die off very rapidly in water of good quality and are not as numerous. The enterococci might well be better indicators of pollution in wells (animal contamination), swimming pools, and sewage treatment plants (high in nutrient, low in oxygen). Similarly, the fecal coliforms might be used in areas other than drinking water, since they are already included in the total coliform count and also die off rapidly in waters of good quality.

Much attention has been given to the genus *Aeromonas*. A review of the data presented by Shotts [*4*] show clearly that *Aeromonas* cannot be considered an indicator of fecal pollution, since they are found infrequently in the intestine of warm-blooded animals, are rarely pathogenic to man, and are a natural inhabitant of water [*4,5*]. *Aeromonas*, when present, inflates the total coliform count on the MF, since approximately 30 percent produce acid from lactose. Generally, they do not interfere with the MPN technique, since gas production from lactose is not usually evident in 48 h. *Aeromonas* has been called a security indicator and, based on the results of the cytochrome oxidase test, might be used to lower total coliform counts while still including them in the total heterotrophic plate count as an indicator of treatment efficiency.

Up-to-date data on the genus *Pseudomonas* presented by Hoadley [*6*] has shown their poor reliability and inconsistency as indicators of pollution in drinking water. While not being considered as an indicator of polluted drinking water in the United States, the additional information presented by Hoadley [*6*] suggests that *P. aeruginosa* determinations along with the fecal coliform test could be of considerable value in the development of criteria for recreational waters.

The consensus on spore formers, *Clostridium* in particular, is that they are not practical as indicators of pollution in drinking water but may be valuable for testing raw waters since their spores may persist unchanged long after coliforms have died out or, in the case of effluents, increased. Minimal importance has been given to *Clostridium* in chlorinated water since their spores are chlorine resistant. The presence of *Clostridium*, however, may be objectionable if the water is used in hospitals or industry [*7*].

Recently, the University of Illinois released information regarding their search for new indicator organisms in water [*8*]. Their work showed many organisms more resistant to chlorine than coliform. The order of resistance to chlorine observed in this study was acid fast organisms, yeasts, *S. typhimurium*, poliovirus, and *E. coli*.

In the University of Illinois study, the yeasts included *Candida parapsilosis, C. krusei, Rhodotorula rubra*, and *Trichosporon fermentans*. The acid-fast organisms were *Mycobacterium fortuitum* and *M. phlei*. So far, these organisms, although consistently found in raw domestic wastewater, have not been found consistently in fecal material. The usefulness of these organisms as indicators of pollution seems better suited to the monitoring of sewage treatment processes or recreational waters rather than drinking water.

These findings agree with work done by Buck [*9*] but disagree with the suggestion made by McCabe [*10*] to use yeasts and acid-fast organisms as indicators of drinking water quality.

The use of chlorine residual measurements in treated water as a substitution for bacterial testing was proposed for the new U.S. Environmental Protection Agency National Interim Primary Drinking Water Regulations in 1975. Ignoring the fact that numerous publications have shown that bacteria are capable of surviving chlorination, McCabe has argued that the coliform test be abandoned in favor of chlorine residuals. The reason given was that 85 percent of the water systems do not collect their minimum number of samples. If this is true, then what assurance is there that these same utilities will substitute four chlorine residual tests for one bacterial test?

An instant and cheap test for drinking water quality apparently is not available at the present time, but many expensive and time-consuming tests are available which would increase the diagnostic capability of the laboratory at a price. The question is how much are we willing to spend to decrease the likelihood of pathogens going unnoticed in drinking water?

One of the simplest, most overlooked, and inexpensive tests is the standard plate count (SPC). It is well recognized that in assessing the potential disease-causing ability of drinking water the SPC is of limited value; however, it is the changes in a plate count which are significant when measuring filter bed efficiency, deterioration in distribution systems, and coliform suppression. The papers presented by Grabow [*11*] and Müller [*12*] support this statement. In reviewing waterborne diseases of the last 30 years, it was found that ground water supplies were responsible for more outbreaks than any other source [*13*]. Bacteriological data from one community water supply study showed the etiological agents to be bacterial (97 percent), viral (2.6 percent), protozoan (0.4 percent), and chemical (0.08 percent). It was also noted that while only 9 percent of the wells showed coliform, 83 percent showed bacteria as determined by the SPC [*13*]. In the absence of a suitable replacement for coliform or multiple parameter testing, the SPC apparently takes on greater importance as a probability statistic or harbinger of pollution. The fact that excessive bacterial populations can suppress coliform detection is immaterial since high or rising SPCs alone should warrant a thorough investigation for the cause. There is no reason to tolerate any organisms in a finished drinking water for which we have adequate preventive measures.

Inasmuch as outbreaks of waterborne disease today are the result of a combination of circumstances connected with treatment and distribution of water, it is probably too much to expect that changes in bacteriological testing and standards will preclude any future outbreaks. Far more important would be the preventive measures taken in eliminating these hazards on the basis of routine sanitary inspections.

In conclusion, it seems logical, based on the data presented, to retain total coliform as the indicator, with restrictions on maximum turbidity units and minimum free chlorine residuals, supplement total coliform counts with SPCs and institute multiple parameter tests when positive total coliform or abnormally high SPCs or both are encountered.

The areas requiring research include:

1. Rapid total coliform and SPC determinations to be used as a screen or presumptive indicator until a conventional bacteriological examination is completed.

2. Maximum allowable SPCs in potable water based on epidemiological data.

3. Massive data input leading to adoption of internationally acceptable criteria and standardized techniques.

References

[*1*] Farmer, J. J. and Brenner, D. J., this publication, pp. 37–47.
[*2*] Mack, W. N., this publication, pp. 59–64.
[*3*] DuFour, A. P., this publication, pp. 48–58.
[*4*] Shotts, E., "Aeromonas," presented at the ASTM International Symposium on Bacterial Indicators of Potential Health Hazards Associated With Water, Chicago, Ill., June 1976.
[*5*] Ptak, D. J., Ginsburg, W. and Willey, B. F. "Aeromonas, The Great Masquerader," *Proceedings*, American Water Works Association Water Quality Technology Conference, 1974.
[*6*] Hoadley, A. W., this publication, pp. 80–114.
[*7*] Cabelli, V. J., this publication, pp. 65–79.
[*8*] Engelbrecht, R. S., Foster, D. H., Masarik, M. T. and Lee, S. H., "Detection of New Microbial Indicators of Chlorination Efficiency," *Proceedings*, American Water Works Association Water Quality Technology Conference, 1974.
[*9*] Buck, J. D., this publication, pp. 139–147.
[*10*] McCabe, L. J., this publication, pp. 15–22.
[*11*] Grabow, W. O. K., this publication, pp. 168–181.
[*12*] Müller, G., this publication, pp. 159–167.
[*13*] Allen, M. J. and Geldreich, E. E., "Bacteriological Criteria for Ground-Water Quality," *Ground Water*, Vol. 13, No. 1, Jan.-Feb. 1975.

V. J. Cabelli[1]

Indicators of Recreational Water Quality

REFERENCE: Cabelli, V. J., **"Indicators of Recreational Water Quality,"** *Bacterial Indicators/Health Hazards Associated With Water, ASTM STP 635*, A. W. Hoadley and B. J. Dutka, Eds., American Society for Testing and Materials, 1977, pp. 222–238.

ABSTRACT: Water quality indicators, in general, and a number of potential health effects, water quality indicators, specifically, are examined with regard to their use in predicting the health hazards associated with the use of recreational waters. The factors considered are the requisite characteristics, constraints upon their use and the data obtained thereby, and their evaluation, especially as regards available epidemiological data. Based upon existing information, two indicator systems, *Escherichia coli* and enterococci, are tentatively recommended as the indicators of choice for this specific application.

KEY WORDS: bacteria, water, coliform bacteria

Ideally, recreational water quality indicators are microorganisms or chemicals whose densities in the water can be quantitatively related to potential health hazards resulting from recreational use therein—particularly those from activities which expose the upper body orifices to the water. In the present context, we will consider only those health hazards due to infectious agents associated with fecal, industrial, nutrient, or thermal pollution. The infectious agents in question and reported outbreaks of recreation associated infectious disease have been reviewed earlier in this publication.[2] The potential sources of the infectious agents reaching water used for recreation include untreated or poorly treated municipal and industrial effluents or sludge, sanitary wastes from seaside residences, fecal wastes from pleasure craft, dredge spoils, drainage from sanitary landfills, storm water runoff, and excretions of lower animals. In addition, the source of the infectious agents may be the aquatic environment itself. The health hazard potentials for infectious disease from each of these sources are not equal, and, of these sources, untreated human fecal wastes present the greatest hazard to human health. If the indicator used reflects these differences, there is no problem.

[1]Chief, Marine Field Station, Health Effects Research Laboratory-Cin., U. S. Environmental Protection Agency, West Kingston, R. I. 02892.

[2]This publication, pp. 15–22.

However, this is not always the case, and alternate or multiple indicators may be required. For example, *Escherichia coli*, a highly specific indicator of fecal wastes, may be an inappropriate indicator of the viral disease hazard in leachates from landfilled sludge high in toxic organics and inorganics, and, when the aquatic environment is the source of the pathogen or it multiplies significantly therein, the pathogen itself becomes the water quality indicator. In some instances, such as small point sources from boat wastes and seaside residences, indicators of fecal pollution would be inappropriate since, in the wastes from a given individual, the pathogen density may equal or exceed the indicator density.

In the final analysis, the best indicator—there is no ideal indicator—is the one whose densities correlate best with health hazards associated with a given (preferably several) type of pollution source. This definition is amenable only to epidemiological analysis. However, one might screen potential indicators for this ultimate test against the following requirements.

1. The indicator should be consistently and exclusively associated with the source of the pathogens.

2. It must be present in sufficient numbers to provide an "accurate" density estimate whenever the level of each of the pathogens is such that the risk of illness is unacceptable.

3. It should approach the resistance to disinfectants and environmental stress, including toxic materials deposited therein, of the most resistant pathogen potentially present at significant levels in the source.

4. It should be quantifiable in recreational waters by reasonably facile and inexpensive methods and with considerable accuracy, precision, and specificity.

These requirements should be considered in the context of the following issues.

Fecal Indicators Versus Water Quality Indicators

Because the discharge of human fecal wastes represents the most hazardous form of pollution with regard to recreational water quality and because this type of pollution is amenable to corrective action, the terms fecal indicator and water quality indicator have been used synonymously. In most cases, this is appropriate. However, there are situations in which this may not be so. Even though the inoculum may have come from fecal wastes, if an infectious agent multiplies in the aquatic environment under the influence of nutrient or thermal pollution, then the source of the organisms is the environment itself. In such situations, an indicator which indexes the human or animal fecal wastes discharged therein will tend to underestimate the potential hazard unless it too multiplies under identical environmental conditions. Potentially, *Aeromonas hydrophila, Vibrio parahaemolyticus, Pseudomonas aeruginosa,* and *Klebsiella* species fall into this category. Obviously, if the source of an infectious agent is not fecal (pathogenic *Naegle-*

ria), a fecal indicator would be of no value in indexing an associated health hazard.

Small Points Sources

The rationale for the use of guidelines and standards based on fecal indicator densities for indexing the health hazards in sewage polluted waters is that, under average conditions of illness in the discharging population, there is a reasonably constant indicator to pathogen ratio in the sewage and its receiving waters. Thereby, an acceptable probability of illness caused by the pathogen can be extrapolated to a given indicator density, which is then recommended as a guideline and promulgated as a standard. Such relationships appear to hold for waters receiving the discharges from relatively large municipal sewage treatment facilities. However, as the number of individuals who contribute to the source of fecal wastes becomes smaller and smaller—ultimately the individuals in a single pleasure boat or the occupants of a single domicile whose sewage is discharged directly or indirectly into a body of water—the indicator-pathogen ratio will vary more and more from the average upon which the guideline or standard is based. In the extreme case where the fecal wastes of a single ill individual or carrier are discharged into the water, the number of pathogens may equal or exceed the number of indicator microorganisms. Routine examination of such waters for fecal indicators would be of no value. Furthermore, the routine examination for pathogens would not be especially useful since the discharges of enteric pathogens will be sporadic, and, in all probability, the first sign of trouble will be illness among recreationists using the waters. The obvious solution is administrative action prohibiting such discharges into recreational waters.

Illness Rates in the Discharging Population

Most epidemiologists and health officers recognize that under epidemic conditions the actual indicator-pathogen ratio may change sufficiently from that upon which the guideline was based so that the acceptable risk of illness will be exceeded unless the guideline is temporarily made more restrictive. The recent swimming associated outbreak of shigellosis on the Mississippi River below Dubuque, Iowa [*1*][3] appears to represent an instance where, although the 200/100 ml fecal coliform guideline was probably exceeded, the outbreak did not occur until there was a large enough number of ill individuals and carriers in the discharging population.

Conversely, if through good public health measures (that is, immunization and the elimination of carriers) there is a significant and consistent decrease of the illness rate in the discharging population over a prolonged period of time, the probability of a specific illness associated with an existing guideline or standard based on a fecal indicator may be considerably lower than predicted. The absence of recreational water associated salmonellosis probably represents a case in point.

[3]The italic numbers in brackets refer to the list of references appended to this paper.

Fecal Indicators Versus Pathogens

The use of fecal indicators such as coliforms or portions of the coliform population, fecal streptococci, and *Clostridium perfringens* for indexing the health hazards in drinking and recreational waters dates back to the late 1800s and early 1900s [*2-4*] shortly after these organisms were first isolated and associated with the fecal wastes of warm-blooded animals. Within the context of the limitations noted previously, such practices were and are sound both on theoretical and practical grounds, since it is recognized that (*a*) there are a large number of pathogenic bacteria and viruses potentially present in municipal sewage [*5*], each with its own probability of illness associated with a given dose; (*b*) monitoring for each of the pathogens on a routine basis would be a herculean task; (*c*) enumeration methods for some of the more important pathogens are unavailable and for the rest are difficult; (*d*) pathogen density data are difficult to interpret because the methodology generally is time-consuming, expensive, imprecise, and inaccurate and because of the meager dose-response data available; and, (*e*) on theoretical grounds, the intent is not to index the presence of the pathogen but rather its potential to be there in sufficient numbers to cause unacceptable health risks. These very sound considerations notwithstanding and because there has been some improvement in enumeration technology for some of the microbial pathogens, some investigators propose to develop criteria and standards based upon the fecal pathogens themselves or even to use one fecal pathogen as an indicator for all the others [*6,7*]. These proposals appear unsound for the reasons stated previously. However, there is ample justification for the examination of recreational waters for pathogens during disease outbreaks and in those instances in which the aquatic environment potentially is the source of the pathogens.

Sewage Versus Feces

There are a number of potential microbial, water quality indicators which are consistently isolated in appreciable but variable numbers from raw as well as treated sewage but which we have infrequently isolated from the feces of normal "individuals" and then only in low numbers (unpublished data). Included are *P. aeruginosa, A. hydrophila, Klebsiella* species, *Citrobacter* species, and *Enterobacter* species. Three of these five organisms are opportunistic pathogens which, under certain rare conditions, could require enumeration because of their potential ability when present at high densities to cause untoward health effects among swimmers. If a fecal indicator, as the name implies, is restricted to microorganisms or chemicals which are specifically, consistently, and exclusively associated with the fecal wastes of warm-blooded animals, then these species cannot be designated as such. "Sewage indicators" would be a more appropriate designation, and they would be expected to variably index the health risks involved depending on the extent of their multiplication in sewage. Four of the organisms noted previously—*A. hydrophila, Enterobacter, Citrobacter*, and *Klebsiella*—meet the definition of coliforms as used in this country, and, as Dr. Dufour pointed up earlier in this publication, a significant portion of the *Klebsiella* population in the aquatic environment satisfies the requirements for fecal coliforms. One consequence of the distinction between sewage and fecal indicators and the rejection of the former would be the abandonment of total and fecal coliforms in favor of *E. coli* for indexing the contamination of surface waters with the fecal wastes of warm-blooded animals.

Human Versus Lower Animal Fecal Wastes

There are several pathogenic microorganisms potentially transmissible to humans by the contamination of water with the fecal or urinary wastes of lower animals [*5*]. With the possible exception of salmonellae, *Leptospira,* enteropathogenic *E. coli*, and some of the intestinal parasites, the transmission of these agents from the fecal wastes of animals to man via recreational water use occurs under limited and unusual circumstances in the United States. In addition, there are no definitive data proving that the potential for transmission of salmonellosis from wild or domestic animals to man via recreational water use can be indexed by fecal indicators or, with the exception of leptospirosis, that there have been outbreaks attributable to this chain of events. Nevertheless, existing guidelines and standards based upon the commonly used recreational water quality indicators make no distinction between pollution arising from human as opposed to lower animal fecal wastes. Furthermore, the major etiological agents of concern from the fecal contamination of recreational waters probably are the enteropathogenic viruses, and significant transmission of these agents to man via animal wastes has not been reported. There are no accurate and precise methods for estimating the proportions of human versus lower animal fecal pollution in environmental waters.

The foregoing discussion is not meant to suggest that there are no health hazards associated with the contamination of recreational waters with the fecal wastes of lower animals. Rather, we would point out that the currently used indicator systems do, in fact, equate the health hazards from human and lower animal fecal wastes, that there is no scientific base for this, and that more information is needed on the relative risks. This issue has relevance to recreational waters subject to contamination from waterfowl and wild animals, runoff from farm lots, and urban storm water runoff when there is no input from municipal wastes.

Fluctuations in Indicator Densities and the Collection of Samples

Generally, guidelines and standards for recreational waters based on microbial indicator systems are expressed (*a*) as a medium or $\log_{10}$ mean value and (*b*) a second value which cannot be exceeded more than 10 percent of the time [*8*]. Fuhs [*9*] (personal communication) points out that these two values are related only for a specific distribution of indicator densities and notes that the distribution will vary from one bathing beach to another.

However, when used as water quality guidelines and standards, these two limits need not be related to each other. Rather, they must be considered as two separate and distinct guidelines each with its own acceptable risk. The 1968 National Technical Advisory Committee guidelines for direct contact recreational waters [*10*] recommends "... waters shall not exceed a log mean of 200/100 ml, nor shall more than 10 percent of total samples during any 30-day period exceed 400/100 ml." The proper interpretation of second limit is that, irrespective of the mean density, the health risk to the bathing population is considered unacceptable when fecal coliform densities in excess of 400/100 ml occur more than 10 percent of the time—this and nothing more. Our own experience and those of others [*9*] has been that the second guideline is usually limiting, particularly at beaches impacted by combined sewer systems which overflow due to rainfall. At least one community [*11*] faced with such a situation has circumvented the problem by applying the existing microbial guidelines to "dry weather" conditions (less than a given amount of rainfall in a given period of time) and temporarily closing the beaches for a number of days following a rainfall in excess of the stated quantity.

In recent years, there has been a tendency to increase the quantity of water examined for pathogens, particularly certain enteroviruses, in recreational waters. The quantity of water examined appears to be a function of available concentration and assay methodology rather than the health risks associated with given viral densities. In the absence of epidemiological data on the risks involved, it may be counterproductive to examine large volumes (10 to 100 gal) of recreational waters for enteroviruses since the notoriety coincident with viral isolations may not be commensurate with the actual risks involved to people using such waters.

Bathers Themselves as a Pollution Source

It is generally accepted that in artificial bathing places, particularly in instances of insufficient water exchange and disinfection, bather density is a major factor in determining the probability of swimming associated illnesses [*12,13*]. However, in natural bathing places pathogenic microorganisms carried by the bathers themselves probably contribute insignificantly to hazards associated with recreational use, and indicator bacteria shed by these individuals contribute little to the quality of the water as perceived by microbiological measurements. The exceptions are small inland bodies of water (ponds, coves, etc.) which are subject to minimal water exchange and at which the bather density is extremely high.

Potential Water Quality Indicators

A number of microorganisms and one chemical, coprostanol, have been or could be considered as potential recreational water quality indicators. These have been reviewed by a number of workers, more recently by Bonde [*14*], Geldreich [*15*], and Cabelli et al [*16,17*]. A list of these indicators is presented in Table 1. Although most of these indicators have been discussed in detail earlier in this symposium some additional comments are in order.

Viruses

A guideline for permissible levels of enteroviruses currently able to be cultured from recreational waters was recently suggested [*6*]. It is unclear whether this guideline was meant to index the overall health hazards from fecal pollution or those specific to the viruses in question. One would suspect the former, since no distinction was made between the attenuated (vaccine) and virulent strains of poliovirus. When this fecal indicator is evaluated against the requirements noted previously or those described by Bonde [*18*], it is deficient on all counts except survival in the environment. Relative to hepatitis A virus, it may also be deficient in its survival properties. *E. coli* does not appear to survive as well as some enteroviruses [*19-23*]; the findings with regard to fecal streptococci are mixed [*20*].

On the other hand, if the suggested guideline was meant to index potential health hazards associated with the specific enteroviruses in question, a number of such guidelines would be required since the infective dose (ID) varies from one agent to another (Table 2). Finally, the author is unaware of any epidemiological data to support the proposed guideline of 1 plaque-forming unit (PFU)/10 gal of water.

A number of bacteriophages, primarily coliphages, have been examined as indicators of virus pollution with mixed results [*21*]. Recently, male specific RNA coliphages such as f-2 [*21-23*] and MS-2 [*24*] have been suggested as water quality indicators because their size, shape, and chemical composition are similar to those of the enteroviruses [*21*]; they mimic the enteroviruses in their resistance to chlorine [*21-25*] and parallel the entero-

TABLE 1—*Potential health effects water quality indicators.*

Indicator
Total Coliforms
Fecal Coliforms
E. coli
Klebsiella
Enterobacter-Citrobacter
Enterococci
C. perfringens
C. albicans
Bifidobcateria
Enteroviruses
Coliphage
Salmonella
Shigella
P. aeruginosa
A. hydrophila
V. parahemolyticus

TABLE 2—*Human ID_{50} for viral agents.*[a]

Microorganisms	Route	Approximate ID_{50}
Viruses		
Poliovirus 3	oral	<10
Rhinovirus	inhalation	<16
Inclusion conjunctivitis	application	10
Parainfluenza 1	inhalation	1 to 10
Coxsackie A_{21}	nasal	20
Influenza	nasal	5 to 100
Parainfluenza 3	nasal	10^3
Adenovirus	nasal	10^3
Parainfluenza 2	nasal	10^3 to 10^4
Salmonellae		
S. typhosa	oral	10^7
Other	oral	10^6 to 10^8

[a]Prepared from data compiled by Mechelas et al [*51*].

virus levels through sewage treatment (H. Wolfe, personal communication). However, because specific assays for these phages under natural conditions in sewage and receiving waters are unavailable, little is known of their densities therein or in human feces. This information as well as epidemiological data are needed in order to assess the value of these specific coliphages as water quality indicators.

Salmonella *and* Shigella

As reported earlier in this symposium, both *Salmonella* and *Shigella* have been responsible for swimming associated outbreaks of disease, but only rarely so. Routine monitoring of recreational waters for their presence seems unwarranted since, in general, they are present sporadically and in low numbers in waters considered acceptable for recreational use by existing guidelines. Furthermore, there are no realistic guidelines against which to evaluate the densities observed.

Most of the data on salmonellae in recreational waters describe the frequency of its isolation, often with sampling methods which do not quantify the amount of water examined. Such data are rather meaningless in assessing the health risk for an organism whose reported human ID_{50} is 10^5 to 10^7 organisms [*26*]. Moreover, the assumption that a single *Salmonella* organism in a litre or more of water reflects a significant health risk to bathers is highly questionable. Correlations between fecal coliform densities and the frequency of *Salmonella* isolations have been described [*15*]. However, when the actual *Salmonella* densities are examined, no correlation is obtained (Fig. 1). Were there a constant relationship, the points would have scattered about a vertical line. Table 3 presents the *Salmonella* recoveries at two bathing beaches, one of which can be regarded as "barely acceptable" and the other as relatively unpolluted according to locally used guidelines. It can be seen that the densities at both beaches were extremely low.

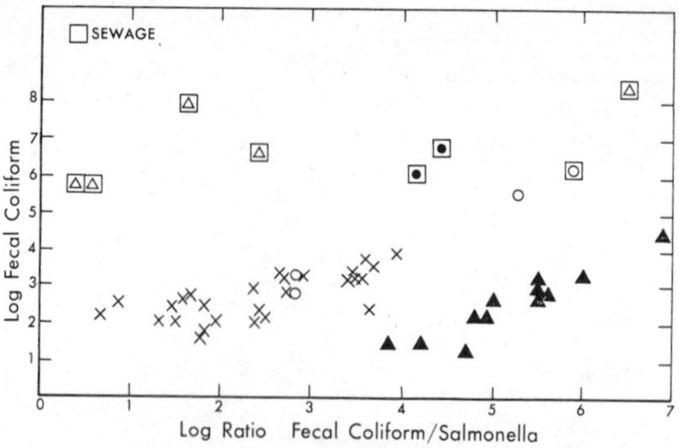

FIG. 1—*Fecal coliform* Salmonella *ratio versus fecal coliform density in sewage and water; prepared from data as follows:* ▲ *Levin, Cabelli, and Dufour (unpublished),* ● *Grabow and Nupen* [52], o *Cheng, Boyle, and Goepeert* [53], x *Smith, Twedt, and Flanigan* [48], △ *Evison and James* [54]; □ *sewage samples.*

P. aeruginosa, A. hydrophila, *and* V. parahaemolyticus

These three organisms are being considered together because all three are found in the aquatic environment and, presumably, have the ability to multiply therein. *V. parahaemolyticus* is not found in fresh water, while *P. aeruginosa* and *A. hydrophila* presumably do not multiply in marine waters. They are, however, isolated from estuarine and coastal waters subject to contamination from municipal sewage wastes [*31,32*]. All three organisms are human pathogens [*33-36*], the first two opportunistically so. All three are fish pathogens [*37*]. *P. aeruginosa* and *A. hydrophila* have been

TABLE 3—*Comparison of* Salmonella *and total coliform densities (per 100 ml) at two beaches arranged in descending order of coliform values* [16].

Relatively Unpolluted Date	Total Coliforms[a]	*Salmonella*[b]
	Barely Acceptable	
11 August	14 500	0.00020
12 August	3 300	0.00045
19 August	1 850[c]	0.00020
18 August	1 550	0.00020
22 July	900	0.00040
29 July	435	0.00020
28 July	360	0.00020
14 July	145	0.00020
18 August	350	<0.00018
22 July	205	<0.00018
29 July	185	0.00040
19 August	90	<0.00018
12 August	70[c]	0.00020
14 July	30	<0.00018

[a]mC estimate of total coliforms from low-tide samples collected concurrently with those for the *Salmonella* assays.
[b]Obtained for examination of 55.51. by S-HVS method.
[c]Estimate obtained by MPN method.

consistently isolated from raw as well as treated sewage effluents at densities of one to three orders of magnitude less than coliforms (unpublished data). However, they are infrequently isolated from the feces of normal humans [*38*], and then only at low densities (unpublished data). Because of this, none of these three organisms appears appropriate as an indicator of potential health hazards coincident with the contamination of recreational waters with fecal wastes. However, examination of recreational waters for these organisms may be warranted under certain special conditions. *P. aeruginosa* is the most common etiological agent of swimmers ear, an otitis externa common among children who swim in small ponds and lakes [*34*]. *V. parahaemolyticus*, or a closely related organism, has been isolated from wound infections in recreationists [*36*] as has *A. hydrophila* [*39*].

Coliforms and Coliform Biotypes

As was pointed out earlier in this publication, *E. coli* is the only coliform biotype consistently and exclusively associated with fecal wastes of warm-blooded animals. As such, it is the coliform indicator of choice for the health hazards associated with the use of recreational wasters polluted with such wastes. The rationale for the enumeration of *E. coli* specifically and that for excluding *Klebsiella* was discussed earlier in this publication by Dr. Dufour. Historically, criteria, guidelines, and standards were based upon total coliform and later fecal coliform densities because facile methods for the enumerating of *E. coli* specifically were not available. With the availability of facile, accurate, and specific most probable number (MPN) [*40*] and membrane filter methods for enumerating *E. coli* in recreational waters [*41*], there is no good reason why *E. coli* should not replace total and fecal coliforms as the coliform indicator of choice. Ideally, one would like to distinguish between those *E. coli* biotypes associated with human as opposed to lower animal fecal wastes and to quantify those strains which are potentially enteropathogenic.

Enterococci

Of all the microorganisms considered as recreational water quality indicators, enterococci is the group which most closely meets the ideal characteristics referred to earlier in this report. Thus, it is something of an enigma that the coliforms and, later, fecal coliforms came into widespread use while enterococci or fecal streptococci are rarely used and then only as an adjunct procedure. Perhaps, the answer lies in the relative ease with which relatively simple enumerative methods were developed for the coliform group of organisms. Facile membrane filter procedures are now available for the routine examination of *Streptococcus faecalis* and *S. faecium* (enterococci) biotypes in water [*42-44*], and this group of organisms should be much more widely used as a recreational water quality indicator. In fact, *S. faecalis* (the streptococcus biotype most closely associated with human fecal wastes) has the advantage over *E. coli* (the coliform biotype most closely associated with fecal wastes) in that it survives better in the aquatic environment [*44*]. One further improvement in the available methodology would appear desirable. It is the development of a facile procedure for differentiating those *S. faecalis* biotypes associated with vegetation and insects [*45*] from those associated with the fecal wastes of warm-blooded animals.

Bifidobacterium *and* Candida albicans

Both these organisms have only in recent years been considered as water quality indicators. Methodological problems will have to be resolved and more information on the distribution of these two organisms in the environment will have to be obtained before they can be seriously considered as potential water quality indicators.

C. perfringens

C. perfringens was one of the earliest organisms considered as a fecal indicator. Because of its widespread distribution in terrestrial and aquatic environments, its use as a water quality indicator is limited to specific instances where (*a*) remote sources of pollution are to be detected, (*b*) the water has come in contact with materials which inordinately destroy the generally used indicators, or (*c*) there is a premium on the survival of the spore, that is, as a conservative tracer in sewage and sludge.

Evaluation of Indicator Systems

Earlier in this publication,[4] I described three approaches towards the evaluation of *C. perfringens* as a water quality indicator. These three approaches bear repetition with regard to recreational water quality indicators in general. The preferable approach is to compare the densities of the various candidate water quality indicators to the final arbiter, illness or symptom rates specifically associated with swimming. After a series of epidemiological-microbiological trials at beaches where the pollution levels vary from trial to trial or beach to beach or both, the indicator system whose densities correlate best with the illness rates is selected. With the exception of the results to be presented, there are no data available using this particular approach. This approach has one weakness—it describes association not causality. However, for the purpose intended, the former should be satisfactory.

The second, and less desirable, approach is to compare the densities of the various candidate indicators to the densities of those pathogens in sewage which have been or are likely to be the etiological agents of recreation water-borne disease outbreaks. This approach, which has been used by a number of workers [*15,46-49*], has a number of problems associated with it. First of

[4]This publication, pp. 65–79.

all, if guidelines and standards are to be based on the indicator selected, this approach begs the question of the pathogen density associated with an unacceptable risk of disease. The response to this question requires good, human dose-response data by the oral route, along with information on indicator-pathogen relationships for all those pathogens in which this route of transmission is significant. Such data are meager. Secondly, it requires good enumerative methods for the pathogen as well as the indicator. As noted earlier, recovery methods for environmental samples are unavailable for three important—possibly the most important—viral agents, hepatitis virus A and the Norwalk and reo-like agents. They are relatively poor for the viral agents that can be grown in tissue culture and for *Shigella*, and they are marginal for salmonellae. This still leaves those pathogens whose source is not, or is not exclusively, fecal wastes; in general, there are no human ID_{50} data for these agents.

The third approach, that of comparing the densities of one indicator to those of another, is a "no win" approach for the reasons stated earlier.

Environmental Protection Agency Epidemiological Study

Over the past three years, our laboratory has been conducting an epidemiological-microbiological investigation to develop recreational water quality criteria for marine bathing beaches. The basic design for this study and the findings from the first two years of the trials have been described in previous publications [*16,50*]. Briefly, the study consisted of a series of discrete trials conducted on weekends in which (*a*) swimmers, rigorously defined as individuals who immerse their heads in the water, and nonswimmers in the same family groups at the beach were enrolled in the study and solicited for information on swimming activity and demography; (*b*) individuals who swam in the midweek before and after a weekend trial were rejected from the study; (*c*) water samples periodically were collected during the time interval when most of the individuals were actually at the beach (between 11:00 a.m. and 5:00 p.m.); and (*d*) illness information in the terms of symptomatology was sought by phone some 7 to 10 days after the weekend trial. Thereby, attack rates for individual symptoms or groups of symptoms (that is, gastrointestinal, respiratory) were obtained for two populations at each beach, swimmers and nonswimmers. Furthermore, the attack rates could be examined by demographic grouping, trial, groups of trials each with a same general indicator density, or beach.

A basic tenet of the experimental design was that no prejudgments would be made as to which is the best indicator. Therefore, measurements were to be made for a number of potential indicators, and the degree of the correlation of indicator densities to symptom rates would be used to select the indicator of choice.

Both in the 1973 pretest (study population about 1300 individuals) and the 1974 trials (study population about 8000 individuals), significant differences in gastrointestinal symptoms (vomiting, nausea, diarrhea, or stomachache) were found among swimmers relative to nonswimmers at a barely acceptable but not at a relatively unpolluted beach. The mean densities for the various bacterial indicators obtained at two locations over three time intervals at each beach are presented in Table 4. It can be seen that, day to day and within day (primarily tidal) variability notwithstanding, significant differences in the mean densities were obtained for a number of indicators at the barely acceptable as opposed to the relatively unpolluted beach. Table 5 shows that attack rates for gastrointestinal symptoms during 1973 and 1974 at each beach for swimmers, nonswimmers, and the differential between the two. A two-way chi square analysis of the data showed that, for each of the two years, the attack rate for swimmers was significantly higher than that for nonswimmers at the barely acceptable beach. This analysis, while showing the appropriateness of the experimental design, did not pro-

TABLE 4—*Mean indicator densities at beaches during time of epidemiological trials.*

		Log Mean Recovery/100 ml			
		1973		1974	
Indicator	Method	Coney Island	Rockaways	Coney Island	Rockaways
Total coliforms	MPN	. . .	. . .	1213[a]	43.2
Total coliforms	mC	983[a]	39.8	. . .	. . .
Fecal coliforms	mC	165[a]	21.5	565[a]	28.4
Escherichia coli	mC	174[a]	24.8	15.3[a]	2.4
Klebsiella	mC	122[a]	13.7	59.2[a]	3.5
Enterobacter-Citrobacter	mC	530[a]	11.1	434[a]	6.6
Fecal streptococci	mSD	91.2	21.8	16.4[a]	3.5
Pseudomonas aeruginosa	mPA	30.4[a]	6.5	45.8[a]	3.1
Aeromonas hydrophila	mA	25.3	26.5	9.6	4.9

[a]Significantly different at 95 percent confidence level.

TABLE 5—*Differential rate of gastrointestinal (GI) symptoms at the Coney Island and Rockaways beaches in 1973 and 1974.*

	Rate of GI Symptoms[a] %					
	Coney Island			Rockaways		
Year	Swim[b]	Nonswim	Δ	Swim	Nonswim	Δ
1973[c]	7.2	2.4	4.8[e]	8.1	4.6	3.5
1974[d]	4.2	2.6	1.6[e]	3.9	3.5	0.4

[a]Vomiting, diarrhea, nausea, or stomachache.

[b]Swimmers—defined a head immersed in water.

[c]Study population—Coney Island: swim 474, nonswim 167; Rockaways: swim 484, nonswim 197.

[d]Study population—Coney Island: swim 1961, nonswim 1185; Rockaways: swim 2767, nonswim 2156.

[e]Significantly higher than nonswimmers (P =0.05).

duce the required information—the comparison of indicator densities to illness rates. When the mean indicator density over the summer at each beach for the 1973 and 1974 data were plotted against their corresponding differential (swimmers minus nonswimmers) gastrointestinal symptom rates, four points were obtained for each indicator. These are shown in Fig. 2. From inspection alone, it can be seen that *E. coli* and enterococci gave the best correlation. The correlation coefficients for each of the indicators are given in Table 6. Not only were extremely high correlation coefficients obtained with *E. coli* and enterococci, but the slopes of the regression lines were significantly different from zero. The analysis of the data by trials also showed that within a given summer, *E. coli* and enterococci were the best indicators.

Summary

It seems appropriate, by way of a summary, to reexamine the two tentative indicators of choice, *E. coli* and enterococci as determined from the epidemiological study, against the characteristics of an ideal indicator as noted earlier in this review. First of all, both are consistently and rather specifically associated with the source of the pathogens, presumably human fecal wastes; however, more specificity with regard to the distinction between human and lower animal wastes would be desirable. Secondly, both organisms are present in the water in sufficient density to index the rather low

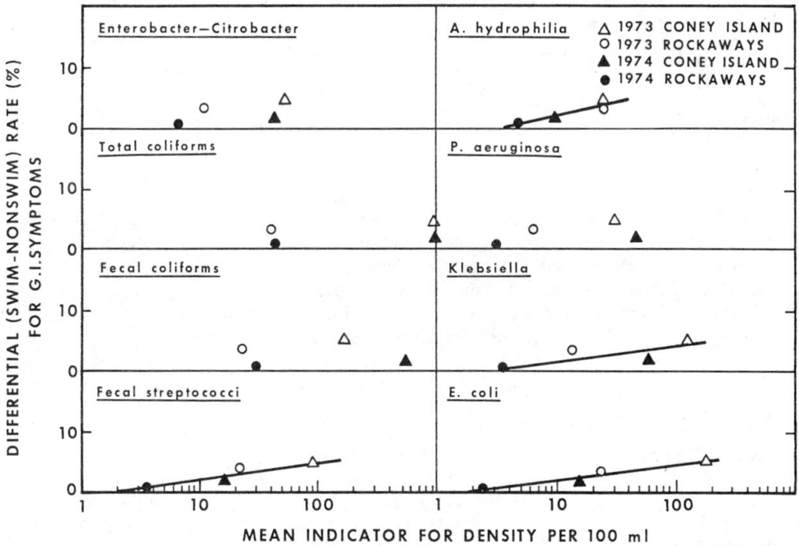

FIG. 2—*Relationship of the differential rate of gastrointestinal symptomatology (swimmer minus nonswimmer) to indicator as obtained from the analysis of 1973 and 1974 data; each point represents the overall gastrointestinal symptom rate and mean indicator density for all the trials conducted at the beach during that summer.*

TABLE 6—*Correlation coefficients for mean density versus gastrointestinal symptoms (1973–1974).*

Indicator	Correlation Coefficient[a]	Slope of Regression Line	
		Value	Significance
Total coliforms	0.33	0.79	NS
Fecal coliforms	0.08	0.24	NS
E. coli	0.95	0.25	. . .[b]
Enterococci	0.95	3.20	. . .[b]
Klebsiella	0.69	1.97	NS
Enterobacter-Citrobacter	0.45	0.87	NS
P. aeruginosa	0.42	1.49	NS

[a]Mean density for all trial per summer per beach versus differential rate (swimmers minus nonswimmers); 4 points.
[b]Significant at 0.95 level.

rates of the rather mild types of gastrointestinal symptoms reported in this study. Neither mortality nor hospitalization was reported. Therefore, one would expect that, except in epidemic situations, more serious but infrequent illnesses such as hepatitis virus A, salmonellosis, shigellosis, and some of the illnesses caused by the recoverable enteroviruses such as polio and Coxsackie virus A would have been indexed at even lower risk levels. Thirdly, it would appear that both *E. coli* and enterococci survive transport from the source (effluent outfall) to the target (bathing beach) well enough to provide a reasonably good correlation to the human health effects (symptomatology) seen at the beaches and presumably caused by pathogenic microorganisms in sewage. Finally, if the problem of satisfactory enumerative methods was a significant factor against the adoption of these two indicator systems, this no longer should be a problem.

References

[*1*] *Morbidity and Mortality Weekly Reports*, Vol. 23, No. 46, 1974.

[*2*] Mathews, A. P., *Technical Quarterly*, Vol. 6, 1893, p. 241.

[*3*] Houston, A. C., "Bactenoscopic Examination of Drinking Water with Particular Reference to the Relations of Streptococci and Staphlococci with Water of This Class," *Twenty-eighth Annual Report of the Local Government Board Containing the Report of the Medical Officer for 1898–99*, Supplement, 1899.

[*4*] Klein, E., "Report on the Morphology and Biology of *Bacillus enteritidis sporogenis*; on the Association of this Microbe with Infantile Diarrhea and with Chlorea Nostras; and on its Relations with Milk, with Sewage, and with Manure," *Twenty-seventh Annual Report of the Local Government Board Containing the Report of the Medical Officer for 1897–98*, Supplement, 1898.

[*5*] McKee, J. E. and Wolf, J. W., *Water Quality Criteria*, 2nd ed. Publication No. 3A, State Water Quality Control Board, Sacramento, Calif., 1963.

[*6*] Berg, G., Bodily, H. L., Lenette, E. H., Melnick, J. L., and Metcalf, T. G., *Viruses in Water*, American Public Health Association, Washington, 1976.

[*7*] Coin, L., Menetrier, M. L., Labonde, J., and Hannoun, M. C., *Proceedings*, International Conference on Water Pollution Research, Vol. 1, 1964, p. 1.

[*8*] Senn, C. L., Berger, B. B., Jensen, E. C., Ludwig, H., Romer, H., and Shapiro, M. A., *Journal of the Sanitary Engineering Division*, Vol. 89, 1963, p. 57.

[*9*] Fuhs, G. W., *The Science of the Total Environment*, Elsievier, Amsterdam, The Netherlands, 1975, pp. 165–175.

[*10*] *Water Quality Criteria*, National Technical Advisory Committee, Federal Water Pollution Control Administration, Dept. of Interior, Washington, D.C., 1968, pp. 7–14.

[*11*] Kupfer, G. S., 1975. "Control of Swimming at Milwaukee's Beaches by Pollution Prediction Formula," Annual Meeting, American Public Health Association, 1975, p. 53.

[*12*] Robinton, E. D. and Mood, E. W., *Journal of Hygiene*, Vol. 64, 1966, p. 489.

[*13*] Brown, J. R., McLean, D. M., and Dixon, M. C., *Canadian Journal of Public Health*, Vol. 54, 1963, p. 121.

[*14*] Bonde, G. J., 1966. Bacteriological methods for estimation of water pollution. *Health Laboratory Science*, Vol. 3, 1966, p. 12.

[*15*] Geldreich, E. E., *Journal of the American Water Works Association*, Vol. 62, 1970, p. 113.

[*16*] Cabelli, V. J., Levin, M. A., Dufour, A. P., and McCabe, L. J., in *Discharge of Sewage from Sea Outfalls*, Pergamon, London, England, 1975, pp. 63–73.

[*17*] Cabelli, V. J., Dufour, A. P., Levin, M. A., McCabe, L. J., and Habermann, P. W., "Relationship of Microbial Indicators to Health Effects at Marine Bathing Beaches," Annual Meeting of the American Public Health Association, Chicago, 1976.

[*18*] Bonde, G. J., 1968. Studies on the dispersion and disappearance phenomena of enteric bacteria in the marine environment. *Rev. Int. Oceanogra. Med. 9*:17.

[*19*] Metcalf, T. G., "Biologic Parameters in Water Transmission of Viruses," *Proceedings*, 13th Water Quality Conference, University of Illinois, Urbana, Ill., 1971.

[*20*] Smith, J., Twedt, R. M., and Flanegan, L. K., *Journal of the Water Pollution Control Federation*, Vol. 45, 1973, p. 1736.

[*21*] Scarpino, P. V., in *International Symposium of Discharge of Sewage from Sea Outfalls*, A. L. H. Gameson, Ed., Pergamon, London, England, 1975, p. 49.

[*22*] Loeb, T. and Zinder, N., "A Bacteriophage Containing RNA," Proceeding of the National Academy of Science, Vol. 49, 1961, p. 857.

[*23*] Kott, Y., Roze, N., Sperber, S., and Betzer, N., *Water Research*, Vol. 8, 1974, p. 165.

[*24*] Dhillon, E. K. S., and Shillon, T. S., *Applied Microbiology*, Vol. 27, 1974, p. 640.

[*25*] Olivieri, V. P., Kruse, C. W., Hsu, Y. C., Griffiths, A. C., and Kawata, K., *Disinfection Water and Waste Water*, J. D. Johnson, Ed., Ann Arbor Science, Ann Arbor, Mich., 1975, p. 145.

[*26*] Hornick, R. B., Greisman, S. E., Woodward, T. E., Dupont, H. L., Dawkins, A. T., and Snyder, M. J., *New England Journal of Medicine*, Vol. 283, 1970, p. 686.

[*27*] Dupont, H. L. and Hornick, R. B., *Medicine*, Vol. 52, 1973, p. 265.

[*28*] Wang, W. L. L., Dunlop, S. G., and Munson, P. S., *Journal of Water Pollution Control Federation*, Vol. 38, 1966, p. 1775.

[*29*] McFeters, G. A., Bissonnette, G. K., Jezeski, J. J., Thomson, C. A., and Stuart, D. G., *Applied Microbiology*, Vol. 27, 1974, p. 823.

[*30*] Cabelli, V. J., Kennedy, H., and Levin, M. A., *Journal of the Water Pollution Control Federation*, Vol. 48, 1976, p. 367.

[*31*] Cabelli, V. J., "The Occurrence of Aeromonads in Recreational Waters," ASM Abtracts, Annual Meeting American Society for Microbiology, 1973, p. 32.

[*32*] Cabelli, V. J., Brezenski, F. T., Dufour, A. P., and Levin, M. A., *Proceedings*, Seminar on Methods for Monitoring Marine Environment, EPA Report 600/4-74-004, U. S. Environmental Protection Agency, 1974, p. 360.

[*33*] Gilardi, G. L., *Applied Microbiology*, Vol. 15, 1967, p. 417.

[*34*] Jones, E. H., *External Otitis, Diagnosis and Treatment*, C. C. Thomas, Springfield, Ill., 1965.

[*35*] Xen-yoji, H., Sakai, S., Terayama, T., Kudo, Y., Ito, T., Benoki, M., and Nagasaki, M., *Journal of Infectious Disease*, Vol. 115, 1965, p. 436.

[*36*] Hollis, D. G., Weaver, R. E., Baker, C. N., and Thornsberry, C., *Journal of Clinical Microbiology*, Vol. 3, 1976, p. 425.

[*37*] Van Duijn, C., Jr., *Diseases of Fishes*, Thomas, Springfield, Ill., 1972, p. 148.

[*38*] Sutter, V. J., *Applied Microbiology*, Vol. 16, 1968, p. 1532.

[*39*] Phillips, J. A., Bernhardt, H. E., and Rosenthal, S. C., *Pediatrics*, Vol. 53, 1974, p. 110.

[*40*] "The Bacteriological Examination of Water Supplies," Reports on Public Health and Medical Subjects No. 71, Her Majesty's Stationary Office, London, England, 1968.

[*41*] Dufour, A. P., Strickland, E. R., and Cabelli, V. J., A membrane filter procedure for enumerating thermotolerant *E. coli*. *Proceedings*, 9th Annual Meeting of the National Shellfish Association, 1975.

[*42*] Levin, M. A., Fischer, J. R., and Cabelli, V. J., *Applied Microbiology*, Vol. 30, 1975, p. 66.
[*43*] Kenner, R. A., Clark, H. F., and Kabler, P. W., *Applied Microbiology*, Vol. 9, 1960, p. 15.
[*44*] Slanetz, L. W., and Bartley, C. H., *Health Laboratory Science*, Vol. 2, 1965, p. 142.
[*45*] Geldreich, E. E., Sanitary significance of fecal coliforms in the environment. F.W.P.C.A. Publ. WP-20-3, 1966.
[*46*] Bonde, G. J., *International Symposium on Discharge Sewage from Sea Outfalls*, Pergamin, London, England, 1975.
[*47*] Slanetz, L. W., Bartley, C. H., and Stanley, K. W., *Health Laboratory Science*, Vol. 5, 1968, p. 66.
[*48*] Smith, J., Twedt, R. M., and Flanegan, L. K., *Journal of Water Pollution Control Federation*, Vol. 45, 1973, p. 1736.
[*49*] Cohen, J. and Shuval, H. I., *Water, Air and Sea Pollution*, Vol. 2, 1973, p. 85.
[*50*] Cabelli, V. J., Dufour, A. P., Levin, M. A., and Haberman, P. W., *Limnology and Oceanography*, Vol. 2, 1976, p. 424.
[*51*] Mechalas, B. J., Hekimian, K. K., Schniazi, L. A., and Dudley, R. H., *Water Quality Criteria Data Book*, Vol. 4, 18040 DAZ 04/72, U. S. Environmental Protection Agency, 1972.
[*52*] Grabow, W. O. K. and Nupen, E. M., *Water Research*, Vol. 6, 1972, p. 1557.
[*53*] Cheng, M. C., Boyle, W., and Goepeert, J. M., *Applied Microbiology*, Vol. 21, 1971, p. 662.
[*54*] Evison, L. W. and James, C., *Journal of Applied of Bacteriology*, Vol. 36, 1973, p. 109.

E. W. Mood[1]

Bacterial Indicators of Water Quality in Swimming Pools and Their Role

REFERENCE: Mood, E. W., "**Bacterial Indicators of Water Quality in Swimming Pools and Their Role,**" *Bacterial Indicators/Health Hazards Associated With Water, ASTM STP 635*, A. W. Hoadley and B. J. Dutka, Eds., American Society for Testing and Materials, 1977, pp. 239–246.

ABSTRACT: The most significant bacterial indicator of the quality of swimming pool water is the total 35°C agar plate count. Specific bacterial indexes such as *Escherichia coli* have been used by various agencies, but these indexes are less sensitive than the total 35°C agar plate count. Also, there is no concensus of opinion as to what constitutes acceptable values for these specific indexes. Properly designed, constructed, and operated swimming pools with an adequate level of residual disinfectant can be maintained with total plate counts which are less than 100 colonies/1.0 ml. The bacterial quality of swimming pool water is usually a measure of the adequacy of the disinfecting process. Microbiological examination of samples of swimming pool water for specific genus or species are needed usually only in conjunction with a suspected outbreak of disease among the users of a swimming pool.

KEY WORDS: bacteria, water, fecal coliforms

From the public health standpoint, it is very important to have a measurement of the quality of swimming pool water which will denote the potential presence of pathogenic microorganisms. In 1923, one of the first reports of the Committee on Bathing Places, Sanitary Engineer Section of the American Public Health Association's tentative standards pertaining to the chemical, physical, and bacterial quality were proposed [*1*].[2] It was recommended that bacterial analyses of swimming pool water include the following two tests on a regular basis; namely, a 37°C, 24 h bacteria count on agar or litmus lactose agar and a partially confirmed test for *Bacterium coli* (later redesignated as *Escherichia coli*). A third test, which was made optional, is the 20°C, 48 h bacteria count on agar. In proposing these three tests, the com-

[1]Associate clinical professor of public health, Department of Epidemiology and Public Health, Yale University School of Medicine, New Haven, Conn. 06516.

[2]The italic numbers in brackets refer to the list of references appended to this paper.

mittee recognized that the methods usually employed for bacterial analyses of potable water did not show the true sanitary condition or quality of the waters in swimming pools and strongly emphasized the need for studies of the bacterial flora of swimming pool water to be used in measuring the quality of the water.

In the more than 50 years which have elapsed since the issuance of that committee report, some studies of the microbiology of swimming pool water have been conducted, but the findings of these researches did not establish quantitative bacterial measurements which were radically different from those used to evaluate potable water. The parameter remained similar; only the values changed.

Present Role of Bacterial Indicators of Swimming Pool Water Quality

The usual role fulfilled today by bacterial indicators of the quality of water in swimming pools is quite different, in many respects, than the role of bacterial indicators for other uses of water. When bacterial indicators are used in swimming pool sanitation activities, the primary purpose is usually to evaluate the accuracy and adequacy of the chemical tests which are used at least daily to regulate the disinfection and chemical conditioning processes of the water. A secondary purpose may be to assist in the evaluation of the effectiveness of the disinfection and filtration of the water and the operation of the recirculation system. A tertiary purpose, which under special circumstances may become a dominant role, is to provide additional information in any epidemiological studies which may be undertaken subsequent to a suspected outbreak of human disease associated with swimming pool water.

In some respects, the three purposes just enumerated are applicable also to the use of bacterial indicators of the quality of water of a public supply. However, the role of indicators of the quality of swimming pool water is different in that the usual source of microbiological contaminants will be from the persons using the swimming pool and, therefore, will have some specific characteristics.

Characteristics of Bacterial Contamination of Swimming Pool Water

A few studies of the types and numbers of microorganisms introduced in swimming pool water by bathers have been conducted by various researchers. Most of the studies show that skin bacteria are the most common bacteria. A study by Robinton and Mood [*2*] using a limited number of young women who teach physical education at a what was then a woman's college showed that the dominant types of organisms shed by bathers into the water were cocci. Members of the genera *Neisseria, Sarcina, Micrococcus,* and *Staphylococcus* were found to be present. Of the 100 samples collected during this study, the last 70 were examined for the presence of staphylococcal organisms. All of the 70 samples were found to be positive for this group of bacteria. Of the 100 samples examined for coliform bacteria, only 61.0

percent showed the presence of these bacteria in 100-ml portions when the multiple dilution tube method was used and only 71.0 percent when the membrane filter technique was used (Table 1). These same samples were examined also for fecal streptococci and *Streptococcus salivarius.* Using 100-ml samples, fecal streptococci were found in only 33.7 percent of the samples when the multiple dilution tube method was used and in 55.0 percent when the membrane filter was used. This study suggested quite clearly that coliform organisms and fecal streptococci are not the indicators of choice for the presence of microorganisms in swimming pool water which may come from the persons using the swimming pool and suggested that members of the genus *Staphylococcus* be used as the principal microbial indicator of swimming pool water quality.

Many other investigators have suggested the use of staphylococci as a measure of swimming pool water quality. Ritter and Treece [*3*] showed that *S. aureus, S. salivarius*, and the enterococci were 5 to 20 times more resistant to chlorine than the coliform bacteria and could be recovered from swimming pool water. Favero and coworkers [*4,5*] reported that *S. aureus* was relatively more resistant to chlorine than *Escherichia coli* and *Pseudomonas aeruginosa.* These investigators suggested that staphylococci be adopted as indicators of pollution in swimming pools. They reasoned that staphylococcal organisms would be valid indicators of pollution as they are derived from the mouth, nose, throat, and skin surfaces and, therefore, indicate the potential presence of pathogens. They are also more resistant to chlorine than coliform bacteria, and the absence of large numbers of staphylococci would imply adequate disinfection and the absence of intestinal bacteria. These researchers proposed an allowable maximum of less than 100 staphylococci per 100 ml of water.

Members of the genus *Streptococci* have been suggested also as indicators of swimming pool water quality, since mouth streptococci are found frequently in swimming pool water when persons were in the pool. Mallmann [*6*] proposed the cocci test as a measurement of swimming pool water contamination but did not recommend a numerical standard. Also Dick et al [*7*] suggested the use of streptococci as indicators of the sanitary quality of swimming pool water.

In 1953, the Public Health Laboratory Service Water Sub-Committee of the United Kingdom examined critically the use of the coliform group of bacteria as indicators of pollution of swimming pool water [*8*]. This group of scientists found that coliforms were seldom isolated from samples of swimming pool water with a free residual chlorine values exceeding 0.10 ppm, hemolytic streptococci and staphylococci were too resistant to chlorination to be of much service as indicator organisms, and plate counts in nutrient agar at 37 and 22°C showed a close correlation with levels of free residual chlorine. This subcommittee recommended that as a standard of microbiological quality of swimming pool water, no sample should contain any coliform organisms in 100 ml of water and that 75 percent of the samples

TABLE 1—*Percent of samples by number of organisms of the results of bacterial examinations of samples of water for specific organisms; samples were collected for special swimming tank used by only one person.*[a]

	Coliform Bacteria		Fecal Streptococci		*S. salivarius*	*S. aureus*	Staphylococci
	MPN/100 ml	Organisms Membrane Filter/100 ml	MPN/100 ml	Organisms Membrane Filter/100 ml	Organisms Membrane Filter/1.0 ml	Organisms Membrane Filter/1.0 ml	Organisms Membrane Filter/1.0 ml
	Number of Samples Examined						
Number of Organisms	100,%	100,%	98,%	100,%	100,%	70,%	70,%
0	39.0	29.0	63.3	45.0	52.0	22.8	0.0
1 to 9	30.0	40.0	10.2	32.0	34.0	51.4	1.4
10 to 99	15.0	28.0	12.2	23.0	14.0	20.0	65.7
100 to 299	13.0	3.0	4.1	0.0	0.0	2.9	24.3
300 to 499	1.0	0.0	1.0	0.0	0.0	0.0	4.3
>500	2.0	0.0	6.2	0.0	0.0	2.0	4.3
Total	100.0	100.0	100.0	100.0	100.0	100.0	100.0

[a]From Ref *2*.

examined should have 37° C agar plate counts not exceeding 10 colonies/1.0 ml with no samples containing more than 100 colonies/1.0 ml.

The "Suggested Ordinance and Regulations Covering Public Swimming Pools" of the Joint Committee on Swimming Pools, which was established in 1964 by the American Public Health Association, recommended similar microbiological standards of swimming pool water quality [*9*]. This model legislation proposed that swimming pool water of acceptable quality should not have more than 15 percent of the samples covering any considerable period of time with more than 200 bacteria per 1.0 ml as determined by the standard (35° C) agar plate count, nor show positive test (confirmed test) for coliform organisms in any five 10-ml portions of a sample nor more than 1.0 coliform organisms per 50 ml when the membrane filter test is used.

Robinton and Mood [*2*] demonstrated that, under test conditions, an average person may shed approximately 2×10^8 organisms into the water while swimming. These are organisms which grow on nutrient agar at 35° C in 24 h under aerobic conditions. Some persons will shed a greater number of organisms, some a fewer number. The important fact is that all swimmers shed a considerable number of microorganisms while swimming, and some of these organisms are easy to quantify using simple microbiological procedures.

Epidemiological Data

Many epidemiological studies have been conducted in which attempts were made to determine if a relationship exists between the quality of swimming pool water as determined by microbiological examinations and the transmission of infectious disease agents to swimmers. Very few studies have been able to show any significant relationship. McCabe [*10*] reported that most of the disease outbreaks which have been associated with swimming pools have involved viruses and mycobacteria. Hoadley [*11*] comprehensively summarized several epidemiological studies which have associated *P. aeruginosa* with outer ear infections among swimmers and skin rashes among users of warm-water whirlpool baths. However, even though there seems to be a significant correlation between outer ear infections and swimming, these data do not imply that the infectious agent was transmitted through the medium of the swimming pool water. *P. aeruginosa* are skin bacteria and may be washed easily from the swimmer's body into the ear channel. Swimming and diving remove some of the protective secretions from the surface of the auditory canal, making it easier for *P. aeruginosa* to cause a localized, inflammatory condition, frequently called otitis externa.

Gallagher [*12*] demonstrated a higher incidence of respiratory illness among swimmers but did not attempt to link these illnesses to the quality of the swimming pool water. Bell et al [*13*] conducted an epidemiological study of pharynoconjunctival fever and suggested that contaminated swimming pool water might have been "a possible source of infection accessory to a person-to-person mode of spread." However, these researchers did not evaluate the quality of the swimming pool water by any microbiological parameter.

Mycobacterium marinum, the causative organism of skin lesions often called "swimming pool granuloma," has been isolated many times from swimming pool water [*14,15*]. This organism is very resistant to chlorine. Park and Brewer [*15*] observed that the organisms which they isolated from swimming pool water were resistant to chlorine concentrations of at least 1.50 ppm free chlorine. These researchers did not report the presence of other concomitant microorganisms found in the swimming pool water samples examined. However, if examinations were made to detect the presence of any of the more common bacterial indicator organisms, the result probably would have been negative as 1.50 ppm of free residual chlorine is bactericidal to most, if not all, of the common bacterial indicator organisms in use today.

Present Bacterial Indicators of Water Quality in Swimming Pools

The most common indicator of the quality of water in swimming pools used by regulatory agencies in the United States is the coliform index derived by using either the multiple dilution tube method or the membrane filter. A few control agencies use the total 35°C agar plate count. It is erroneous to assume that fecal pollution of swimming pool water is the only or most important type of pollution of public health concern. Epidemiological data summarized earlier do not support the importance which has been placed on the role of fecal organisms as indicators of public hazards associated with swimming pools.

The 1969 recommendations of the Joint Health Committee on Swimming Pools [*9*] specifies minimum water quality conditions to be maintained in swimming pools. As discussed earlier, two parameters—the coliform count and the total agar plate count—are used. At present, these recommendations are under revision, and there appears to be no urgency to change these parameters either quantitatively or qualitatively.

While there may be some benefit to the use of such bacterial indicator organisms as *S. aureus, P. aeruginosa*, and streptococci, there is a major deterrent to the use of any one of these three indexes; namely, there are no uniform methods for detecting or enumerating such organisms or both. Many various procedures have been recommended, but, to date, there is no agreement concerning a "standard method" to be used in evaluative procedures of swimming pool water quality involving one of the other potential indicator organisms.

Except for the temperature of incubation, the use of the total plate count as a measurement of the bacterial quality of swimming pool water appears to be consistent with the concept of sanitary control of swimming pools in the Federal Republic of Germany as outlined by Müller [*16*]. The use of the total plate count as the basic bacterial indicator of the quality of swimming pool water has merit. A properly designed, constructed, and maintained swimming pool which is operated such that when in use the recommended minimum levels of residual disinfectant and the proper pH are maintained in all areas of the swimming pool should consistly yield samples of water which will have virtually no coliform organisms and a total 35°C agar plate count of less than 100 colonies/1.0 ml. The total 35°C agar plate count is simple, quick, and an economical microbiological test for monitoring the bacterial content of swimming pool water.

In most studies of swimming pool water disinfection that have included the total plate count as one of the indexes of water quality with other evaluative methods, the results have indicated clearly that the 35°C agar plate count is a sensitive measure of the presence of bacteria in water. When the other indexes have suggested that the quality of the swimming pool water was below recommended values, invariably the total plate counts were high. In some studies, it has been demonstrated that the total plate count method is a more sensitive evaluative procedure than the methodology utilizing specific organisms or groups of organisms as a measurement of disinfection of swimming pool water. When a swimming pool is in use and the level of free available residual chlorine is below recommended minimums, the total plate count method may be the only practical biological test which may reflect the impairment of the quality of the water.

The use of the total plate count as a bacterial indicator of quality of swimming pool water—and hence of the potential health hazards— is consistent with the recommendations of Farmer and Brenner [*17*] that water quality standards avoid all genus and species terms. A total plate count involves a methodology which can be precisely defined and easily used.

Summary

There appears to be only a limited need for the extensive use of a bacterial indicator of the potential health hazards associated with swimming pool water. Chemical tests, which measure the disinfecting process, are usually all that is needed for daily evaluations of swimming pool water quality. Such chemical tests are indirect indicators of the bacterial quality of the water.

When on occasion a direct test of the bacterial content of swimming pool water is needed or desired to check the disinfecting action of the chemical agents employed, the total 35°C agar plate count method is recommended as being effective, rapid, simple, and economical. Bacterial measurements utilizing selected genus or species need only to be made in conjunction with a suspected outbreak of disease among the users of a swimming pool.

References

[*1*] Committee on Bathing Places, *American Journal of Public Health*, Vol. 14, No. 7, July 1924, pp. 597–602.

[*2*] Robinton, E. D. and Mood, E. W., *Journal of Hygiene* (Cambridge), Vol. 64, 1966, pp. 489–499.

[*3*] Ritter, C. and Treace, E. L., *American Journal of Public Health*, Vol. 38, No. 11, Nov. 1948, pp. 1532–1538.
[*4*] Favero, M. S., Drake, C. H., and Randall, G. B., *Public Health Reports*, Vol. 79, No. 1, Jan. 1964, pp. 61–70.
[*5*] Favero, M. S. and Drake, C. H., *Public Health Reports*, Vol. 79, No. 3, March 1964, pp. 251–257.
[*6*] Mallman, W. L., *American Journal of Public Health*, Vol. 52, No. 12, Dec. 1962, pp. 2001–2008.
[*7*] Dick, E. C., Shull, I., and Armstrong, A. S., *American Journal of Public Health*, Vol. 50, No. 5, May 1960, pp. 689–695.
[*8*] Public Health Laboratory Service Sub-Committee, *Monthly Bulletin of the Ministry of Health and the Public Health Laboratory Service*, Dec. 1953, pp. 254–267.
[*9*] Joint Committee on Swimming Pools, *Suggested Ordinance and Regulations Covering Public Swimming Pools*, American Public Health Association, New York, N. Y., 1964.
[*10*] McCabe, L. J., this publication, pp. 15–22.
[*11*] Hoadley, A. W., this publication, pp. 80–114.
[*12*] Gallagher, J. R., *New England Journal of Medicine*, Vol. 238, 24 June 1948, pp. 899–903.
[*13*] Bell, J. A., Rowe, W. P., Engler, J. I., Parrott, R. H., and Huebner, R. J., *Journal of the American Medical Association*, Vol. 157, 1955, pp. 1083–1092.
[*14*] Cleere, R. L., *The Sanitarian*, Vol. 23, No. 2, Sept.-Oct. 1960, pp. 105–107.
[*15*] Park, V. K. and Brewer, W. S., *Journal of Environmental Health*, Vol. 38, No. 6, May-June 1976, pp. 390–392.
[*16*] Müller, G., this publication, pp. 159–167.
[*17*] Farmer, J. J., III, and Brenner, D. J., this publication, pp. 37–47.

E. M. Clausen,[1] *B. L. Green,*[1] *and Warren Litsky*[1]

Fecal Streptococci: Indicators of Pollution

REFERENCE: Clausen, E. M., Green, B. L., and Litsky, Warren, **"Fecal Streptococci: Indicators of Pollution,"** *Bacterial Indicators / Health Hazards Associated With Water, ASTM STP 635,* A. W. Hoadley and B. J. Dutka, Eds., American Society for Testing and Materials, 1977, pp. 247–264.

ABSTRACT: The significance of fecal streptococci as indicators of fecal pollution is reviewed. Consideration is given to the classification and definition of these streptococci, methods and media suitable for their enumeration, and the incidence and persistence of fecal streptococci in clean and polluted enviornments.

KEY WORDS: bacteria, water, coliform bacteria

Fecal streptococci are present in the feces of humans and warm-blooded animals. Although consistently recovered from waters known to receive fecal contamination, there is no indication that they multiply in natural or fecally polluted waters or soils. They appear able to survive longer than bacterial pathogens and are not considered to be of pathogenic significance themselves. For these reasons, fecal streptococci have been long regarded as potential indicators of fecal pollution. A more complete evaluation of reported sources and survival of fecal streptococci in nature is required, however, in order to establish the reliability of these organisms as indicators of pollution.

Classification and Definition of the Fecal Streptococci

Fecal streptococci are defined as those species of streptococci which are recovered from feces in significant quantities. Originally, the term referred only to the group D streptococci. Now, however, *Streptococcus mitis*, *S. salivarius*, and *S. avium* are also included in this group (Fig. 1).

Group D includes only those fecal streptococci which possess the group D antigen. This is a carbohydrate antigen composed of glycerol teichoic acid

[1]Research assistant, research assistant, and Commonwealth professor and chairman, respectively, Department of Environmental Sciences, University of Massachusetts, Amherst, Mass. 01003.

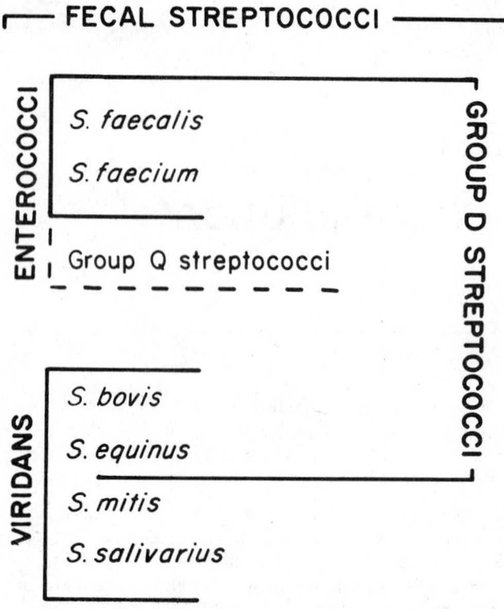

Courtesy of American Society for Microbiology.

FIG. 1—*Definition of the terms "enterococci," "group D streptococci," and "fecal streptococci" based on* Streptococcus *species belonging to each group.*

[*1,2*][2] which is located between the cell membrane and the cell wall [*3*]. The group D streptococci include *S. faecalis, S. faecium, S. durans, S. bovis*, and *S. equinus*.

The term enterococci is used to describe those species of fecal streptococci which conform to the Sherman criteria. These criteria include the ability to grow at both 10 and 45°C and survive at 60°C for at least 30 min. Growth should occur at a pH of 9.6 and in the presence of 6.5 percent sodium chloride. These organisms also should be able to reduce 0.1 percent methylene blue in milk. The enterococcus group includes *S. faecium, S. faecalis, S. durans*, and related biotypes.

Fecal streptococcal species are characterized largely by their biochemical reactions. Because speciation is employed for the identification of pollution sources, a brief description of the distinguishing characteristics of each strain is warranted.

Enterococcus Group

S. faecalis is characterized by its resistance to tellurite and its ability to reduce 2, 3, 5-triphenyl tetrazolium chloride, ferment sorbitol, and anaerobically ferment glycerol (Fig. 2). Three subspecies are recognized. *S. faecalis*

[2]The italic numbers in brackets refer to the list of references appended to this paper.

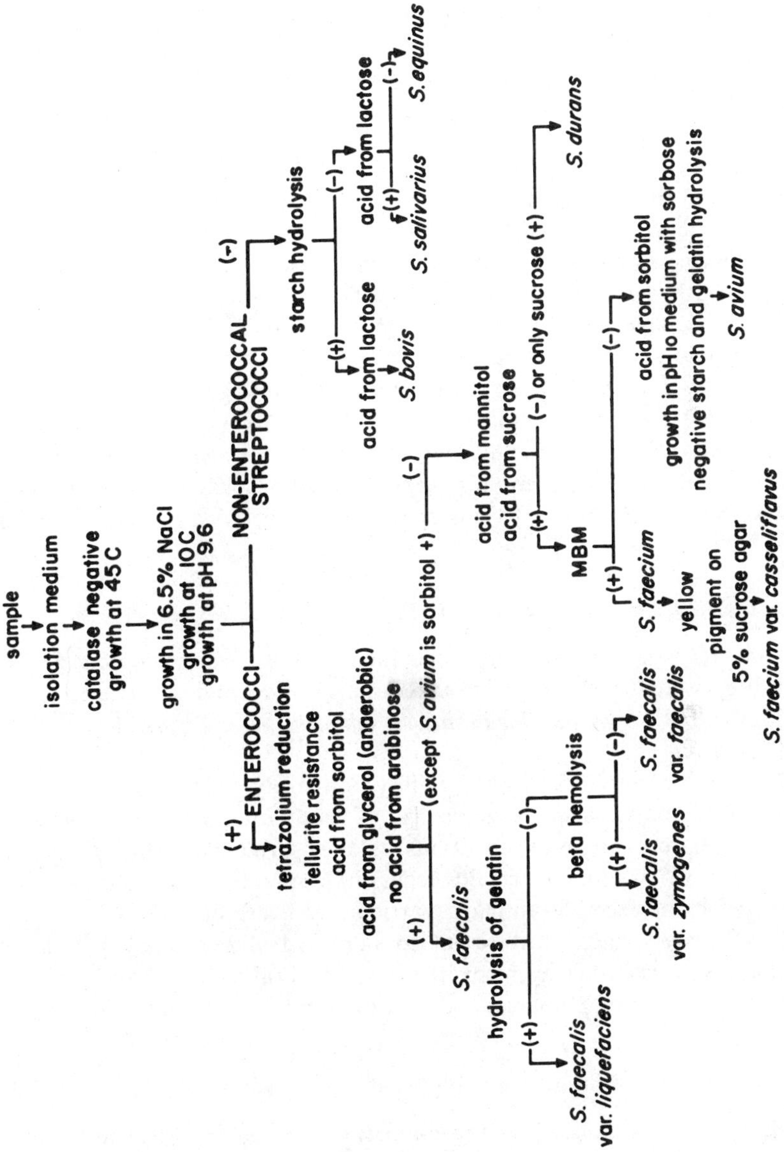

FIG. 2—*Biochemical identification of fecal streptococci.*

subspecies *faecalis* is distinguished both by its inability to liquefy gelatin and produce beta hemolysis. This organism is quite common to human feces. *S. faecalis* subspecies *liquefaciens* is also nonhemolytic but is capable of gelatin liquefication. It is believed by some to be a ubiquitous strain and, therefore, not indicative of fecal contamination. *S. faecalis* subspecies *zymogenes* is recognized by its beta-hemolytic reaction [*4*].

S. faecium is characterized by its ability to ferment arabinose, mannitol, and sucrose. It does not reduce tetrazolium and is not resistant to tellurite. *S. faecium* subspecies *casseliflavus* is recognized by the production of yellow pigment when grown on 5 percent sucrose agar. Its reactions to tellurite, tetrazolium, and the catalase test are variable [*4*]. *S. faecium* subspecies *casseliflavus* is usually recovered from plants.

S. durans is sometimes regarded as analogous to *S. faecium* [*5*] and sometimes described as a separate species [*6*]. Its biochemical reactions are similar to those described for *S. faecium*, except that mannitol is not fermented and utilization of sucrose is variable [*4*].

Although *S. avium* is not considered a member of the enterococci, its biochemical reactions resemble those of *S. faecium*. It is distinguished by its inability to grow in methylene blue milk [*4*]. It also ferments sorbose, grows in pH 10 medium containing sorbose, and lacks the ability to hydrolyze starch and gelatin [*7*]. This species is recovered principally from the feces of fowl.

Nonenterococcal Fecal Streptococci

Nonenterococcal fecal streptococci include those fecal streptococci which do not grow at 10°C, in 6.5 percent sodium chloride, in methylene blue, or at a pH of 9.6 [*8*]. *S. bovis, S. equinus, S. salivarius,* and *S. mitis* are members of this group.

S. bovis is characterized by its ability to hydrolyze starch, grow in litmus milk, and produce acid in lactose [*4*]. *S. equinus* is unable to hydrolyze starch and does not grow in litmus milk or ferment lactose [*4*]. Neither *S. bovis* nor *S. equinus* normally occurs in human feces, but both are associated with the feces of animals, particularly livestock.

S. salivarius ferments lactose but does not hydrolyze starch [*5*]. Although this organism is primarily recovered from the oral cavity, it may be found also in the feces of humans. It does not occur in animal feces.

Methods for the Enumeration of Fecal Streptococci

Methods recommended for the enumeration of fecal streptococci in water and wastewater include membrane filtration, multiple tube dilution, and direct pour plating. Detailed instructions for the execution of these tests are included in a number of publications and will not be repeated here. Rather, the reader is advised to refer to *Standard Methods for the Examination of Water and Wastewater* [*6*] for such information. This discussion shall instead emphasize evaluation of the various media and procedures which have been recommended for the recovery of fecal streptococci.

Membrane Filtration

The membrane filter procedure is recommended for enumerating fecal streptococci in fresh waters, marine waters, and unchlorinated sewage [*6*]. The test is performed by filtration of a known volume of liquid sample through a 0.45 μm membrane filter. Bacteria present in the liquid are retained on the membrane surface. The filter is then transferred to a selective medium for colony development. Colonies are counted to determine the number of organisms present per volume of sample. This procedure has a number of advantages over the multiple tube or most probable number (MPN) technique. Not only are results available sooner, but a larger volume of sample may be tested and organisms are recovered as isolated colonies which may be immediately subjected to biochemical identification. Filtration may not be used, however, for highly turbid samples.

A number of media have been suggested for the enumeration of fecal streptococci by membrane filtration, but KF agar is most widely used. On this medium fecal streptococci appear as pink to dark red colonies after incubation for 48 h at 35°C. Brodsky and Schiemann [*9*] have observed that, although autoclaving is recommended by the manufacturer, it may be detrimental to the selectivity of the medium. They recommended boiling for 5 min as a preferable method of preparation.

Pavlova et al [*8*] reported that KF medium recovered high proportions of fecal streptococci with low numbers of false positives. The fact that *S. bovis* and *S. equinus* are recovered on KF medium is supported by the work of Geldreich and Kenner [*10*], Pavlova et al [*8*], Daoust and Litsky [*11*], Kenner et al [*12*], and Brodsky and Schiemann [*9*]. Recovery of *S. mitis* and *S. salivarius* on KF medium also has been demonstrated [*11,12*]. It must be noted, however, that a few reports have indicated failure of this medium to support the growth of *S. bovis* and *S. equinus* [*13,14*]. False positives which have been recovered include *Pediococcus cerevisiae* [*12*], *Lactobacillus plantarum* [*12*], and *Staphlococcus aureus* [*15,16*].

Pfizer selective enterococcus (PSE) agar is not generally employed for membrane filtration, but some investigators advocate its use. The major advantage of this medium is that plates may be read following an 18 to 24 h incubation, whereas other media require 48 h. Fecal streptococcus colonies are distinguished by dark brown halos indicative of esculin hydrolysis. Use of PSE with membrane filters is complicated by the fact that esculin hydrolysis is only visible in the agar beneath the membrane. Daoust and Litsky [*11*] have demonstrated that this medium can be adapted for membrane use by overlaying the filters with PSE agar. This modification makes esculin hydrolysis readily apparent. Brodsky and Schiemann [*9*] suggested that esculin hydrolysis is not a necessary criterion for proper identification. They reported that 95 percent of the total number of colonies which developed on

PSE nourished membranes were identified as group D streptococci. Their methods of speciation are suspect, however, in that higher percentages of group D than of fecal streptococci were consistently reported.

PSE recovers *S. bovis* and *S. equinus* [*8,11,17,18*] and is also reported to support the growth of *S. salivarius* and possibly *S. avium* [*11*]. Recovery rates on PSE have been found to be equivalent to recovery rates on KF agar [*8,9,11,19*]. Pavlova et al [*8*] reported, however, that PSE produced a higher percentage of false positive reactions than were observed on KF medium.

M-enterococcus medium [*20*] was the first selective medium designed for the enumeration of enterococci by membrane filtration. Plates are incubated at 35°C for 48 h, after which pink or red colonies are counted. Crofts [*21*] reported that M-enterococcus counts are comparable to azide dextrose-ethyl violet azide MPN values. *S. bovis* and *S. equinus* are recovered with low efficiency on this medium [*11,13,22*].

Switzer and Evans [*14*] reported improved recovery of *S. bovis* using enterococcosel broth. This medium recovered almost 50 percent more *S. bovis* than any of the media described previously. Enterococcosel broth, which is a product of another manufacturer, has the same formulation as PSE except that the agar is omitted. The broth was dispensed on absorbent pads for membrane use.

Recently Levin et al [*19*] proposed a two-step technique for the enumeration of enterococci from marine water. Filters are first placed on a selective medium. Following a 48 h incubation, the membrane is transferred to an esculin-iron agar. Within 30 min black spots indicative of esculin hydrolysis develop in the agar beneath enterococcus colonies. The authors reported that 90 percent of the esculin hydrolyzing colonies as well as 12 percent of the nonhydrolyzing colonies were confirmed as enterococci. *S. bovis, S. equinus, S. mitis*, and *S. salivarius* were not recovered. Verified recovery by this method exceeded recovery by KF and PSE by one order of magnitude.

Membrane filtration is not recommended for chlorinated effluent because recovery by this method is found to be consistently less efficient than MPN determinations. Lin [*23*], however, succeeded in increasing membrane counts so that MPN values were equalled. He reported that this may be accomplished either by enrichment with bile broth or by extending the incubation period from 48 to 72 h.

Multiple Tube Dilution or Most Probable Number Method

The multiple tube dilution method may be used for detection of enterococci in water, sewage, or feces. Because MPNs are more time consuming and less convenient, the membrane filter procedure is usually preferred. The MPN technique is recommended for chlorinated effluents [*6*] or situations in which turbidity, high background numbers, and the presence of metallic compounds or coagulants prohibit use of membrane filtration [*7*]. MPNs are not recommended for the analysis of seawater [*6*].

The MPN procedure involves inoculation of a minimum of three decimal dilutions of sample into three or five replicate tubes of azide dextrose (AD) broth. Positive tubes are recognized by the appearance of turbidity following 24 to 48 h incubation at 35°C. These tubes are confirmed by transfer to ethyl violet azide (EVA) broth. In this medium, a positive reaction is indicated by formation of a purple button or dense turbidity after 48 h incubation at 35°C. The estimated number of enterococci is computed according to the number of positive EVA tubes and is calculated from an MPN table. Use of azide dextrose broth was first recommended by Mallmann and Seligmann [*24*]. Litsky et al [*25*] proposed the use of EVA as a confirmatory medium for the purpose of eliminating certain false positive reactions which were known to occur in azide dextrose. It must be stressed that the AD-EVA MPN is a method specific for the quantification of enterococci and that other fecal streptococci are not recovered.

Other media have been recommended for use with the MPN procedure. Kenner et al [*12*] reported high recoveries using KF broth. This medium recovers not only enterococci but other fecal streptococci as well. The use of PSE broth has also been recommended. Buck [*26*] found that seawater samples produced frequent false positive reactions when AD-EVA media were used. He reported, however, that agreement between apparent and real MPN values was greatly improved when PSE broth was substituted. According to Litsky [*27*], higher MPN values were recorded for freshwater samples when positive azide dextrose tubes were confirmed on PSE agar.

Pour Plates

Plate counts are recommended as an alternative procedure to the membrane filter technique when chlorinated sewage effluent and waters of high turbidity are encountered. This method is included as a tentative procedure in the 14th edition of *Standard Methods* [*6*]. Sample aliquots are pipetted into petri dishes and mixed with KF or PSE agar tempered to 45°C. Use of these media already has been discussed. The principal limitation of this method is that only small volumes of sample may be plated.

Fluorescent-Antibody Technique

Use of fluorescent-antibody (FA) techniques for the identification of group D streptococci was long hindered by the lack of a potent, specific group D antiserum. In 1972, however, Pavlova et al [*28*] produced such an antiserum by pooling sera produced by a variety of immunizing strains. Cross reactions with staphylococci and nongroup D streptococci were eliminated by treating smears with trypsin prior to staining with fluorescent-antibody. Trypsin presumably digests nonspecific proteins on the cell surface which otherwise react with the antiserum.

Pavlova's work was confirmed by Pugsley and Evison [*29*] using commercially available group D antisera. They observed that the antisera reacted with a high proportion of group D isolates and that cross reaction with other organisms was largely eliminated by trypsinization. Thus, there is presently available commercial group D antisera suitable for FA identification of fecal streptococci.

Abshire and Guthrie [*30*] have reported a slide technique specific for the detection of *S. faecalis* in water. The antiserum developed for this test gave positive fluorescence when reacted with most strains of *S. faecalis* but did not react with *S. faecalis* subspecies *liquefaciens* or with nongroup D organisms. A field study was performed to evaluate this antiserum. Water samples were inoculated into azide dextrose broth and incubated for 3 h. Smears were then prepared, and organisms were isolated for biochemical identification. It was reported that 96.5 percent of the organisms which were identified biochemically as typical *S. faecalis* also produced a positive fluorescent reaction.

For a more complete discussion of the preparation and use of fluorescent antisera, reference should be made to the writings of Pavlova et al [*28*], Cherry [*31*], and Thompson and Wells [*32*].

Speciation

Biochemical tests useful for the speciation of fecal streptococci have been briefly reviewed. For a more complete discussion of the methods and media required for speciation, *Microbiological Methods for Monitoring the Environment: 1 Water and Wastewater* [*7*] is recommended.

Occurrence of Fecal Streptococci in Nature

Since fecal streptococci may be regarded as indicators of fecal pollution, it is important to determine whether they occur in the environment only in association with feces or whether they persist and multiply independent of fecal pollution. The distribution of fecal streptococci in feces, plants, insects, soil, and water is therefore of great concern.

In Feces

The ratio of enterococci to other streptococci occurring in feces is known to differ among vertebrate species. This fact is of particular importance in distinguishing human from animal sources of pollution. Human feces are characterized by a great predominance of enterococci. Although exact speciation of the enterococci from this source has been generally neglected, it is apparent that *S. faecalis* subspecies *liquefaciens* constitutes about 25 percent of the total streptococcus population [*10,33*]. Cooper and Ramadan [*33*] reported that typical *S. faecalis* comprised about 40 percent of the streptococcus population and that *S. durans* accounted for 4 percent. A number of fecal streptococci have been reported to be unique to human feces. These include *S. faecalis* [*34*], *S. durans* [*33*], and *S. mitis* and *S. salivarius* [*12*].

The latter two species are buccal streptococci which enter the digestive tract and populate feces in low numbers. They are not invariably present but if detected indicate human fecal pollution. *S. bovis* and *S. equinus* are rarely recovered from human stools.

Feces from livestock are distinguished by significant proportions of *S. bovis* and *S. equinus*. Reported combined percentages of these species range from 18.9 percent for pigs to 66.2 percent for cattle [*10,12*]. The remaining population is primarily enterococci, of which up to 18 percent have been identified as *S. faecalis* subspecies *liquefaciens* [*24*]. Streptococci recovered from feces of various other vertebrates have been speciated, and data are reported for dogs, cats, and rodents [*10*], fish [*35*], and wild animals, including small mammals, birds and reptiles [*36*].

The feces of fowl are characterized by the presence of group Q streptococci as well as group D. Nowlan and Deibel [*18*] indicate that group Q densities range from 100 fold less than corresponding densities of enterococci up to equivalent populations. Other workers [*10,12*] have not attempted isolation of group Q, but indicate high proportions of atypical enterococci. Kenner et al [*12*] reported that 38 percent of the streptococci recovered were designated as biotypes, a category which may have included *S. avium*. *S. bovis* and *S. equinus* were not detected in significant quantities.

Human feces may be distinguished not only by the speciation of fecal streptococci but more simply by determination of the ratio of numbers of fecal coliforms to fecal streptococci present. Animal feces are characterized by a higher density of streptococci, such that the ratio of fecal coliforms to fecal streptococci is always less than 0.7. Human feces contain a greater number of coliforms, causing the fecal coliform-fecal streptococcus ratio to exceed 4.0 [*10*]. This relationship has proved extremely useful in determining sources of pollution. However, the ratio is altered by die-off and so it only can be successfully applied to recent fecal pollution. Geldreich and Kenner [*10*] suggest that this ratio is only valid during the first 24 h following discharge of fecal matter into a body of water.

On Plants

The incidence of fecal streptococci on plants is of concern for two reasons. First, streptococci present on plants may be introduced into soil and water. Detection in water of populations of group D streptococci originating from plants would invalidate use of these organisms as indicators of fecal pollution. Secondly, if streptococci are natural residents of plants, recovery of enterococci from processed vegetables cannot be considered indicative of fecal contamination. Careful consideration of the incidence and origin of streptococci on plants is therefore warranted.

Numerous studies have reported the occurrence of streptococci on plants. Mundt reported enterococcus recovery from 58.5 percent of 106 samples involving 63 species of plants from agricultural areas [*37*], 34 percent recovery from flowers of nonagricultural plants, 32.2 percent recovery from flowers of agricultural plants, 10.4 percent recovery from grasses and cereals [*38*], and 14 percent recovery from nearly 2200 flowers sampled from a variety of species in a wild environment [*39*]. Recovery of entercococci from plants also has been reported by Geldreich et al [*40*]. Enterococci have been found to be present in silage [*41*] and the seeds of corn and peas [*42*].

Geldreich and Kenner [*10*] speciated fecal streptococci recovered from plants and concluded that 43 percent of their isolates were enterococci, 9 percent were *S. bovis* or *S. equinus*, 34.9 percent were atypical *S. faecalis*, and 13 percent were *S. faecalis* subspecies *liquefaciens*.

There has been some confusion as to the origin of enterococci isolated from plants. Three possibilities have been discussed.

1. Streptococci may be present as a result of direct contamination from warm-blooded animals, insects or both. Support for this theory is provided by a study of plants in a wild environment [*39*]. Low incidence of enterococci and similarity between species recovered from plants and from animal feces suggest that enterococci may be chance contaminants. Subsequent studies indicate that this explanation is not sufficient.

2. Enterococci may be temporary residents on plants, capable of limited reproduction. Work done by Mundt [*38*] in 1961 suggests this to be the case. Fecal streptococci were isolated from a variety of plant species. In most instances, they were found to be neither invariably present nor invariably absent on a given species. Simultaneous occurrence or absence of enterococci on plants and in the surrounding soil suggests that counts in the soil result from spreading of the plant population, probably by means of insects, rain, wind, and gravity.

3. Finally, enterococci may exist on plants in a truly epiphytic relationship. Mundt defines a commensal as an organism which occurs in seed or soil, moves from the seed to emergent parts, and is able to reproduce on the growing plant. In a series of greenhouse experiments he has demonstrated that a strain of *S. faecalis* subspecies *liquefaciens* was able to meet the foregoing criteria in association with bean, rye, corn, and cabbage plants [*43*]. Further evidence of commensal growth includes the widespread occurrence of fecal streptococci in the seeds of corn and peas [*42*] and the recovery of substantial numbers of enterococci from cornflowers [*38*].

Large numbers of enterococci isolated from vegetation have been reported to be physiologically different from recognized species common to warm-blooded animals. Mundt [*44*] isolated 130 *S. faecium*-like strains which differed in one or more biochemical reactions from *S. faecium*. Additionally, he isolated 375 strains which were heterogeneous in regard to the Sherman criteria, as well as to tellurite, tetrazolium, and sugar fermentation reactions. While the guanine-cytosine content for this group fell within the range characteristic of streptococci, DNA/DNA homology indicated only distant relationship to *S. faecium*, *S. faecalis*, and *S. lactis*.

Geldreich et al [*40*] detected starch hydrolysis in 37.7 percent of 646 cultures of enterococci isolated from plants. This characteristic was uncommon to strains recovered from water, warm-blooded animals, cold-blooded animals, insects, or soil. Facklam [*4*] also failed to recover enterococci capable of starch hydrolysis from among 250 group D strains of human origin.

Mundt [*45*] reported a consistent difference between litmus milk reactions of *S. faecalis* isolated from plant and human sources. Ninety-eight percent of the human isolates produced a reduced acid curd or acid proteolysis. Eighty-six percent of the plant isolates showed a reduced rennet reaction. Eighty-two percent of animal isolates and 94 percent of insect isolates also produced a rennet reaction.

S. faecium subspecies casseliflavus was originally isolated from plants and has been recovered from a wide variety of leaves and vegetables [*46,47*]. Population levels range from zero to 10^6 organisms/g. Mundt observed that *S. faecium* subspecies *casseliflavus* is frequently the most numerous member of the lactic acid producing bacteria isolated from plants. Martin et al [*48*] have reported a high frequency of *S. faecium* subspecies *casseliflavus* among insects. Few organisms of this strain have been reported in association with humans. Bartley and Slantez [*34*] mention the isolation of a yellow pigmented streptococcus from human feces, but its biochemical characteristics were not reported. Isenberg [*49*] reported 23 presumptive *S. faecium* subspecies *casseliflavus* cultures from a group of 250 clinical group D isolates, but confirmatory tests were not performed. Facklam [*4*] failed to detect this subspecies while speciating a total of 262 clinical isolates. Clearly *S. faecium* subspecies *casseliflavus* is much more prevalent on plants and insects than in human feces. No attempts to isolate this organism from animals or soil have been reported.

Comparisons have been made between numbers of fecal streptococci, total coliform, and fecal coliform organisms isolated from a variety of plant samples [*40*]. Fecal streptococcus counts per gram of plant material were observed to be significantly higher than fecal coliform counts or confirmed total coliform counts. Three possible explanations are offered. (1) Counts may have resulted from plant contact with insects; insects were also found to have high densities of fecal streptococci and low densities of fecal coliforms. (2) Counts may have been affected by commensal growth. (3) The AD-EVA procedure used for quantification of fecal streptococci may have recovered *S. lactis*, resulting in a significant number of false positive reactions.

In summary, it appears that streptococcus populations are not residual to all vegetation but are most likely carried to individual plants by insects. Certain species appear able to grow and multiply on plants at least under some conditions. The observation that fecal streptococci are found considerably less frequently on plants growing in wilderness areas would suggest that plant populations originate from a fecal source. The unique biotypes characteristic of plants suggest, however, that residual populations on certain plants provide a streptococcus reservoir which is spread to other vegetation. Clearly more work is necessary before the relationship between streptococci and plants is fully understood.

In Insects

The fecal streptococcus population of insects is of particular significance because of the possible involvement of insects in transferring streptococci to plants. Fecal streptococci have been recovered from a variety of insects including Orthoptera (grasshoppers), Hemiptera (true bugs), Lepidoptera (moths and butterflies), Coleoptera (beetles), Hymenoptera (bees, wasps, and ants), Diptera (flies), and Homoptera (leaf hoppers) [*40,48,50,51*]. Geldreich et al [*40*] report that the fecal streptococcus population recovered from among five orders of insects is composed of 52 percent enterococci and 48 percent *S. faecalis* subspecies *liquefaciens*. Neither *S. bovis* nor *S. equinus* was recovered. Martin and Mundt [*48*] speciated streptococci recovered from eight insect orders. *S. faecalis* was recovered from 32 percent, *S. faecium* from 22 percent, and *S. faecalis* subspecies *casseliflavus* from 44 percent of the insects examined.

It is apparent from studies done by Geldreich et al [*40*] and Eaves and Mundt [*52*] that fecal streptococci cannot be recovered consistently from a given species. In many instances, not all individuals of a species harbor the organisms. This random recovery suggests that fecal streptococci are not consistent residents of insects but are present instead from chance contacts with streptococci in the environment. The similarity of strains and biotypes recovered from insect and plant populations is indicative of a frequent exchange of organisms between these sources. Differences between these strains and the streptococci occurring in human feces suggest infrequent intermingling of these populations.

In Soil

The presence of fecal streptococci in natural waters is known to be directly related to the incidence of streptococci in soil. This is due primarily to stormwater runoff which washes soil organisms into streams, rivers, and lakes. In evaluating fecal streptococci as indicators of water pollution, it is therefore appropriate to consider the frequency with which streptococci occur in soil.

Significant populations of enterococci in unpolluted soils have not been reported. Medreck and Litsky [*53*] recovered enterococci from only 2.2 percent of 369 soil samples taken in watershed areas. Total coliforms were detected in 71.8 percent and fecal coliforms occurred in 1.1 percent of the samples. It would be interesting to determine whether this low frequency of streptococci reflects a low incidence of streptococci on the watershed plants. Mundt [*39*] observed that streptococcus densities on plants in a wilderness region were much lower than in agricultural areas. Correlations between plant and soil populations have been observed [*39*].

Speciation of strains isolated from agricultural soils produced the following distribution: 63 percent enterococci other than *S. faecalis* subspecies *liquefaciens*, 35 percent *S. faecalis* subspecies *liquefaciens*, and 2 percent *S. bovis* and *S. equinus* [*10*].

Persistence of enteric bacteria in the soil following contamination is also of significance in the evaluation of indicator organisms. The survival of *S. faecalis* and *Escherichia coli* in soils has been compared. Mallmann and Litsky [*54*] tested the survival of these organisms in a number of soils which were contained in metal cylinders. Temperatures during the study ranged from 25 to 28°C. *S. faecalis* persisted for about 40 days, whereas *E. coli* were still recovered after eleven weeks. *Salmonella typhi* survived for a maximum of 18 days.

Van Donsel et al [*55*] studied survival in two outdoor test areas which were periodically inoculated with test strains of *E. coli* and *S. faecalis*. Significant seasonal variation in survival was noted. During the summer months, fecal coliforms outlived fecal streptococci. Fecal streptococci survived 2.7 days, while fecal coliforms persisted for 3.3 days. During the autumn, survival time was about 13 days for both species. Throughout the winter and spring, fecal streptococci survived much longer than fecal coliforms, persisting for as long as 20 days.

In Water

Fecal streptococci are of importance primarily as indicator organisms. Comparison of densities of streptococci, total coliforms, and fecal coliforms is therefore central to this discussion. Frequencies of these indicators in reservoirs, natural bodies of water, storm waters, and sewage effluent merit consideration.

Reservoirs—Because reservoirs are protected from most fecal contamination, indicator organisms are expected to occur in low numbers or not at all. Studies of three reservoirs in Washington and Oregon demonstrate that fecal streptococci are no more numerous than fecal coliforms in clean water and occur less frequently than total coliforms [*56*].

Natural Waters—In rivers and lakes fecal streptococci are generally less numerous than fecal coliforms and are present in lower numbers than total coliforms. Litsky et al [*57*] reported the ratio of fecal coliforms to fecal streptococci in the Connecticut River to be seven to one. Leninger and McCleskey [*58*] compared fecal streptococci to total coliforms in surface waters of varying quality and concluded that clean and polluted conditions were best indicated by enterococci. Although total coliforms were recovered from sources presumably uncontaminated by human pollution, enterococci were not detected. Furthermore, fecal streptococci have been recovered from polluted wells and springs in which fecal coliforms were absent [*59*].

Stormwater Runoff—Urban runoff is characterized by greater numbers of fecal streptococci than fecal coliforms [*10,60-62*]. Van Donsel et al [*55*] observed that the recovery rates of fecal coliforms and fecal streptococci are proportional to their densities in the soil. Clearly, however, proportions of these organisms in stormwater are also related to their survival rates in water.

Sewage—Total and fecal coliforms are present in sewage in much higher densities than fecal streptococci [*10,59,63*]. Susceptibility to sewage treatment and aftergrowth of indicator organisms will be discussed later.

Critical to the evaluation of indicator organisms is the length of time they survive in the environment. A proper indicator of fecal pollution should persist for as long as intestinal pathogens but not a great deal longer. Survival of fecal streptococci, fecal coliforms, total coliforms, and pathogens is therefore of importance.

Enterococci generally appear to be more persistent than either fecal or total coliforms. Survival studies conducted with laboratory-seeded stormwater [*10*] indicate that *S. faecalis* and *S. faecalis* subspecies *liquefaciens* persist significantly longer than either fecal coliforms or *Enterobacter aerogenes*. Other fecal streptococci are not as resistant. *S. bovis* died off rapidly in stormwater, and survival of *S. equinus* in the environment is known to be extremely limited [*10*]. Survival of *Salmonella* paralleled that of the fecal coliforms. After six to twelve days, however, fecal coliform numbers showed more rapid decline so that by the fourteenth day the percent survival of *Salmonella* was exceeded only by that of the enterococci [*10*]. Survival studies in marine water also indicate a more rapid die-off of *Salmonella* than of *E. coli* [*64*].

It is apparent that survival of indicator organisms, and probably pathogens as well, is considerably influenced by water temperature and by the amount of soluble organic matter present. In the previous study, all of the test organisms died off more rapidly at 20 than at 10°C. Similar conclusions can be drawn from the soil survival studies of Van Donsel et al [*55*]. High nutrient concentrations have been shown to prolong the survival of *S. faecalis* [*65*]. Other indicator organisms are also affected by nutrient levels, but these effects shall be considered later.

The use of indicator organisms for the evaluation of sewage effluent involves problems specific to sewage treatment. The selection of an indicator organism must be based on two criteria: (*a*) whether the die-off of the organism during sewage treatment reflects the rate of removal of pathogens and (*b*) whether numbers of indicator organisms in the nutrient-rich effluent are a correct indication of the sanitary quality of the water. Cohen and Shuval [*59*] have compared survival of total coliforms, fecal coliforms, fecal streptococci, and viruses in sewage. Samples were taken before and after primary settling and biological filtration. These treatments removed viruses with the least efficiency. Removal of fecal streptococci was considerably less than that of coliforms and, thus, more closely paralleled virus survival. In an open sewage channel receiving effluent, viruses again showed the least die-off followed by fecal streptococci.

Because sewage effluent is chlorinated to reduce bacterial densities, susceptibility of indicators to chlorine is also important. Some researchers have reported that chlorination reduces numbers of total coliforms, fecal coliforms, and fecal streptococci with equal efficiency [*60*], while others have found fecal streptococci to be more resistant than coliform organisms [*66*]. These variations are most likely related to differences in the chlorine residual and contact time.

Subsequent to chlorination regrowth of bacteria is known to occur in the nutrient-rich effluent. Evans et al [*60*] studied aftergrowth of indicator organisms in laboratory-chlorinated stormwater. Total coliform counts were observed to increase as much as 10 000 fold within 24 h following chlorination. Fecal coliforms and fecal streptococci did not demonstrate significant aftergrowth. Shuval et al [*22*], however, have detected regrowth of both total and fecal coliforms in chlorinated wastewater effluent. Slanetz and Bartley [*64*] also reported regrowth of total and fecal coliforms in sewage effluent and pure cultures which were placed in dialysis tubing and suspended in seawater. Regrowth of fecal streptococci was not observed. Furthermore, laboratory studies indicated that *E. coli* multiplied in as little as 0.28 ppm of organic matter in solution, whereas much higher concentrations were required for growth of *S. faecalis* [*65*]. Viral pathogens do not reproduce in water. Although there is some indication that *Salmonella* are capable of aftergrowth in effluent waters [*64*], adequate chlorination has been shown to successfully remove *Salmonella* from sewage effluent [*67*]. Thus, fecal streptococci must be considered the preferred indicator of sanitary quality in waters which receive chlorinated effluent.

Controversy regarding the distribution of *S. faecalis* subspecies *liquefaciens* cannot be ignored. Geldreich et al [*10*] have reported the recovery of high numbers of *S. faecalis* subspecies *liquefaciens* from two unpolluted wells and also high percentages of these organisms in insects and soil. In that there are significant numbers of *S. faecalis* subspecies *liquefaciens* in the feces of certain animals, the percentages reported for insects and soil are not conclusive proof of the ubiquitous nature of this organism. Furthermore, there is evidence that *S. faecalis* subspecies *liquefaciens* is not generally a predominant species in water. Bartley and Slanetz [*34*] have speciated streptococci isolated from sewage, seawater, rivers, ponds, and wells. They observed that *S. faecalis* subspecies *liquefaciens* comprised only 5 to 14 percent of the streptococci isolated from each of these sources. At present, the two wells reported by Geldreich are an isolated case, but they cannot be disregarded. The possibility exists that this species is ubiquitous, and more work must be done to resolve the issue.

In summary, fecal streptococci are not only abundant in feces but also have been detected in significant quantities on plants and insects. Because fecal streptococci appear capable of reproduction in the nutrient loaded wastewater of vegetable processing plants [*68*], these organisms are not recommended as indicators of sanitation in fruit and vegetable processing or in waters known to receive this effluent. Thus far, however, there is no evidence that plant and insect populations have significant effects on the bacterial population of other waters.

Fecal streptococci generally have been found to be more persistant than fecal coliforms and are therefore a safer indicator of pollution. Whereas fecal coliforms may better parallel die-off of enteric bacterial pathogens such as salmonellae, fecal streptococci better indicate the possibility of viral contamination. Streptococci do not multiply in polluted waters, which gives them a significant advantage over coliform indicators in detecting recent pollution. In addition, speciation of the group D streptococci can provide information as to the source of pollution.

Although fecal streptococci may not prove to be an ideal indicator in all circumstances, the use of streptococci may be advised in the following situations: (1) when assessing the quality of reservoirs, drinking water, and other waters in which viral contamination is particularly undesirable, (2) when determining the quality of organically rich water, especially those waters receiving chlorinated sewage effluent, and (3) when attempting to locate the sources of fecal pollution.

References

[*1*] Wicken, A. J., Elliott, S. D. and Baddiley, J., *The Journal of General Microbiology*, Vol. 31, No. 2, May 1963, pp. 231–239.

[*2*] Wicken, A. J. and Baddiley, J., *Biochemical Journal*, Vol. 87, No. 1, 1963, pp. 54–62.

[*3*] Smith, D. G. and Shattock, P. M. F., *The Journal of General Microbiology*, Vol. 34, No. 1, Jan. 1964, pp. 165–175.

[*4*] Facklam, R. R., *Applied Microbiology*, Vol. 23, No. 6, June 1972, pp. 1131–1139.

[*5*] Buchanan, R. E. and Gibbons, N. E., ed., *Bergey's Manual of Determinative Bacteriology*, 8th ed., The Williams and Wilkins Co., Baltimore, 1974, pp. 490–509.

[*6*] American Public Health Association, *Standard Methods for the Examination of Water and Wastewater*, 14th ed., American Public Health Association, New York, 1976.

[*7*] Bordner, R. and Winter, J., ed., *Microbiological Methods for Monitoring the Environment. 1 Water and Wastes.* U.S. Environmental Protection Agency, Environmental Monitoring and Support Laboratory, Cincinnati, Ohio, 1977.

[*8*] Pavlova, M. T., Brezenski, F. T. and Litsky, W., *Health Laboratory Science*, Vol. 9, No. 4, Oct. 1972, pp. 289–298.

[*9*] Brodsky, M. H. and Schiemann, D. A., *Applied and Environmental Microbiology*, Vol. 31, No. 5, May 1976, pp. 695–699.

[*10*] Geldreich, E. E. and Kenner, B. A., *Journal of the Water Pollution Control Federation*, Vol. 41, No. 8, Part 2, Aug. 1969, pp. R336–R352.

[*11*] Daoust, R. A. and Litsky, W., *Applied Microbiology*, Vol. 29, No. 5, May 1975, pp. 584–589.

[*12*] Kenner, B. A., Clark, H. F. and Kabler, P. W., *American Journal of Public Health*, Vol. 50, No. 10, Oct. 1960, pp. 1553–1559.

[*13*] Slanetz, L. W. and Bartley, C. H., *American Journal of Public Health*, Vol. 54, No. 4, April 1964, pp. 609–614.

[*14*] Switzer, R. E. and Evans, J. B., *Applied Microbiology*, Vol. 28, No. 6, Dec. 1974, pp. 1086–1087.

[*15*] Mossel, D. A. A., *Journal of the Science of Food and Agriculture*, Vol. 15, No. 6, June 1964, pp. 349–362.

[*16*] Mossel, D. A. A., von Diepen, H. M. and de Bruin, A. S., *Journal of Applied Bacteriology*, Vol. 20, No. 2, Nov. 1957, pp. 265–272.

[*17*] Sabbaj, J., Sutter, V. L. and Finegold, S. M., *Applied Microbiology*, Vol. 22, No. 6, Dec. 1971, pp. 1008–1011.

[*18*] Nowlan, S. S. and Deibel, R. H., *Journal of Bacteriology*, Vol. 94, No. 2, Aug. 1967, pp. 291–296.

[*19*] Levin, M. A., Fischer, J. R. and Cabelli, V. J., *Applied Microbiology*, Vol. 30, No. 1, July 1975, pp. 66–71.

[20] Slanetz, L. W. and Bartley, C. H., *Journal of Bacteriology*, Vol. 74, No. 5, Nov. 1957, pp. 591–595.

[21] Crofts, C. C., *American Journal of Public Health*, Vol. 49, No. 10, Oct. 1959, pp. 1379–1387.

[22] Shuval, H. I., Cohen, J. and Kolodney, R., *Water Research*, Vol. 7, No. 4, April 1973, pp. 537–546.

[23] Lin, S., "Evaluation of Methods for Detecting Coliforms and Fecal Streptococci in Chlorinated Sewage Effluents," Report Investigation 78, Illinois State Water Survey, Urbana, Ill., 1974.

[24] Mallmann, W. L. and Seligmann, E. B. Jr., *American Journal of Public Health*, Vol. 40, No. 3, Mar. 1950, pp. 286–289.

[25] Litsky, W., Mallmann, W. L. and Fifield, C. W., *American Journal of Public Health*, Vol. 43, No. 7, July 1953, pp. 873–879.

[26] Buck, J. D., *American Journal of Public Health*, Vol. 62, 1972, pp. 419–421.

[27] Litsky, W., personal communication.

[28] Pavlova, M. T., Beauvais, E., Brezenski, F. T. and Litsky, W., *Applied Microbiology*, Vol. 23, No. 3, Mar. 1972, pp. 571–577.

[29] Pugsley, A. P. and Evison, L. M., *Water Research*, Vol. 8, No. 10, Oct. 1974, pp. 725–728.

[30] Abshire, R. L. and Guthrie, R. K., *Water Research*, Vol. 5, No. 11, Nov. 1971, pp. 1089–1097.

[31] Cherry, W. B., in *Manual of Clinical Microbiology*, Blair, J. B., Lennette, E. H. and Truant, J. P., ed., American Society for Microbiology, Bethesda, Md., 1970, pp. 693–704.

[32] Thompson, B. M. and Wells, J. G., *Applied Microbiology*, Vol. 22, No. 5, Nov. 1971, pp. 876–884.

[33] Cooper, K. E. and Ramadan, E. M., *The Journal of General Microbiology*, Vol. 12, No. 2, Apr. 1955, pp. 180–190.

[34] Bartley, C. H. and Slanetz, L. W., *American Journal of Public Health*, Vol. 50, No. 10, Oct. 1960, pp. 1545–1552.

[35] Geldreich, E. E. and Clarke, N. A., *Applied Microbiology*, Vol. 14, No. 3, May 1966, pp. 429–437.

[36] Mundt, J. O. *Applied Microbiology*, Vol. 11, No. 2, Mar. 1963, pp. 136–140.

[37] Mundt, J. O., Johnson, A. H. and Khatchikian, R., *Food Research*, Vol. 23, 1958, pp. 186–193.

[38] Mundt, J. O., *Applied Microbiology*, Vol. 9, No. 6, Nov. 1961, pp. 541–544.

[39] Mundt, J. O., *Applied Microbiology*, Vol. 11, No. 2, Mar. 1963, pp. 141–144.

[40] Geldreich, E. E., Kenner, B. A. and Kabler, P. W., *Applied Microbiology*, Vol. 12, No. 1, Jan. 1964, pp. 63–69.

[41] Mieth, H., *Zentralblatt für Bakteriologie, Parasitenkunde, Infektionskrankheiten, und Hygiene*, Abt. 1 Orig., Vol. 183, No. 1, Sept. 1961, pp. 68–89.

[42] Fitzgerald, G. A., *American Journal of Public Health*, Vol. 37, No. 6, June 1947, pp. 695–701.

[43] Mundt, J. O., Coggins, J. H. Jr. and Johnson, L. F., *Applied Microbiology*, Vol. 10, No. 6, Nov. 1962, pp. 552–555.

[44] Mundt, J. O., *International Journal of Systematic Bacteriology*, Vol. 25, No. 3, July 1975, pp. 281–285.

[45] Mundt, J. O., *Journal of Milk and Food Technology*, Vol. 36, No. 7, July 1973, pp. 364–367.

[46] Mundt, J. O. and Graham, W. F., *Journal of Bacteriology*, Vol. 95, No. 6, June 1968, pp. 2005–2009.

[47] Mundt, J. O., Graham, W. F. and McCarty, I. E., *Applied Microbiology*, Vol. 15, No. 6, Nov. 1967, pp. 1303–1308.

[48] Martin, J. D. and Mundt, J. O., *Applied Microbiology*, Vol. 24, No. 4, Oct. 1972, pp. 575–580.

[49] Isenberg, H. D., Goldberg, D. and Sampson, J., *Applied Microbiology*, Vol. 20, No. 3, Sept. 1970, pp. 433–436.

[50] Steinhaus, E. A., *Journal of Bacteriology*, Vol. 42, No. 6, Dec. 1941, pp. 757–790.

[51] West, L. S., *The Housefly*, Comstock Publishing Associates, Ithica, N.Y., 1951.

[*52*] Eaves, G. N. and Mundt, J. O., *Journal of Insect Pathology*, Vol. 2, No. 3, Sept. 1960, pp. 289–298.
[*53*] Medrek, T. F. and Litsky, W., *Applied Microbiology*, Vol. 8, No. 1, Jan. 1960, pp. 60–63.
[*54*] Mallmann, W. L. and Litsky, W., *American Journal of Public Health*, Vol. 41, No. 1, Jan. 1951, pp. 38–44.
[*55*] Van Donsel, D. J., Geldreich, E. E. and Clarke, N. A., *Applied Microbiology*, Vol. 15, No. 6, Nov. 1967, pp. 1362–1370.
[*56*] Lee, R. D., Symons, J. M. and Robeck, G. G., *Journal of the American Water Works Association*, Vol. 62, No. 7, July 1970, pp. 412–422.
[*57*] Litsky, W., Mallmann, W. L. and Fifield, C. W., *American Journal of Public Health*, Vol. 45, No. 8, Aug. 1955, pp. 1049–1053.
[*58*] Leninger, H. T. and McCleskey, C. S., *Applied Microbiology*, Vol. 1, No. 3, May 1953, pp. 521–524.
[*59*] Cohen, J. and Shuval, H. I., *Water, Air and Soil Pollution*, Vol. 2, No. 1, March 1973, pp. 85–95.
[*60*] Evans, F. L. III, Geldreich, E. E., Weibel, S. R. and Robeck, G. G., *Journal of the Water Pollution Control Federation*, Vol. 40, No. 5, Part 2, May 1968, pp. R162–R170.
[*61*] Geldreich, E. E., Best, L. C., Kenner, B. A. and Van Donsel, D. J., *Journal of the Water Pollution Control Federation*, Vol. 40, No. 11, Part 1, Nov. 1968, pp. 1861–1872.
[*62*] Weibel, S. R., Anderson, R. J. and Woodward, R. L., *Journal of the Water Pollution Control Federation*, Vol. 36, No. 7, July 1964, pp. 914–924.
[*63*] Burm, R. J. and Vaughn, R. D., *Journal of the Water Pollution Control Federation*, Vol. 38, No. 3, March 1966, pp. 400–409.
[*64*] Slanetz, L. W. and Bartley, C. H., *Health Laboratory Science*, Vol. 2, No. 3, July 1965, pp. 142–148.
[*65*] Allen, L. A., Pasley, S. M. and Pierce, M. A. F., *The Journal of General Microbiology*, Vol. 7, Nos. 1 and 2, Aug. 1952, pp. 36–43.
[*66*] Silvery, J. K. G., Abshire, R. L. and Nunez, W. L., *Journal of the Water Pollution Control Federation*, Vol. 46, No. 9, Sept. 1974, pp. 2153–2162.
[*67*] Brezenski, F. T., Russomanno, R. and DeFalco, P. Jr., *Health Laboratory Science*, Vol. 2, No. 1, Jan. 1965, pp. 40–47.
[*68*] Mundt, J. O., Larsen, S. A. and McCarty, I. E., *Applied Microbiology*, Vol. 14, No. 1, Jan. 1966, pp. 115–118.

A. K. Highsmith,[1] *J. C. Feeley,*[1] *and G. K. Morris*[1]

Isolation of *Yersinia enterocolitica* From Water

REFERENCE: Highsmith, A. K., Feeley, J. C., and Morris, G. K., **"Isolation of *Yersinia enterocolitica* From Water,"** *Bacterial Indicators/Health Hazards Associated With Water, ASTM STP 635*, A. W. Hoadley and B. J. Dutka, Eds., American Society for Testing and Materials, 1977, pp. 265–274.

ABSTRACT: Only in the last few years has *Yersinia enterocolitica* been recognized as an etiologic agent. An increasing awareness of this organism is evidenced by the number of recorded cases and isolations each year from a variety of sources throughout the world. Previous failure to isolate *Y. enterocolitica* may be related to lack of familiarity rather than absence of the organism. Presence in the animate and inanimate environments does provide opportunity for transmission by person to person, animals, foodstuffs, and water, but vehicles of disease transmission are not fully delineated. General hygienic techniques in regard to food and water sanitation should apply in the methods for controlling the disease caused by *Y. enterocolitica*. The bacteriology of *Y. enterocolitica* is reviewed and laboratory methodology is described.

KEY WORDS: bacteria, water, coliform bacteria

The genus *Yersinia* was named in honor of the French bacteriologist A. J. E. Yersin who first isolated the causative agent of plague in 1894. This genus contains three species—*Y. pestis, Y. pseudotuberculosis,* and *Y. enterocolitica*. The first two species were formerly in the genus *Pasteurella* and considered members of the family Brucellaceae. However, the eighth edition of the *Bergey's Manual of Determinative Bacteriology* classifies the *Yersinia* genus as a member of the family Enterobacteriaceae.

During the 1960s many investigators described the *Y. enterocolitica* organism as *P. pseudotuberculosis, P. pseudotuberculosis*-like organisms, *Pasteurella*, type X or *Pasteurella*, type Y. In 1964 Frederiksen [*1*][2] recognized that these strains were similar to the *Bacterium enterocoliticum* organism isolated by Schleifstein and Coleman in 1939 [*2*]. Hence, he proposed the name *Y. enterocolitica*.

[1]Microbiologist, Hospital Infections Laboratory Section; chief, Special Pathogens Laboratory Section; and chief, Epidemiologic Investigations Laboratory Branch, respectively; Bacterial Diseases Division, Bureau of Epidemiology, Center for Disease Control, Public Health Service, U. S. Department of Health, Education, and Welfare, Atlanta, Ga. 30333.

[2]The italic numbers in brackets refer to the list of references appended to this paper.

Y. enterocolitica is a gram-negative ovoid or rod-shaped organism, measuring 0.8 to 3.0 μm by 0.8 μm. It is a facultative anaerobe. While a few strains of *Y. enterocolitica* have been reported motile at 35 to 37°C, most are motile only at temperatures below 30°C. *Y. enterocolitica* is comprised of a varied group of organisms representing multiple serotypes, biotypes, and phage types. Much discussion has been generated about what is a true *Y. enterocolitica*. Recently, Brenner et al [*3*] have described four DNA relatedness groups within this species. This may be of help in clarifying true species designation.

Clinical Aspects

Human infection may exhibit several clinical manifestations. The most prevalent is acute gastrointestinal illness; other forms of illness are polyarthritis [*4*], erythema nodosum [*4*], Reiter's syndrome [*5*], septicemia [*6*], and meningitis [*7,8*].

The type of clinical illness may be influenced by the age of the patient [*9*]. Gastroenteritis, mesenteric lymphadenitis, acute terminal ileitis, and pseudoappendicitis are observed in children and young adults, whereas acute enteritis with or without fever is observed more often in middle aged and old people; arthritis and erythema nodosum may be complications, especially in older individuals [*9*].

Diagnosis is usually by bacteriological examination of the stool. The recovery of the organism from the stool is best accomplished during the acute enteritis phase. An alternative method, commonly practiced in European countries, where serotypes 3 and 9 prevail, is the examination of patients' sera for antibodies to *Y. enterocolitica.* Currently in the United States, it is recommended that serological examinations be performed only in epidemic situations using the antigen prepared from the epidemic strain.

Pathogenicity

The mechanism of pathogenicity of this organism has not been elucidated. Early investigation by Mollaret and Guillon [*10*] resulted in no appropriate animal model being established. Subsequently, they injected over 100 strains of *Y. enterocolitica* into 15 species of animals by various inoculation routes and found no animal mortality. However, recent work by Quan et al [*11*] has shown that the gerbil is a sensitive animal for pathogenicity testing of *Y. enterocolitica* isolates. In addition, Carter [*12*] has shown that the mouse can be used, provided that it is an appropriate inbred line. Alonso et al [*13*] have shown athymic mice to be of value for pathogenicity testing.

Epidemiology

World Prevalence—In 1966, only 23 cases of *Y. enterocolitica* had been reported. This number increased to 642 in 1970, 1000 in 1972, and now is greater than 4000 [*14,15*]. Most of these reports are from European countries. Only a few cases have been reported in the United States even though the first isolates of this organism were made from 12 cases that occurred in humans between 1923 and 1947. In 1972, Weaver and Jordan [*16*] reported that 29 isolates had been made in the United States since 1966. They stated that serotype 8 was the predominant type and that types 3 and 9 were absent. In contrast, types 3 and 9 are very prevalent in Europe, and type 8 is absent. It appears that the epidemiologies of *Y. enterocolitica* infections in the United States and Europe are different. An explanation of the low frequency of isolation in the United States may be the absence of laboratory screening for *Y. enterocolitica* in the routine analysis of enteric specimens. In one hospital in Canada which initiated routine screening for this organism, many isolates have been made subsequently [*14*].

Presence in Water—Y. enterocolitica has been isolated from drinking water [*7*] and from nonchlorinated well water [*17,18*]. The earliest suggestion of water as a source of infection was made in 1956 when Coleman [*20*] reported that two sisters had become ill with an organism he called *Bacterium enterocoliticum* and attributed the source of their infections to be some "water that was unsafe unless boiled."

In 1974, Keet reported a case of septicemia in which mountain stream water was identified as the vehicle of transmission [*21*]. The isolates made from both the patient and the stream were of the same biotype and serofactor. Most of the strains that have been isolated from water in the United States are rhamnose positive and are serologically untypable or possess multiple serofactors. *Y. enterocolitica* in water for human consumption regardless of serotype or biotype should be of public health concern.

Y. enterocolitica also has been isolated from foodstuffs, such as milk [*22*], ice cream [*23*], mussels [*24*], oysters [*19*], specimens from swine slaughter houses, samples of market meat, and vacuum-packed beef [*25*]. Water may have been the ultimate source of contamination.

Isolation and Enumeration of *Y. enterocolitica* in Water

Investigation of various waters for the presence of *Y. enterocolitica* has been limited in the past for three reasons. These bacteria were not thought to have public health significance. They differ ecologically from other common enteric pathogens, for example, they survive and grow better at low temperatures and compete poorly with other bacteria when grown at 37°C. There are no standard laboratory methods for the isolation and enumeration of this organism.

Collection of Samples

All water samples are collected according to standard methods.

Primary Isolation Procedures

Membrane filtration methods have been used successfully for the isolation of *Y. enterocolitica* from contaminated waters either by direct culture on M Endo broth [*18*] or through preenrichment of the membrane in cooked meat broth (CMB) [*19*]. Highsmith et al [*26*] incubated their specimens for 72 h at 25 and 36°C. *Y. enterocolitica* colonies were dark red and very small. A dissecting scope or hand lens was used to observe colonies for picking. These same authors also found that membrane filters grown on BHI and MacConkey agar did not allow *Y. enterocolitica* colonies to be distinguished from those of other enteric organisms.

Toma [*19*] found that CMB incubated at 25°C yielded six out of seven isolates as contrasted with preenrichment in Selenite F which yielded only one out of seven isolates.

Cold enrichment techniques have been used primarily for the enrichment of fecal material. Optimum recovery is made at three weeks with specimen suspended in 0.067 M phosphate-buffered saline pH 7.6 and incubated at 4°C [*14,19,27-29*].

Recommended Methods

The recommended procedure combines the membrane filter, CMB, and cold enrichment techniques (Fig. 1).

Plating Media

Inoculation

Y. enterocolitica has a tolerance to high concentrations of bile salts. This characteristic is useful for its selective isolation. The two plating media recommended are MacConkey agar and Salmonella-Shigella agar (SS). The latter medium has inhibited the growth of some strains of *Y. enterocolitica*. Recently, new plating media have been developed that may be of value [*17,30*].

Some of the common enteric media must not be used. Eosine methylene blue agar (EMB) containing sucrose is an example. *Y. enterocolitica* colonies will ferment the sucrose and have green metallic sheens similar to *Escherichia coli* colonies. In addition, media containing brilliant green dye should be avoided because it is toxic to *Y. enterocolitica*.

Incubation

Although 25°C is the optimum temperature for incubation of plate media, isolation of the organism can be made relatively well from plate media incubated in the temperature range of 22 to 30°C. At higher temperatures such as 36°C, *Y. enterocolitica* does not compete well with other enteric bacteria; therefore, it is recommended that plates be incubated at 25°C for 48 h.

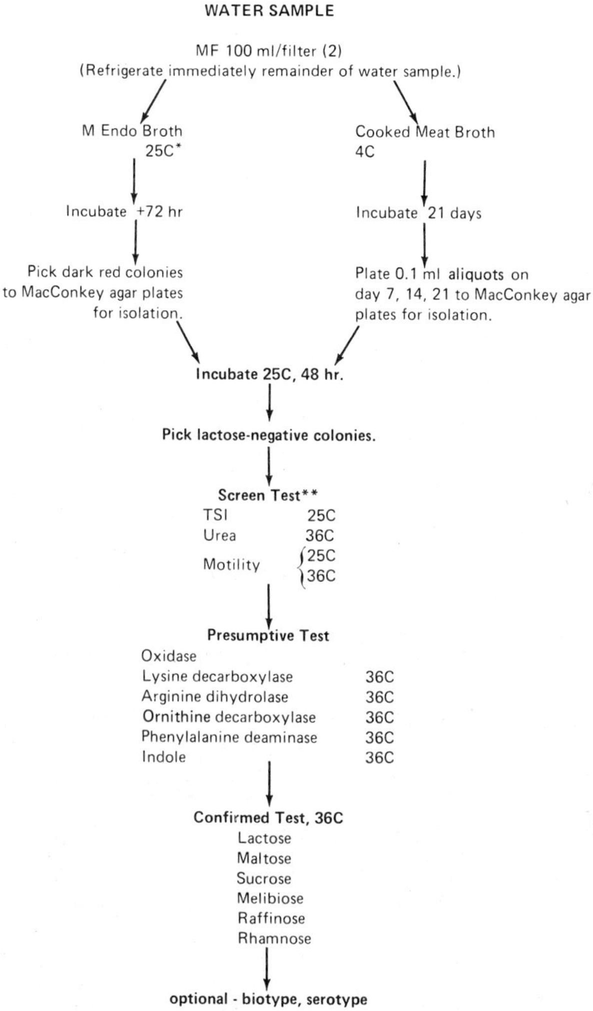

Y. enterocolitica can also be isolated on MMF when incubated at 36C, (18, 26).
**See biochemical characterization section.

FIG. 1—*Method for the isolation of* Y. enterocolitica *from water.*

Colony Selection

Colony size can vary with media and temperature. At 24 h the colonies will be pinpoint sized; at 48 h they will be large enough to pick. On MacConkey agar, colonies that have not fermented lactose and are 2 to 3 mm in diameter (with or without entire margins) should be picked for identification. Colonies on SS agar are similar but slightly smaller in diameter. A hand lens or dissecting scope may be of help in this examination.

TABLE 1—*Biochemical reactions of* Y. enterocolitica *and other related bacteria (tests performed at 36° C unless otherwise noted).*

Tests	*Y. enterocolitica*	*Y. pestis*	*Y. pseudotuberculosis*	*V. cholerae*	*A. hydrophila*	*Serratia*	*Enterobacter*	*C. diversus*	*Klebsiella*	*P. morganii*	*P. rettgeri*	*Providencia*	*C. violaceum*
Oxidase	−	−	−	+	+	−	−	−	−	−	−	−	−(W+)−
Christensen's urea	+	−	+	−	−(+)	−(+)	−(+)	+,−	+,−	+	+	−	−(L+)
Lactose	N	−	−	+(L)	−(+)	−(+)	+(−)	+,−	+(−)	−	−	−	−
Maltose	+(L)	+	+	+	+	+	+(L)	+	+	−	−	−	−
Sucrose	+	−	−	+	+(−)	+	+(−)	+,−	+(−)	−	−(+)	+L(−)	+,−
Melibiose	−(+)T	−	+	−	−	−	+	−	+	−	−	−	−
Raffinose	−(+)T	−	−	−	−	−	+(−)	−	+	−	−	−	−
Rhamnose	−(+)T	−	+	−	−	−	+	+	+	−	+(−)	−	−
Motility, 22°C	+	−	+	+	+	+	+	+	−	+	+	+	+
Motility, 36°C	−	−	−	+	+	+	+	+	−	+	+	+	+
Arginine dihydrolase	−	−	−	−	+	−	−(+)	+	−	−	−	−	+
Lysine decarboxylase	−	−	−	+	−(+)	+	+(−)	−	+,−	−	−	−	−
Ornithine decarboxylase	+	−	−	+	−	+	+(−)	+	−	+	−	−	−
Phenylalanine deaminase	−	−	−	−	−	−	+,−	−	−	+	+	+	−
Indole	−(+)	−	−	+	+	−	−	+	−	+	+	+	−

NOTE—+ = positive within 7 days.
( ) = minority of reactions.
W = weak.
L = late.
N = Majority are negative with enteric base medium, positive with Hugh-Leifson O-F medium..
T = Test performed at 25°C.

Biochemical Characterization

Screen Tests

Suspected colonies picked from the plate media should be inoculated into the media indicated in Fig. 1 and incubated as directed. The TSI reaction should be acid slant, and acid butt containing no hydrogen sulfide (H_2S) or gas bubbles. Some strains that ferment sucrose rapidly may cause the slant to revert to alkaline after 24 h of incubation. The urea should be positive, while the motility should be positive only at 25°C and negative at 36°C.

Presumptive and Confirmed Tests

Isolates that show reactions on the screening media indicative of *Y. enterocolitica* should be further examined by inoculating the presumptive tests stipulated in Fig. 1, and, if still characteristic, these isolates should be inoculated finally into the confirming test media [*31*]. Table 1 lists the reactions for *Y. enterocolitica* and other closely related bacteria. The three biochemicals (rhamnose, raffinose, and melibiose) should be tested. Brenner et al [*3*] have determined these isolates to be of a different DNA relatedness group within the *Yersinia* genus. Although most of the water isolates in the United States to date have been rhamnose positive, only 53 out of 4800 isolates of *Y. enterocolitica* studied at the Institute Pasteur, Paris, France, till 1975 were rhamnose positive. None of these isolates were isolated from patients having gastroenteritis. They were from water or isolates from patients with an illness other than gastrointestinal [*41*].

Typing Systems

Biotype

A biotyping system for *Y. enterocolitica* was first described by Nilehn [*32*]. Subsequently Wauters [*23*] modified this system with the hope of simplifying it. However, the substitution of lecithinase for salicin and esculin resulted in some American isolates having different biotypes for each respective biotyping system. Therefore, it is suggested that both of these systems be utilized. Table 2 indicates the tests to be performed, temperatures of incubation, and common tests. One of the authors proposed that isolates be described according to a combined Nilehn-Wauters (NW) schema. For ease of designation, strains might be described as NW 12 (Nilehn 1, Wauters 2) or NW 21 (Nilehn 2, Wauters 1). This should be done by the best fit method. Test exceptions should also be noted for possible epidemiologic markers. An example would be NW 11 (salicin negative), etc.

Serotype

Serologically there are 34 "O" factors and 19 "H" factors recognized [*22,33,34*]. Some antigens are shared in common with other bacteria; for

TABLE 2—*Biotype schema for* Y. enterocolitica.

Tests	Biotypes: Wauters' 1	2	3	4	5	Nilehn's 1	2	3	4	5
Salicin[a]	...[b]	...	...	...	...	+	−	−	−	−
Esculin[a]	...	...	...	...	...	+	−	−	−	−
Lecithinase	+	−	−	−	−	...	...	...	...	...
Indole[c]	+	+	+	−	−	+	+	+	−	−
Lactose (O/F)	+	+	+	−	−	+	+	+	−	−
Xylose[a]	+	+	+	−	−	+	+	+	−	−
Nitrate[a]	+	+	+	+	−	+	+	+	+	−
Trehalose[a]	+	+	+	+	−	+	+	+	+	−
B galactosidase	+	+	+	+	−	+	+	+	+	−
Ornithine decarboxylase	+	+	+	+	−	+	+	+	+	−
Voges-Proskauer	...	...	...	...	...	+	+	+[d]	+	−
Sorbose[a]	...	...	...	...	...	+	+	+	+	−[d]
Sorbitol[a]	...	...	...	...	...	+	+	+	+	−
Sucrose	...	...	...	...	...	+	+	+	+	−[d]

[a]All biochemicals are incubated at 25°C except those marked with an *a*; they are incubated at 36°C

[b]Dots indicate tests that are not done in the indicated schema.

[c]For practical purposes, results should be recorded after seven days even though the authors cited may have incubated tests longer; indole is incubated at 25 and 36°C even though Wauters' incubated it at 29°C.

[d]Reactions may vary for specific strains.

example, *Y. enterocolitica*, serotype 9 has cross reactions with *Brucella* species [*35*]. Other crosses are with *Vibrio* and *Salmonella* [*36*].

Phage type

A phage typing system has been developed in Europe [*37*]; however, most of the isolates from the United States are insensitive to the European typing.

Conclusion

Although the epidemiology of *Y. enterocolitica* is not fully understood, water has been implicated as a vehicle of transmission for human illness. *Y. enterocolitica* have been isolated from a variety of waters, including potable. For this reason laboratories should be prepared to examine water for this organism. The membrane filter technique with M Endo broth has been used effectively in recovery of *Y. enterocolitica* from water. The combination of this method and cold enrichment in CMB is recommended as the procedure for isolating this organism from water.

References

[*1*] Frederiksen, W., *Proceedings*, 14th Scandinavian Congress of Pathology and Microbiology, Norwegian University Press, Oslo, Norway, 1964, pp. 103–104.
[*2*] Schleifstein, J. and Coleman, M. B., *New York State Journal of Medicine*, Vol. 39, 1939, pp. 1749–1753.
[*3*] Brenner, D. J., Steigerwalt, A. G., Falcao, D. P., Weaver, R. E., and Fanning, G. R., *International Journal of Systematic Bacteriology*, Vol. 23, 1976, pp. 205–216.
[*4*] Ahvonen, P., *Annals of Clinical Research*, Vol. 4, 1972, pp. 39–48.
[*5*] Solem, J. H. and Lassen, J., *Scandinavian Journal of Infectious Diseases*, Vol. 3, 1971, pp. 83–85.
[*6*] Mollaret, H. H., Omland, T., Henriksen, S. D., Baeroe, P. R., Rykner, G., and Scavizzi, M., *La Presse Medicine*, Vol. 9, 1971, pp. 345–348.
[*7*] Keet, E. E., *New York State Journal of Medicine*, Vol. 74, 1974, pp. 2226–2230.
[*8*] Sonnenwirth, A. C., *Annals of New York Academy of Science*, Vol. 174, 1970, pp. 488–502.
[*9*] Winblad, S., *Microbiology and Immunology*, Vol. 2, 1973, pp. 129–132.
[*10*] Mollaret, H. H., and Guillon, J. C., *Annales De L'Institute Pasteur*, Vol. 109, 1965, pp. 608–613.
[*11*] Quan, T. J., Meek, J. L., Tsuchiya, K. R., Hudson, B. W., and Barnes, A. M., *Journal of Infectious Disease*, Vol. 129, 1974, pp. 341–344.
[*12*] Carter, P. B., *Infection and Immunity*, Vol. 11, 1975, pp. 164–170.
[*13*] Alonso, J. M., Bercovier, H., Destombes, P., and Mollaret, H. H., *Annales of Microbiology*, Vol. 126B, 1975, pp. 187–189.
[*14*] Toma, S. and LaFleur, L., *Applied Microbiology*, Vol. 28, 1974, pp. 469–473.
[*15*] Mollaret, H. H., *Annales De Biologie Clinique*, Vol. 30, 1972, pp. 1–6.
[*16*] Weaver, R. E. and Jordan, J. G., *Microbiology and Immunology*, Vol. 2, 1973, pp. 120–125.
[*17*] Saari, T. N. and Quan, T. J., Abstracts of the Annual Meeting, ASM Abstract C119, American Society of Microbiology, Atlantic City, N. J., 1976.
[*18*] Highsmith, A. K., Feeley, J. C., Wood, B. T., Skaliy, P., Wells, J. G., and Rosenberg, M. L., Abstracts of the Annual Meeting, ASM Abstract C120, American Society of Microbiology, Atlantic City, N. J., 1976.
[*19*] Toma, S., *Canadian Journal of Public Health*, Vol. 64, 1973, pp. 477–487.
[*20*] Coleman, M. B., "Annual Report of the Division of Laboratory Research," New York State Department of Health, 1956, p. 91.

[21] Lassen, J., *Scandinavian Journal of Infectious Diseases*, Vol. 4, 1972, pp. 125–127.
[22] Weaver, R. E., personal communication, 1976.
[23] Wauters, G., "a l'etude de *Yersinia enterocolitica*," These d'agregation, Vander, Louvain, France, 1970.
[24] Spardo, M., and Infortuna, M., *Bulletin*, Society of Italian Biologists, Vol. 44, 1968, pp. 1896–1897.
[25] Morris, G. K. and Feeley, J. C., *Bulletin of the World Health Organization*, Vol. 54, 1976, pp. 79–85.
[26] Highsmith, A. K., Feeley, J. C., Skaliy, P., Wells, J. G., and Wood, B. T., 1977, to be published.
[27] Zen-Yoji, Maruyama, T., Sakai, S., Kimura, S., Mizuno, T., and Momose, T., *Japanese Journal of Microbiology*, Vol. 17, 1973, pp. 220–222.
[28] Wilson, D., McCormick, J. B., and Feeley, J. C., *Journal of Pediatrics*, Vol. 89, 1974, pp. 767–769.
[29] Greenwood, J. R., Flanigan, S. M., Pickett, M. J., and Martin, W. J., *Journal of Clinical Microbiology*, Vol. 2, 1975, pp. 559–560.
[30] Lee, W. H., Abstracts of the Annual Meeting, ASM Abstract P16, American Society of Microbiology, New York, N. Y., 1975.
[31] Feeley, J. C., Lee, W. H., and Morris, G. K., *Compendium of Methods for the Microbiological Examination of Foods*, American Public Health Association, 1976.
[32] Nilehn, B., *Acta Pathologica Et Microbiologica Scandinavica*, Supplement, Vol. 206, 1969, pp. 1–46.
[33] Wauters, G., Le Minor, L., and Chalon, A. M., *Annales De L'Institute Pasteur*, Vol. 120, 1971, pp. 631–642.
[34] Corbel, M. J., *Journal of Hygiene, Cambridge*, Vol. 75, 1975, pp. 151–171.
[35] Maeland, J. A., and Digranes, A., *Acta Pathologica Et Microbiologica Scandinavica*, Vol. 83, 1975, pp. 382–386.
[36] Nicolle, P., Mollaret, H. H., and Brault, J., *Microbiology and Immunology*, Vol. 2, 1973, pp. 54–58.
[37] LaFleur, L., Martineau, B., and Chicoine, L., *Union Medicale Du Canada*, Vol. 101, 1972, p. 2407.
[38] Wauters, G., *Microbiology and Immunology*, Vol. 2, 1973, pp. 68–70.
[39] Weaver, R. E., Tatum, W. W., and Hollis, D. G., "The Identification of Unusual Pathogenic Gram Negative Bacteria (Elizabeth O. King)," Center for Disease Control, Atlanta, Ga., 1975.
[40] Darland, G., Ewing, W. H., and Davis, B. R., "The Biochemical Characteristics of *Yersinia enterocolitica* and *Yersinia pseudotuberculosis*," Center for Disease Control, Atlanta, Ga., 1974.
[41] Alonso, J. M., Bejot, J., Bercovier, H., and Mollaret, H. H., *Medicines et Maladies Infectieuses*, Vol. 5, No. 10, 1975, pp. 490–492.

L. T. Vlassoff[1]

Klebsiella

REFERENCE: Vlassoff, L. T., ***"Klebsiella,"*** *Bacterial Indicators/Health Hazards Associated With Water, ASTM STP 635*, A. W. Hoadley and B. J. Dutka, Eds., American Society for Testing and Materials, 1977, pp. 275–288.

ABSTRACT: *Klebsiella pneumoniae* is well distributed in the environment and found to be associated primarily with wastes high in organic matter and human activity.

K. pneumoniae is a serious opportunistic pathogen of increasing concern in hospitals, where precolonization seems to be a prerequisite to infection. Epidemiological data are lacking to link infection with high *K. pneumoniae* concentrations in the environment. Presently, it cannot be enumerated reliably by primary plating methods.

The presence of *K. pneumoniae* in water indicates degraded quality, and it is probably as significant as finding *Escherichia coli* and a good indicator of pollution from certain organic wastes.

KEY WORDS: bacteria, water, coliform bacteria, klebsiella

The coliform *Klebsiella* is an opportunistic pathogen, becoming prominent in hospital infections, and is found in high concentrations in organic wastes. Because of its prominence in these areas, it is worthwhile to consider its epidemiological role in nature and in the hospital and its origin and distribution. *Klebsiella*'s role as an opportunistic pathogen is well established although its etiology is vague and its taxonomy confused.

Should clinical and environmental *Klebsiella* be grouped in the same genus? Are the high densities of *K. pneumoniae* in organically polluted environments a public health hazard? Can this genus be used to determine the extent of influence of certain organic wastes, and is it the best organism for this purpose? Does *Klebsiella* have the same health significance as *Escherichia coli* in aquatic environments?

If methods can be developed readily to identify and enumerate *Klebsiella* with certainty, these questions should be answered. Only some of the data can be examined here; there are no clear-cut answers available today.

The preponderance of environmental literature available is based on the Edwards and Ewing nomenclature [*1*];[2] therefore, *K. pneumoniae* and *Enterobacter aerogenes* will be used here. Percent positive reactions listed by

[1]Manager, Microbiology Section, Laboratory Services Branch, Ministry of the Environment, Rexdale, Ontario, Canada.

[2]The italic numbers in brackets refer to the list of references appended to this paper.

these taxonomists is also useful to determine doubtful reaction. This deserves closer examination but is beyond the scope of this report.

Only *K. pneumoniae* will be discussed here, because 95 percent of isolates in the United States fall into this species [*2*]. *K. ozaenae* is found infrequently, and *K. rhinoscleromatis* is isolated rarely. Bascomb, Seidler, Knittel and Dufour (personal communications) agree that *K. pneumoniae* is the species most likely to be present in aquatic environments.

In Environment

The distribution of *Klebsiella pneumoniae* in the environment has been studied only recently [*3*]. Because of its former confusion with "*Aerobacter aerogenes*," a saprophyte with little health significance except that it was a coliform, *Aerobacter* has been reported for some time in soil, water, and in high-organic wastes [*4*]. With more recent taxonomic clarification [*1*] that renamed *Aerobacter* to *Enterobacter* and that provided cultural reactions that consistently separate *Enterobacter* and *Klebsiella*, reports of *K. penumoniae* in environmental samples have emphasized concern over its role as an opportunistic pathogen [*5*].

It appears that *Klebsiella-Enterobacter* dominate the coliforms in soil samples, and *Klebsiella* has been reported to be as high as 71 percent of the isolates (Table 1) [*6-8*]. *E. coli*, however, is an infrequent (less than 10 percent) isolate from "uncontaminated" soil and plant samples [*8*].

Market produce present a different distribution of coliforms [*7-9*]. Market fruit and vegetables show an appreciable density of *E. coli* (14 percent) with *Klebsiella* populations similar to environmental samples (Table 2).

Water samples seem to present still another coliform distribution pattern [*3,8*]. In varied geographical locations, *E. coli* was shown to dominate the aquatic coliforms (57 percent isolates and 90 percent samples) with *Klebsiella* less frequently isolated (38 percent isolates and 28 percent samples) [*3,10*]. It would appear that *Klebsiella* may be ubiquitous in nature, except for the water environment (Table 3) [*3*].

TABLE 1—Klebsiella, *coliform, fecal coliforms, and* E. coli *reported in environmental samples.*

Source	Percent of Coliform Isolates			
	K. pneumoniae	*K. penumoniae,* 44.5°C	Fecal coliforms or *E. coli*	Reference
Soil, tree needles, and bark	71	35%	2% infrequent isolates	[*8*]
Uncontaminated	50 *Klebsiella-Enterobacter* (− − + +), ubiquitous	. . .	small percent	[*6*]
Uninhabited areas (Greece)	ubiquitous	. . .	. . .	[*9*]

TABLE 2—Klebsiella, *coliform, fecal coliforms, and* E. coli *reported in market produce.*

Source	Percent of Coliform Isolates: *K. pneumoniae*	Percent of Coliform Isolates: *K. pneumoniae*, 44.5°C	Fecal coliforms or *E. coli*	Reference
Seeds and vegetables	10^3/g	...	...	[*50*]
Market fruit and vegetables	64	25	400/100 g (MPN) 14%	[*8*]

Industrial organic wastes have been shown to generate high numbers of *Klebsiella*, and their presence in pulp and paper effluents and their treatment facility effluents have been well documented (Table 4). Sulfite paper mill effluents show 10^4 and 10^6/100 ml [*11,12*], while kraft mill effluents generate lower (< 10^3/100 ml) *Klebsiella* densities, as do bleaching process effluents for both mills (< 2 to 2700/100 ml) [*11*]. The *Klebsiella* population dominates other organisms [*13*] and forms from 50 to > 90 percent of the coliform population [*11,12,14-16*]. Recent surveys have shown that waters receiving pulp and paper mill wastes—whether treated, segregated, or not—are a major source of coliforms.

Textile, sugarcane, sugar refining, and kelp processing wastes high in carbohydrates show much the same pattern of coliform, *Klebsiella*, and *E. coli* distribution as for paper mill wastes [*11,14,17*]. *K. pneumoniae* comprise 80 to 90 percent of the 10^5/100 ml coliform with a small portion being *E. coli* (Table 5).

Reports [*8,17,18*] show that 18 to 45 percent of environmental isolates can grow at 44.5°C (see Tables 1 through 4). Concern that the fecal coliform test includes colonies of thermotolerant *Klebsiella* has lead to reevaluations of that test's validity [*5*]. It has been suggested that thermotolerant *K. pneu-*

TABLE 3—Klebsiella, *coliform, fecal coliforms, and* E. coli *reported in water.*

Source	Percent of Coliform Isolates: *K. pneumoniae*	Percent of Coliform Isolates: *K. pneumoniae*, 44.5°C	Fecal coliforms or *E. coli*	Reference
Raw water	...	38%	57%	[*10*]
	16	...	30%	[*59*]
Drinking water	59	...	13%	[*10*]
	19	...	15%	[*59*]
Surface water	...	...	2–2/100 ml (MPN-FC)	[*8*]
Surface waters various geographic areas	28% of samples; not ubiquitous in waters	...	90% of samples; *E. coli* dominated isolates	[*3*]

TABLE 4—Klebsiella, *coliform, fecal coliforms, and* E. coli *reported in pulp and paper wastes.*

Source	Coliforms per 100 ml	Coliform Isolates *K. pneumoniae*, %	Fecal coliforms or *E. coli*, %	Reference
Pulp and paper wastes and receiving waters		>90	<10	[*15*]
		80		
		69 to 100	3.3	[*11*]
		>90		[*8*]
		predominate		[*13*]
	10^5 to 10^6/100	80		[*39*]
Pulp and paper wastes	...	60 (MPN)	34 (MPN)	[*12*]
Sulfite process	...	10^4 to 10^6/100	...	[*11*]
Kraft process	...	<10^3	...	[*11*]
Bleached sulfite and kraft process	...	<2 to 2700/100	...	[*11*]
Woodroom effluents	10^5 to 10^6 N^2-fixing coliforms per ml	...	...	[*31*]

moniae should be considered as significant as *E. coli* to indicate deteriorated water quality and an increased health hazard [*18*]. Others have concurred on the grounds that *Klebsiella* are coliforms and pathogens [*19*].

In soils and on plants, *Klebsiella* appear to be ubiquitous, but, in unpolluted water, they are more difficult to find. In drinking water, *E. coli* is usually present with *Klebsiella* and is a more frequent isolate [*3,10*]. There are high populations of *Klebsiella* in the environment [*3,8,20*] which are associated with degraded water and human activity. It remains to be shown whether or not these opportunistic pathogens are linked with human disease and infection, but their presence denotes deteriorated water quality. In the quantities present in waters below certain effluent discharges, they are of concern, as similar concentration of any bacteria would be, and could create a health hazard. The origin of *Klebsiella* in wastes, the environment, and its epidemiology need to be studied. The sooner more information is gathered, the sooner *Klebsiella* will find its niche in environmental studies.

Pathogenicity

Water pollution microbiologists formerly have misclassified *Klebsiella* isolates as *Aerobacter aerogenes*, because classification was limited to IMViC patterns. They were considered primarily nonpathogenic, normal soil inhabitants. More recent reports, differentiating these organisms by more than their IMViC characteristics, have shown that indeed *K. penunomiae* dominate the coliform population in many environmental samples, prompting an evaluation of the pathogenicity and virulance of environmental isolates.

That *K. pneumoniae* is an opportunistic pathogen is a matter of record

TABLE 5—Klebsiella, *coliform, fecal coliforms, and* E. coli *reported in other sources.*

Source	Coliforms/100 ml	Percent of Coliform Isolates		Fecal coliforms or *E. coli*, %	Reference
		K. pneumoniae	*K. pneumoniae*, 44.5°C		
Textile wastes	. . .	10^7/100 ml 80	45	. . .	[*17*]
receiving water	10^6/100 ml	50	. . .	. . .	. . .
Sugarcane	10^5 to 10^8/ml	86	. . .	. . .	[*14*]
Sugar refining wastes	10^3 to 10^4/100 ml	90 *Klebsiella-Enterobacter*	. . .	. . .	[*11*]
Kelp process wastes	10^7/100 ml	. . .	. . .	25	[*11*]
Clinical samples	. . .	. . .	96	. . .	. . .
Mastitis samples	. . .	. . .	100	. . .	[*18*]
Environmental samples	. . .	. . .	18	. . .	. . .

[*21*]. It was first isolated from infected lungs and has caused genito-urinary infections, bacteremia, osteomyelitis, and meningitis [*16,21,22*]. *K. pneumoniae* also has been implicated in bovine mastitis [*18,23,24*]. Although it is found in the environment [*3,8*], it appears to be related more to humans [*19,25*]. It comprises a small portion of the intestinal flora of 20 to 30 percent of humans, where it is generally not found to the extent that *E. coli* is found [*26,27*]. Twenty percent of incoming hospital patients have been found to be colonized with *K. pneumoniae* [*25*].

Colonization of the human intestinal tract has been related to nosocomial infections [*28*]. Precolonization, in fact, may be preliminary to hospital acquired clinical infection [*21,29*]. Symptomless colonized patients have been shown to be four times more susceptible to hospital acquired infections than noncolonized patients [*19*].

Infection of the upper respiratory tract by *K. pneumoniae* can be quite serious. They cause only 2 percent of the bacterial pneumonia cases but are the cause of 60 to 70 percent of deaths by all bacterial pneumonia [*2,27*].

Klebsiellae are next in importance to *E. coli* in urinary tract infections but are more often the cause of septicemia, pneumonia and postoperative infections [*25,27*].

Although the epidemiology of hospital infections is poorly understood [*22,29*], there are indications that infections due to "aerobic gram-negative" bacteria are becoming more prominent. Gram-negative bacteria as causative agents of infectious diseases have been noted [*30,31*], and *Klebsiella-Enterobacter-Serratia* infection frequency has more than doubled in the last ten years [*32*]. Prior to the 1960s, staphylococci were the major agents in nosocomial infections in hospitals, whereas *Pseudomonas*, *Klebsiella-Enterobacter-Serratia*, *E. coli*, and anaerobes such as *Bacteroides* are of primary concern today. In a German hospital [*33*], the death rate for children due to secondary pneumonia rose to 21.4 percent during the period 1969 to 1973 and from 16.8 percent during the period 1954 to 1958. The decrease in *Staphylococcus* infections and the increase in patients with "massively" lowered resistance have been cited as possible causes. Estimates of 300 000 episodes and more than 100 000 deaths annually are ascribed to bacteremia in which *Klebsiella-Enterobacter* are some of the bacteria implicated. The patient mortality rate in a hospital study was reported to be 12 percent due to *Klebsiella-Enterobacter-Serratia*. Five of six patients died when *Klebsiella* bacteremia was associated with a respiratory tract infection [*22*].

Environmental and clinical isolates have been submitted to mouse pathogenicity tests in several investigations [*14,25,34,35*]. No obvious relationships were seen between environmental (fresh and saline water, sugarcane, and textile isolates) and clinical *K. pneumoniae* isolates, but both reacted similarly to the mouse test. The validity of this test requires some clarification, since where mice were killed, high numbers of bacteria were required [*5*]. Even though it is important to reiterate that no difference in pathogenicity was noted when their source was compared, *K. pneumoniae* of similar pathogenicity could be found in environmental sources as in hospitals isolates.

K. pneumoniae strains from geographically different waters appeared essentially similar in broad serotype distribution. Sixty of the 72 types were represented although 49 percent fell within eleven serotypes [*36*]. Although capsular serotypes 1 and 6 are historically related to respiratory tract infections [*22,37*], serotypes 1, 3, 4, and 5 are found rarely in hospital infections. Many of these were found in textile finishing plant wastes [*34*]. No clear relation was found between serotype and virulence.

A combination of numerically coded biochemical biotypes and serological typing has been used to further differentiate clinical isolates [*38*]. Over 100 *Klebsiella* types were clearly distinguished using this system, and its use was suggested to study nosocomial *Klebsiella* associated hospital epidemiological infections. Such a system may be useful to further divide environmental biotypes when comparing them to clinical isolates and determine the relevance of environmental *Klebsiella* to health hazard.

Water strains have a greater susceptibility to antibiotics than human isolates [*36*]. Multiple resistance from human strains and a similar resistance of nonhuman strains suggest a significant percentage of resistance transfer factor possession (that is, noninherited antibiotic resistance). However, water strains have a greater susceptibility to antibiotics than human isolates [*36*]. This appears to be the only significant difference between water and clinical strains. However, nothing is known of the rapidity with which R factors are lost in nature or how this influences environmental isolates. Clinical and environmental *Klebsiella* isolates have been analyzed for genetic differences [*3,39,40*]. DNA-DNA duplex experiments show that pulp and paper mill and human infection isolates are similar.

The realization that all − − + + IMViC types should not be lumped together as *Aerobacter aerogenes*; some of them are *K. pneumoniae*, can occur in nature, and are pathogenic; some can occur in large quantities in the environment due to proliferation in polysaccharidic wastes; some are coupled with the reported worldwide concern over increased nosocomial infections; knowledge of the antibiotic resistance of this organisms; and reported similar pathogenicity of environmental and clinical isolates suggests the importance of recognizing and mapping *K. pneumoniae* in the environment and considering it as a serious indication of pollution.

The complicated route by which *Klebsiella* infections occur, the lack of epidemiological data, the reported illness in the young, the aged, and those under stress, and their lack of invasive potential [*60*] require careful consideration in determining the environmental significance of *Klebsiella*.

Identification

K. pneumoniae has been of concern in hospitals where it is mainly associated with respiratory and urinary infections. It was first known as Friedlander's bacillus (1882) and later named *K. pneumoniae*. Capsules of these organisms were shown to possess serological specificity which formed the basis for typing procedures in the 1920s and 1930s. In the late 1940s and 1950s, it became possible to classify the major types serologically, and 72 serotypes are now recognized using the antigenic specificity of the capsular polysaccharides and a quellung reaction with specific antisera.

Unfortunately because of the close physiological characteristics of *Klebsiella* and *Enterobacter* at that time, they were taxonomically differentiated according to isolation source. Sewage and environmental isolates were designated *Enterobacter* formerly *Aerobacter*, and clinical isolates were named *Klebsiella*. *Bergey's Manual of Determinative Bacteriology* [*41*] continues to classify these bacteria in this way, and some confusion still exists. Edwards and Ewing [*1*] and Cowan and Steel [*42*] published classifications that have clarified the distinction between *Klebsiella* and *Enterobacter* although some environmental microbiologists have yet to adopt these criteria.

Three species are now recognized in the United States [*1*]—*K. pneunomiae, K. rhinoscleromatis*, and *K. ozaenae*.

However, the Cowan and Steel [*42*] classification recognizes six *Klebsiella* retaining *E. aerogenes* as a separate genus. They are *K. aerogenes, K. pneumoniae, K. rhinoscleromatis, K. ozaenae, K. oxytoca*, and *K. edwardsii*.

Bascomb [*43*] in Britain, using 50 biochemical groupings and numerical taxonomic techniques, suggests six taxonomy groupings, three of Cowan's species, adding a new "unnamed" group, and including *E. aerogenes*. They are *K. aerogenes/oxytoca/edwardsii, K. pneumoniae, K.* (unnamed group), *E. aerogenes* (proposes *K. mobilis*), and *K. rhinoscleromatis*.

Recently, Johnson et al [*44*] used 216 characters and numerical taxonomy to regroup the tribe Klebsielleae into five species, retaining the genus *E. aerogenes*. They are *K. aerogenes, K. pneumoniae, K. rhinoscleromatis, K. ozaenae*, and *K.* species.

While the taxonomists improve their methods and regroup genera of the tribe Klebsielleae, the current environmental and clinical literature becomes clearer, although some confusion still exists.

Some of the variations in biochemical reactions reported in the four principal taxonomic classifications can be seen in Tables 6 and 7, in which only the major differences of opinion are reported. Growth of *K. pneumoniae*, in KCN medium, for example, is reported 40 percent [*44*], 97.9 percent [*1*], 0 to 15 percent or 85 to 100 percent [*42*], and 0 to 10 percent [*43*] positive. By contrast, there is greater similarity reported for *E. aerogenes* by these authors (Table 7) although agreement is still lacking.

Current North American literature uses the Edwards and Ewing [*1*] classification which separated *K. pneumoniae* and *E. aerogenes* for both environmental and clinical isolates. However, the recent Johnson scheme differentiates a portion of the *K. pneumoniae* of Edwards and Ewing into *K. pneumoniae* and *K. aerogenes*, and some European [*45*] clinical microbiologists are using Cowan's *K. aerogenes*. To avoid further confusion, it would seem necessary to meticulously confirm isolates with a sufficient number of physi-

TABLE 6—*Variation in biochemical reactions reported.*

	K. pneumoniae				
			[*42*][a]		
Character	[*44*]	[*1*]	sensu stricto	sensu lato[b]	[*43*][c]
Growth in KCN medium	d(2/5)	+(97.9%)	−	+	−
Acid from dulcitol	d (3/5)	±(31.5%)	+	d	+
Methyl Red	d (3/5)	±(13.3%)	+	−	+ (RT, 37°C)
Voges-Proskauer	d (2/5)	±(91.1%)	−	+	− (RT, 37°C)

[a] + = 85 to 100% strains positive;
− = 0 to 15% strains positive;
d = 16 to 84% strains positive.
[b]Sensu lato strain equivalent to *K. aerogenes.*
[c] + = 90 to 100% strains positive;
− = 0 to 10% strains positive;
(RT) = room temperature.

ological tests and report results so that, awaiting taxonomic agreement, European and American literature can be compared.

Through the taxonomic haze, however, *K. pneumoniae* and *E. aerogenes* have emerged reasonably separate, but the subdivision of *K. pneumoniae* biotypes certainly requires clarification. An example of this confusion is the genus *K. aerogenes* of Cowan and Bascomb (*K. aerogenes/oxytoca/edwardsii*), placed in Edwards and Ewing's *K. pneumoniae* by Johnson. Cowan separates *K. aerogenes* on the basis of fimbriation [*46*], while Bas-

TABLE 7—*Variations in biochemical reactions reported.*

	E. aerogenes			
Character	[*44*]	[*1*]	[*42*][a]	[*43*][b]
Acid from dulcitol	−	− (4%)	d	−
Salicin	d (3/8)	+ (98.7%)	+	+
Malonate	. . .	±(74.7%)	+	+
Methyl Red	−	− (0%)	−	−(RT) d (37°C)
Voges-Proskauer	+	+ (100%)	+	d (RT) −(37°C)
Gelatin hydrolysis	−	− (0%) (+) (77.3%)[c]	d	−

[a] + = 85 to 100% strains positive;
− = 0 to 15% strains positive;
d = 16 to 84% strains positive.
[b] + = 90 to 100 % strains positive;
− = 0 to 10% strains positive;
d = 11 to 89% strains positive;
(RT) = room temperature.
[c](+) = positive reaction after three or more days.

comb uses growth on KCN. Neither test is done routinely by environmental nor clinical laboratories. The variable Methyl Red and Voges-Proskauer results reported by taxonomists [*47*] are of no help in this dilemma. The importance here is that *K. aerogenes* [*42*] is considered a nonpathogenic saprophyte [*45*], but, as a member of *K. pneumoniae* [*1*], it is considered an opportunistic pathogen.

Undoubtedly, due to some of the frustrations of this taxonomic confusion, researchers [*38*] investigating a hospital epidemic devised a numerically coded group of biochemical tests and serotyping scheme where over 100 *K. pneumoniae* biotypes were recognizable. Such a scheme undoubtedly will be useful in future work to be done to unravel the environmental-clinical-taxonomic dilemma.

It is accepted by all four authors [*1,42-44*], however, that *K. pneumoniae* and *E. aerogenes* can be distinguished by motility and the decarboxylation of ornithine. Other tests, of course, would include those necessary to place them in the tribe Klebsielleae.

K. pneumoniae [*1*] are gram-negative, rod-shaped, nonmotile, capsulated bacteria, not producing indolphenol oxidase, not capable of producing indol (only 6 percent +), not possessing phenylalanine diaminase, ornithine decarboxylase, or arginine dihydrolase, slow to produce urease, give a positive Voges-Proskauer test (that is, produce acetymethylcarbinol), utilize citrate and sodium alginate as sole sources of carbon, and produce lysine decarboxylase. *K. pneumoniae* also produces acid from adonitol, inositol, and lactose, and utilizes malonate.

Practical confirmation tests would include motility (−), ornithine decarboxylase (−), urease (+), and sodium alginate (+/− 88 percent) to differentiate *K. pneumoniae* from *E. aerogenes*. The latter gives +, +,−,− reactions to these tests, respectively.

Primary isolation and identification methods are not available. Purified coliform isolates from membrane filter (MF) colonies or most probable number (MPN) tubes are confirmed with a battery of physiological tests. Cultures from elevated temperature tests restrict the number of *Klebsiella* recovered, so 35°C incubation should be used.

Primary isolation and enumeration medium has been proposed capitalizing on the ability of some *K. pneumoniae* to fix nitrogen. This nitrogen-deficient medium [*48*] frequently used for *Azotobacter* isolation [*31*] was aerobically incubated at 37°C, and the large, convex gummy (mucoid), glistening colonies were considered *Klebsiella* or *Enterobacter* or both. Although Eller [*49*] used it successfully with fecal samples and frozen meats, a modification by Brown [*50*] failed to isolate *Klebsiella* from vegetable peels.

Another modification of the Hino medium, with 2 percent mannitol and 5 μg yeast extract per millilitre incubated in an atmosphere of 80 percent nitrogen (N_2) and 20 percent argon, was used to enumerate *Enterobacter* and *Klebsiella* from aquarium water where experimental pet green turtles were held for several months [*20*]. High densities (10^3 to 10^4/ml) of *Klebsiella* were isolated in association with *Aeromonas, Enterobacter*, and *Salmonella*. However, no comment was made regarding the medium's performance.

In nitrogen fixing studies of pulp and paper effluents [*31*] another modification of the Hino medium (25 μg of yeast extract) verified MPN coliform counts that were able to fix nitrogen. The coliforms were not specified as *Klebsiella*.

Bile esculin agar [*51*] has been used successfully to differentiate pure cultures of *E. coli* and *K. pneumoniae* [*52*]. The prepared bile esculin agar was found slightly inhibitory to *Klebsiella* when modified with 1 percent lactose. Esculin is hydrolyzed to 6,7-dihydroxycoumarin which reacts with the iron in the medium to form a black compound. Ninety-nine percent of the *E. coli* tested were negative, and 99 percent of the *Klebsiella-Enterobacter-Serratia* were positive. It is not known whether this medium has been tried on environmental samples.

With streptococci work, blood agar is used to indicate pathogenicity; with staphylococci, it is the coagulase reaction. DNase medium was suggested for *K. pneumoniae* [*51*], but was reported unsuccessful [*15*].

A citrate agar, successful for separating *Klebsiella-Enterobacter* from *E. coli* by suppressing the latter, was not effective in environmental isolates [28,34].

A medium with simple growth substances employing antibiotics to suppress other environmental organisms is being developed [*34*] and tested on a wide variety of environmental samples in geographically diverse areas. It is reported to be able to detect and enumerate *Klebsiella* in a two-step procedure.

Methyl violet 2B medium [*47,53*] using the ability of a 1:1000 dye dilution to inhibit other naturally associated bacteria (*E. coli, Pseudomonas aeruginosa, Aeromonas, Proteus, Salmonella,* and *Citrobacter freundii* [*54,55*] was developed recently. With pure cultures, this medium produced colonies that were generically and morphologically distinct. In our laboratory [*56*], only two of 38 coliform isolates from a clean river and a sewage polluted bay were confirmed as *Klebsiella*. *Klebsiella* stock cultures streaked on this medium failed to grow. However, a further evaluation of double violet agar by the authors of this medium, using river waters, has indicated that it confirms 80 percent of primary isolates [*53*].

Another medium using acriflavine and violet red bile agar was reported recently [*55*] to separate *K. pneumoniae* successfully from other organisms in cow bedding. It is reported to produce large, mucoid, golden colonies of *K. pneumoniae*, easily differentiated from small brown *E. aerogenes* and dark-brown *P. aeruginosa* colonies in pure cultures. This media remains to be further tested with environmental and clinical samples.

While many investigators have indicated the need for primary isolation, differentiation, and enumeration procedures to identify *Klebsiella* from both environmental and clinical samples, there are few published methods. The few evaluations reported have not been conducted on a wide enough variety of samples to allow any of the methods to be proposed for wider or specific uses. Presumably, the current interest in Klebsielleae will prompt more experimentation.

Environmental microbiologists were wrongly classifying *K. pneumoniae* as *Aerobacter aerogenes* and attributing no significance to it. Recently, some clarification of taxonomy has shifted attention in North America to *K. pneumoniae*, and environmental isolates are now considered opportunistic pathogens of some significance. These organisms are reported in high numbers in organic (carbohydrate) polluted waters. Emerging from recent numerical taxonomic studies [*43,44*] is concern regarding proper grouping. Environmental isolates in North America identified as *K. pneumoniae* and *Enterobacter aerogenes* are considered pathogenic and saprophytic, respectively. In Britain, six *Klebsiella* species are identified [*42*], and numerical taxonomy has verified them [*45*]. One has to wonder where Edwards and Ewing's three *Klebsiella* species fit into this six species taxonomic scheme and what differences this makes to the environmental work where, for example, 90 percent of the coliform isolates from paper mill effluents were shown to be *K. pneumoniae*. The work of Fallon [*45*] with *K. pneumoniae*, for example, suggests that the old *Aerobacter aerogenes* connotations of "soil bacteria" with no health significance rears its ugly head once again.

The suggestion of many investigators [*15,49,57,58*] that the ornithine decarboxylase and the motility tests, be used as a basis of identification until the taxonomic question is resolved is strongly endorsed.

References

[*1*] Edwards, P. R. and Ewing, W. H., *Identification of Enterobacteriaceae*, 3rd ed., Burgess Publishing Company, Minneapolis, Minn., 1972.

[*2*] Lennette, E. H., Spaulding, E. H., and Truant, J. P., *Manual of Clinical Microbiology*, 2nd ed., American Society for Microbiology, Washington, D.C., 1974.

[*3*] Knittel, M. D., *Applied Microbiology*, Vol. 29, 1975, pp. 595–597.

[*4*] Parr, L. W., *Bacteriological Reviews*, Vol. 1, No. 3, 1939.

[*5*] Bordner, R. H. and Carroll, B. J., EPA Technical Report 3, Environmental Protection Agency, 4–5 May, 1972.

[*6*] Geldreich, E. E., Water Pollution Control Research Series Publication No. WP-20-3, U.S. Dept. of the Interior, Cincinnati, Ohio, 1966.

[*7*] Brown, C. and Seidler, R. J., *Applied Microbiology*, Vol. 25, 1973, pp. 900–904.

[*8*] Duncan, D. W. and Razzell, W. E., *Applied Microbiology*, Vol. 24, 1972, pp. 933–938.

[*9*] Papavassiliou, et al, *Journal of Applied Bacteriology*, Vol. 30, 1967, p. 219.

[*10*] Ptak, D. J., Ginsburg, W., and Willey, B. F., *Journal of the American Water Works Association*, September 1973, p. 604.

[*11*] Bauer, R. R., see Ref *5*.

[*12*] Huntley, B. E., Jones, A. C., and Cabelli, V. J., *Journal of the Water Pollution Control Federation*, Vol. 48, July 1976, p. 1766.

[*13*] Dutka, B. J., Bell, J. B., Collins, P., and Popplow, J., Manuscript Report No. KR-69-1, Division of Public Health Engineering, Dept. of National Health and Welfare, Canada, 1969.

[*14*] Nunez, W. J. and Colmer, A. R., *Applied Microbiology*, Vol. 16, No. 12, December 1968, pp. 1875–1878.

[*15*] Knittel, M. D., see Ref *5*.
[*16*] Knittel, M. D., Technicial Report 660/2-75-024, U.S. Environmental Protection Agency, June 1975.
[*17*] Dufour, A. P., personnel communication, 1975.
[*18*] Bagley, S. T. and Seidler, R. J., ASM Abstract, American Society of Microbiology, 1975.
[*19*] Geldreich, E. E., see Ref *5*.
[*20*] McCoy, R. H. and Seidler, R. J., *Applied Microbiology*, April 1973, pp. 534–538.
[*21*] Eichoff, T. C., see Ref *61*.
[*22*] Steinhauer, B. W., Eickhoff, T. C., Kislak, J. W., and Finland, M., *Amn. Inst. Med.*, Vol. 65, 1966, p. 1180.
[*23*] McDonald, T. J., McDonald, J. S., and Rose, D. L., *Journal of Veterinary Research*, Vol. 31, 1970, p. 1937.
[*24*] Carroll, E. J., *American Journal of Veterinary Research*, Vol. 32, 1970, pp. 689–699.
[*25*] Matsen, J. M., Spindler, J. A., and Blosser, R. O., *Applied Microbiology*, Oct. 1974, pp. 672–678.
[*26*] Thom, B. T., *Lancet*, Vol. 2, November 1970, pp. 1033.
[*27*] Guarraia, L., see Ref *5*.
[*28*] Montgomerie, J. Z., Doak, P. B., Taylor, D. E. M., and North, J. D. K., *Lancet*, Oct. 1970, p. 787.
[*29*] Duncan, I. B. R., *International Journal of Clinical Pharmacology, Therapy and Toxicology*, Vol. 11, No. 4, 1975, pp. 277–282.
[*30*] Wundt, W., *Fortschritte der Medizin*, Vol. 93, No. 8, 1975, pp. 379–381; Microbiology Abstract, 1975.
[*31*] Knowles, R., Neufelo, R., and Simpson, S., *Applied Microbiology*, Vol. 28, 1974, pp. 4–608.
[*32*] Matsen, J. M., *Applied Microbiology*, Vol. 19, No. 3, March 1970, pp. 438–440.
[*33*] Simon, C., Lange, C., and Harms, D., *Deutsche Medizinische Wochenschrift*, Vol. 100, No. 18, 1975, pp. 990–995.
[*34*] Dufour, A. P. and Cabelli, V. J., *Journal of Water Pollution Control Federation*, Vol. 48, May 1976, p. 872.
[*35*] Menon, A. S. and Bedford, W. K., Surveillance Report EPS 4–AR–73–1, Water Pollution Control Directorate, Department of Environment, Ottawa, June 1973.
[*36*] Matsen, J. M., see Ref *5*.
[*37*] Edwards, P. R. and Fife, M. A., *Journal of Bacteriology*, Vol. 70, 1955, pp. 382–390.
[*38*] Rennie, R. P. and Duncan, I. B. R., *Applied Microbiology*, Vol. 28, No. 4, October 1974, pp. 534–539.
[*39*] Knittel, M. D. and Seidler, R. J., ASM Abstracts, American Society of Microbiology, 1974, p. 51.
[*40*] Seidler, R. J., ASM Abstract, American Society of Microbiology, to be published.
[*41*] Buchanan, R. E. and Gibbons, N.E., Eds., *Bergey's Manual of Determinative Bacteriology*, 8th ed., Williams and Wilkins, Baltimore, Md., 1975.
[*42*] Cowan, S. T., *Manual for the Identification of Medical Bacteria*, 2nd ed., Cambridge University Press, England, 1974.
[*43*] Bascomb, S., and Gower, J. C., *Journal of General Microbiology*, Vol. 86, 1975, pp. 93–1020.
[*44*] Johnson, R., Colwell, R. R., Sakazaki, R., and Tamura, K., *International Journal of Systematic Bacteriology*, Vol. 25, No. 1, Jan. 1975, pp. 12–37.
[*45*] Fallon, J., *Journal of Clinical Pathology*, Vol. 26, 1973, p. 523.
[*46*] Duguid, J. P., *Journal of General Microbiology*, Vol. 21, 1959, pp. 271–286.
[*47*] Campbell, L. M. and Roth, I. L., *Applied Microbiology*, Vol. 30, No. 2, Aug. 1975, pp. 258–261.
[*48*] Hino, S. and Wilson, P. W., *Journal of Bacteriology*, Vol. 75, 1968, p. 408.
[*49*] Eller, C. and Edwards, F. F., *Applied Microbiology*, Vol. 16, June 1968, p. 896.
[*50*] Brown, C., Seidler, R. J., Latimer, J., and Brown, L. R., ASM Abstracts G94, American Society of Microbiology, 1973.
[*51*] Vasconcels, G. J., see Ref *5*, p. 95.
[*52*] Wasilauska, B. L., *Applied Microbiology*, Vol. 21, No. 1, Jan. 1971, pp. 162–163.
[*53*] Campbell, L. M., Roth, I. L., and Klein, R. D., *Applied and Environmental Microbiology*, Vol. 31, No. 2, 1976, p. 213.

[54] Fung, D. Y. C. and Miller, R. D., *Applied Microbiology*, Vol. 25, No. 5, May 1973, pp. 793–799.
[55] Fung, D. Y. C., ASM Abstracts, American Society of Microbiology, 1976.
[56] Rokosh, D., Ontario Ministry of the Environment, Toronto, Ontario, personal communication, 1976.
[57] Campbell, L. M. and Roth, I. L., *International Journal of Systematic Bacteriology*, Vol. 25, No. 4, Oct. 1975, pp. 386–387.
[58] Matsen, J. M. and Blazevic, D. J., *Applied Microbiology*, Vol. 18, No. 4, Oct. 1969, pp. 566–569.
[59] Clark, J. A., personal communications.
[60] Darrell, J. H., *Journal of Clinical Pathology*, Vol. 26, 1973, p. 894.
[61] Gellman, I., Technical Bulletin 254, National Council of the Paper Industry for Air and Stream Improvement, March 1972.
[62] Gellman, I., Technical Bulletin 279, National Council of the Paper Industry for Air and Stream Improvement, March 1975.

G. I. Barrow[1]

Bacterial Indicators and Standards of Water Quality in Britain

REFERENCE: Barrow, G. I., "**Bacterial Indicators and Standards of Water Quality in Britain,**" *Bacterial Indicators/Health Hazards Associated With Water, ASTM STP 635,* A. W. Hoadley and B. J. Dutka, Eds., American Society for Testing and Materials, 1977, pp. 289–336.

ABSTRACT: Health hazards vary greatly with different waters and their uses as well as with geography, climate, populations and their living standards. Only by defining practicable water quality objectives and priorities in the light of current knowledge and likely cost benefit can the risks of actual and potential disease be reduced to acceptable levels. These, however, will still vary in different parts of the world. Universal standards, in contrast to objectives, are still a long way ahead. In this paper, the British approach to the choice and use of bacterial indicator organisms in relation to potential health hazards and water quality criteria is reviewed briefly. Although no mandatory standards have ever been imposed, drinking water supplies in Britain are among the best in the world in quality and safety, even though about two thirds of them are derived directly from sewage polluted surface waters. However polluted these may be, in principle, they can be adequately treated, and British philosophy has always relied on this rather than on the more elusive targets of disinfecting sewage effluents and purifying water sources. For almost 50 years, coliform organisms and *Escherichia coli* have been used as the main indicators of fecal pollution, and, in this way, the efficiency of treatment and thus the safety of drinking water supplies have been effectively monitored. Despite increased water usage and recycling, there is no valid reason to change this view. Although viruses can be isolated readily from sewage effluents, there is no epidemiological evidence of transmission to man by the water route, and the current position is one of surveillance and research. Similarly, despite intensive investigation, there is no authenticated evidence of significant disease attributed to sewage polluted bathing beaches or efficiently maintained swimming pools. Aesthetic values, however, are now playing a greater role in environmental pollution control. In addition, compliance with microbiological and other standards for water for abstraction as well as for bathing recently accepted by the European Economic Community will be necessary in the future.

KEY WORDS: bacteria, water, coliform bacteria, pollution, disease

[1]Consultant bacteriologist and director, Public Health Laboratory, Royal Cornwall Hospital (City), Truro, Cornwall; also, chairman, Standing Subcommittee on the Bacteriological Examination of Water Supplies, Public Health Laboratory Service, England and Wales.

In the beginning, man the hunter did not harm the aquatic environment. Like the animals on which he preyed, he had to live in the vicinity of water. His excrement, as did theirs, added to the nutrients in the environment. Like the animals he hunted, it probably contained numerous parasites, some of which presumably spent part of their life cycle in water. During progression from nomadic herder to primitive farmer, the same sort of biological balance prevailed and pollution as we know it today did not occur. It was not until man settled in groups and lived together in close proximity to others that the problems of waterborne disease arose. As water is one of the prime necessities of life, its presence governed the siting of communities. This resulted in the ironic situation in which the water source which attracted a community in the first instance became polluted by the same community. The term pollution is used as quoted by Davies [*1*],[2] that is, the introduction of material or effects at a harmful level by man. As the community became greater in size and the population more dense, the level of contamination of the environment rose. This increased the possibilities of individuals becoming infected with specific parasites, which then proliferated and further contaminated the local environment, thus perpetuating the host-parasite cycle. Although the causal organisms were not known, a relation gradually became recognized between drinking water and sewage pollution.

That ingestion of polluted water may be associated with certain health hazards has been known since the beginning of history. The classical triad of waterborne infections—enteric fever, bacterial dysentery, and cholera—were later identified. To these, bacterial and viral gastroenteritis, leptospiral infections, and infectious hepatitis may be added. Many other waterborne diseases are caused by certain protozoa, worms, and flukes whose ova, larvae, or cysts, excreted in feces or urine, may gain access to surface waters which act directly or indirectly as the vehicle of transmission. These can be, and still are, a major problem in some parts of the world but, fortunately, not in Britain, and they will not be considered further. Even with modern water technology, there are still risks, the degrees of which vary with geography, climatic conditions, and the economic circumstances and living standards of communities as well as with the different sources and uses of different waters. In theory, with unlimited resources for water pollution control, potential health hazards to man or beast should never result in infection let alone actual disease. But we live in a real world where resources are usually limited, some more than others, so that this ideal is unlikely to be achieved in practice even if it were possible to eliminate human error and mechanical failures. In addition, the applications of scientific knowledge are advancing so rapidly that more areas of environmental pollution will inevitably come to light. However, by defining practicable water quality objectives as well as priorities in the light of current knowledge and likely cost benefit, the risks of actual and potential health hazards may be reduced

[2]The italic numbers in brackets refer to the list of references appended to this paper.

to acceptable levels. These, however, will still vary with different waters, regions, and circumstances. Universal standards, in contrast to objectives, are still a long way ahead.

In this paper, the British attitude towards water quality criteria in relation to microbial health hazards and safety is briefly outlined with particular reference to potable supplies, recreational uses, and indicator organisms.

Water Cycle

This subject is vast and the literature now voluminous. The way in which water resources are used and exploited in Britain are reviewed briefly by Holden [*2*], Downing [*3*], and Martin [*4*]. Rivers have always been, and probably always will be, used to carry domestic and waste effluents to the sea. Not only are they the cheapest means of disposal, but without such discharges many rivers in Britain would become mere trickles or even dry up. In the same way, most of these sewage polluted rivers are used also for abstraction for drinking water. Indeed, some 60 percent of public supplies of drinking water in England and Wales are derived directly from sewage polluted surface waters, and the high standard of safety is a tribute to the water authorities and all concerned. Few outbreaks of infection attributed to public supplies of drinking water have occurred during this century, and all have been due to accidental failures of one kind or another. After the serious outbreak of typhoid fever at Croydon in 1937, recommendations about the protection of gathering grounds, the detection of typhoid carriers among personnel, and the hygienic practices to be used in daily management of waterworks were made in a report [*5*]. There has been no typhoid attributed to drinking water since and only occasional incidents due to other causes [*7,8*]. This satisfactory state of affairs would probably have continued perfectly well were it not for the rapid developments in the field of chemical technology with their hazardous wastes—often of ill-defined substances whose effects on the aquatic environment are unknown. Quite apart from ecological considerations, it is right that this should cause concern and that the possible effects, direct and indirect as well as short and long term, of the continued ingestion of small amounts of toxic substances in water should be thoroughly investigated and, if necessary, remedial action taken. However, so far as is known, bacterial pollution of the environment in Britain is probably no worse now and, indeed, may well be better than it was before the reports of the Royal Commission on Sewage Disposal during 1898–1915. Its well known recommendations [*9*] that sewage effluents discharged to rivers should contain not more than 30 mg/litre of suspended solids and a five-day biochemical oxygen demand (BOD) of not more than 20 mg/litre were formulated to prevent the creation of nuisances and the destruction of animal, plant, and fish life. These recommendations were related to both the volume and the BOD of the water receiving the discharge. If the dilution of the effluent after discharge was low, a specially stringent standard might be prescribed. If very great, the general standard might be relaxed or ignored altogether. Where the dilution exceeded 500 volumes, it was suggested that all tests might be dispensed with and that crude sewage could be discharged without treatment other than simple screening and the removal of grit. These recommendations have served well despite the pollution associated with increase of population and industry. All effluents ultimately reach either inland lakes or the sea. The factors concerned in the subsequent ecology of the microflora and their numbers are complex and have been extensively investigated by Gameson et al [*10*]. However, despite changes in the nature and extent of environmental pollution, the concept and role of bacterial indicators for the quality and safety of water in Britain remain essentially unchanged.

Source of Pathogenic Organisms

The function of sewage treatment is to remove organic matter and thus produce an effluent satisfying a physical and chemical standard. It is essential that a sharp distinction be made between the chemical and the bacteriological aspects of sewage pollution. This is necessary because many industrial wastes, regarded chemically as highly polluting, are free from bacteria, while crude sewage and many effluents satisfying the recommended chemical standards, and so regarded as nonpolluting, contain vast numbers of intestinal bacteria derived from both man and animals. The distinction is also necessary because the essential function of sewage treatment is to remove as much as possible of the putrescible organic matter present. The greater part, however, of the organic matter in human and animal feces is present in the form of the bacteria from the intestine. Crude sewage contains almost all the agents capable of causing infectious disease in man—bacteria, viruses, protozoa, and other organisms excreted through the intestinal tract. Of these, the salmonella group, which are widely distributed in man and animals, are by far the most common cause of infectious intestinal disease in temperate climates. Salmonellae of human origin are derived from acute and convalescent cases, symptomless excreters, and chronic carriers in the local population. Salmonellae of animal origin stem not only from animals themselves but from the associated trades which process their products for human or animal use. These include abattoirs and the meat products industry, poultry packing and processing plants, tanneries, knackers' premises, meatmeal, bonemeal, fertilizer, and animal feeding stuffs plants. There is always a background salmonella level in sewage effluents derived from these sources as well as from apparently healthy populations. For example, when Harvey et al [*11*] monitored untreated sewage from a housing estate over a period of 15 months, they isolated a surprising number of salmonella serotypes with no apparent illness among the residents, presumably reflecting the excretion of various salmonellae ingested in noninfective doses. The extensive ramifications and recycling of such microbial pathogens between man and animals—directly in food and indirectly from the environment due to pollution at some stage of the water cycle—are illustrated in Fig. 1 adapted from Fennell [12]. Theoretically, such infections in man and animals may be reduced by breaking the cycle at more than one point by disinfection or other means.

Survival and Fate of Pathogenic Bacteria

Efficient sewage treatment does not remove all bacteria. It only reduces their numbers, and the resulting effluent contains the same types of bacteria, normal and pathogenic, in about the same proportion but diminished numbers [13]. Under normal flow conditions where effluents are discharged into rivers, sewage organisms have often disappeared or at least can be isolated only with difficulty a few miles downstream. However, it must be realized that they are being discharged practically all day and every day, and the fate of pathogens such as salmonellae and viruses is clearly important in relation to potential health hazards. With even limited survival, the contamination of receiving waters with either treated or untreated sewage presents a poten-

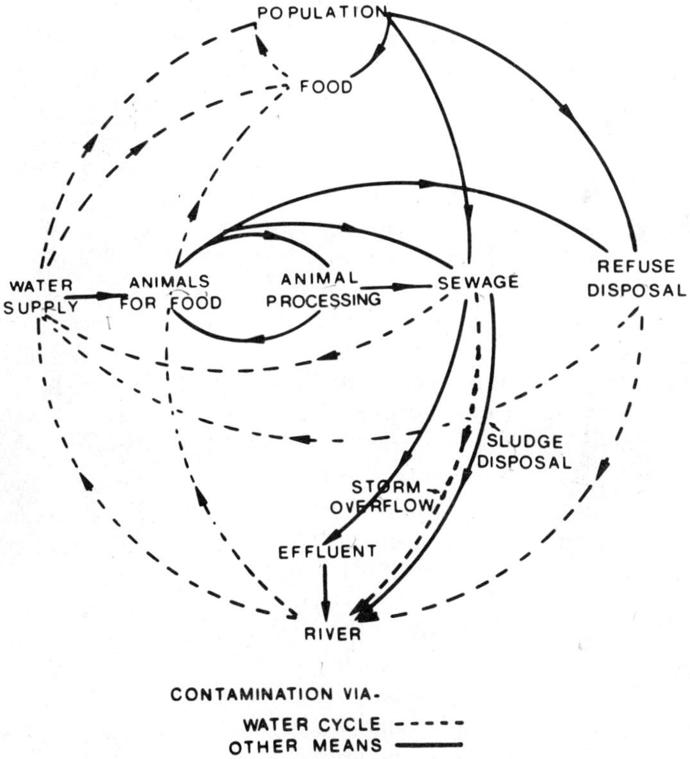

FIG. 1—*The* Salmonella *cycle.*

tial risk to the health of the community. In general, the risks are least when sewage effluent after efficient treatment is discharged to natural waters where sufficient dilution allows the natural processes of self-purification to take place; they are greatest where crude sewage is discharged to natural waters of limited volume. They arise either directly from the ingestion of pathogenic organisms or from the consumption of foods contaminated by sewage or sewage-polluted water during growth or preparation. The risks are well documented and have been long recognized. They include enteric fever and infective hepatitis from shellfish grown or stored in sewage-polluted water, infections transmitted by vegetables either fertilized with crude sewage or sludge or irrigated or sprayed with polluted water, by raw milk similarly contaminated, and by canned foods cooled after sterilization with sewage polluted water. These methods of transmission are limited but recurrent.

That pathogens such as salmonellae can be recovered from sewage effluents invariably and that most of them remain viable for some time in natural waters leads to questions about the disinfection of effluents. Apart from possible recreational risks, this has a number of practical drawbacks. With effluents typical of those produced at conventional sewage works in Britain, the dosage of chlorine required, and the costs thus involved, would be comparatively high especially where substantial concentrations of ammonia occur. The toxic effect of chlorine on fish would require the provision of facilities to remove any excess, and the possible ecological consequences of inevitable accidental failures would need careful consideration. Cyanogen chloride, which is also highly toxic to fish, may be produced by chlorination of certain industrial effluents. The decrease in numbers of nonpathogenic organisms after disinfection may itself reduce the rate of self-purification of receiving waters [*3*]. In addition, rapid bacterial aftergrowths may occur, creating an increased oxygen demand and resulting in less dissolved oxygen in the water, thus also hindering the processes of self-purification. Further undesirable effects of chlorinating sewage effluents where receiving waters are subsequently used as a source of drinking water include the formation of chlorophenols with their attendant taste problems, as well as potentially carcinogenic organochlorine derivatives which are relatively resistant to biodegradation. Also disinfection of effluents can not prevent contamination of receiving waters with pathogens from other sources. For example, the report of the Metropolitan Water Board (1969-1970) [*14*] and recent investigations in Yorkshire by Fennell et al [*15*] have shown that seagulls can cause significant contamination of filter beds and storage reservoirs with salmonellae by carrying infected material from their feeding grounds at refuse tips and sewer outfalls to the water on which they roost.

In Britain, disinfection is not generally used, and there are no bacterial standards for sewage effluents discharged after conventional treatment to surface waters. Instead, stringent criteria are recommended for potable waters [*13*]. To ensure absence of pathogenic organisms, the criteria accepted for potable supplies is the absence of *Escherichia coli* from 100 ml of water. As the numbers of *E. coli* in wet feces may approach 10^9/g, this standard implies a dilution of one part of crude sewage to 10^9 parts of water. Such a dilution is rarely obtained or necessary. The numbers of intestinal organisms in effluents are reduced greatly after discharge to natural waters by sedimentation, exposure to light, starvation through lack of nutrients in adequate concentration, and by various predatory organisms, including Myxobacteria, *Bdellovibrio*, Protozoa, Rotifera, *Daphnia*, *Cyclops*, and other zooplankton. Intestinal organisms thus survive in water for a period but do not usually multiply. This is to be expected as they are highly specialized parasites, growing in a high concentration of nutrients in darkness at a constant temperature of 37°C. In fresh water, their survival time may be measured in weeks, depending largely on the initial numbers discharged; in seawater, the period is measured in hours. Extensive work by Gameson et al [*16*] on the survival and distribution of coliform organisms in coastal waters in relation to the discharge of crude sewage revealed wide variations between samples taken at different times. More recently, Gameson [*17*] has shown that at least in one area maximum pollution seemed to be related to droppings from seagulls during the nesting season. This would invalidate the application of coliform standards for sea outfalls.

Monitoring and Surveillance

Before discussing the role and application of indicator bacteria to environmental pollution, it may be useful to define the terms monitoring and surveillance. Investigations are often made with different objectives. For example, short-term studies may be required to determine the microbial content or effects of certain pollutants in particular circumstances to allow prompt action—as in the control of drinking water supplies. In contrast, long-term studies may be needed to ascertain general trends. Thus, monitoring may be regarded as the regular and systematic measurement of specified microbial populations in the environment to ensure that particular criteria are not only being attained but maintained. Surveillance, on the other hand, follows the incidence and distribution of microbial pollutants over a period of time, and it is often accompanied by other measurements which may aid interpretation of data. The aim of surveillance is thus to determine trends over a long period, and it rarely requires regulatory action. Routine or continuing measurements of environmental parameters made either for the purpose of monitoring or surveillance are often costly in terms of laboratory facilities and other resources, and the observations required should be the minimum consistent with the objectives. To achieve this in a new situation, a preliminary investigation or baseline study is usually needed to decide whether monitoring or surveillance is required as well as the best kind of samples, how they should be collected, the location and frequency of sampling, the choice of organisms to be sought, and the isolation methods to be used. Different organisms may be used as indicators in different situations, both for monitoring and surveillance, and these criteria must be considered in relation to the objectives. Whatever indicator organisms are chosen, ease and accuracy of identification, as well as maximum recovery, reproducibility, and comparability of results are all important.

Indicator Bacteria and Health Hazards

Indicator organisms can be defined only in terms of what they are intended to indicate. Thus, the presence of any pathogens clearly indicates a health risk, however remote, just as the presence of smoke denotes fire. Since seeking pathogens in water is usually impracticable for monitoring purposes, other bacteria are used to indicate the presence and extent of pollution. Ideally, such indicator bacteria should cover the possible presence of all pathogenic organisms. They also should be abundant in the substance for which their presence serves as an indicator, absent or, at least, very few in number from all other sources, capable of easy isolation and numerical estimation, and unable to grow in the aquatic environment [*18*]. To these can be added greater resistance than pathogens to disinfectants and the aqueous environment [*19*]. In practice, these criteria cannot always be met, but, in Britain, they are virtually fulfilled by the use of *E. coli* as an indicator of pollution by fecal material of human or animal origin. *E. coli* is so abundantly present in the feces of man and warm-blooded animals that its presence in water may be used to detect recent fecal or sewage pollution—its numbers indicating the degree of pollution, and its absence the probable freedom from pathogenic organisms excreted through the intestinal tract—and, thus, the saftey of the water. Its presence, however, denotes no more than the potential presence of pathogenic organisms.

Although "fecal coliforms" also are described for this purpose, the term as currently defined comprises all coliform organisms which are capable of producing acid and gas from lactose at 44°C. These include many coliforms which are not fecal in origin and which can multiply in the aquatic environment under suitable conditions. *E. coli*, in contrast, is an obligate intestinal parasite of man and animals. It fails to multiply in water in which its survival time is limited, and, in sewage, it vastly outnumbers any pathogens. In Britain, the use of *E. coli* is therefore considered preferable to fecal coliforms as an indicator of fecal pollution.

The disadvantages of its use are few. Although impossible to distinguish between *E. coli* of human or animal origin, probably because they are identical, it is not essential because animals are known to excrete many human pathogens. The presence of *E. coli* in a natural water therefore requires topographical examination of the source to determine the origin and nature of the pollution. The microbiological laboratory must not be used instead of field inspection and commonsense engineering hygiene. When supplementary information is needed for interpretation of equivocal data, the use of additional bacterial indicators, such as the presence and types of fecal streptococci, may be helpful. Apart, however, from surveillance and special investigations including appraisal of new water sources, examination for fecal streptococci and *Clostridium perfringens (welchii)* are not often used in routine practice in Britain.

Environmental Health Hazards from Microbial Pollution

In Britain now, despite the increasing need for recycling wastewater as well as changing attitudes towards recreational uses and environmental pollution, infections associated with water are rare—indeed, this is taken for granted [*20*]. When considering potential health hazards, however, it is important to appreciate that the presence of a pathogen does not mean that disease will necessarily ensue or is even likely to occur unless a significant epidemiological relationship with the incidence of a specific infection has been either observed or can be established. The role of indicator organisms in the control of potential health hazards from water should be considered in the context of the questions posed by Moore [*21*]. First, how should one assess whether a polluted environment is causing or, is likely to cause, infection and, second, having established suitable criteria, what are the health hazards from microbial pollution of an environment, and how should they be tackled?

As Moore has indicated, a cause and effect relation between pollution and disease is not always simple or easily determined. The possibility usually starts from an observation that a particular disease occurs more often than "normal" under certain circumstances—such as exposure to a polluted environment—and a common source is then sought. This does not, of course, mean that all those exposed will suffer disease, otherwise it would be self-evident. Indeed, it may well require retrospective or prospective surveys to yield a statistically significant association. Even then, it is not necessarily one of cause and effect, and critical examination in the light of criteria cited by Bradford Hill [*22*] is required. The more of these satisfied, the greater the probability of a cause and effect being related to each other. These criteria include the strength of the association, frequency of previous observations, nonassociation with other environmental factors or of the suspect pollutant with other diseases, occurrence of the disease only following and not preceding exposure to the suspect pollutant, increasing incidence of disease with increasing exposure, compatibility of the association (or at least nonconfliction) with the known natural history of the disease, supporting experimental evidence, and close analogy with some other similar situation.

These criteria are equally applicable to any environmental hazard, but as Moore [*21*] has so ably pointed out the assessment of infection risks requires even further consideration.

Epidemiology and the Host-Parasite Relationship

Microbial infection is a complex host-parasite reaction. There is no absolute relation between the presence of a pathogen and the risks of contracting the associated disease, even in susceptible hosts. Bacteriologists face pathogens every day, but simple practical codes of safety usually prevent acquisition of the associated diseases. Uncritical bacteriology assumes that a so-called pathogen isolated from a clinical specimen is the cause of the patient's current illness. This is often true, but the actual conclusion is rarely justified without additional information. Each case must be considered on merit, and no general statement can cover all the combinations of possible or relevant variables. This complex biological situation, however, is very important when assessing the significance of pathogens in the environment. For example, Perry et al [*23*] showed that haemolytic streptococci liberated into the environment rapidly lost their virulence and became irrelevant in the epidemiology of streptococcal infection. *Staphylococcus aureus*, although seriously cited by Brisou [*24*] as a possible seawater hazard, is normally present in the nose or skin of almost half the population; the risk of developing staphylococcal infections therefore depends on personal susceptibility rather than simply on exposure to this organism in the environment. Similarly, *C. perfringens (welchii)* is ubiquitous, but clostridial disease will only occur if devitalized tissues provide anaerobic conditions necessary for growth, and, even then, the outcome cannot be predicted. Wild poliovirus can be isolated from polluted waters only when the disease is present in the population, but the chances of actual infection occurring are vastly greater through personal contact than from pollution of the environment. Basic knowledge of the reservoir of a pathogen, its origin, mode of spread, route of infection—oral, inhalation, cutaneous, or via mucous membranes—and the susceptibility of the community or individual are all essential for the control of communicable disease. Indeed, they are so important, that not only does each health hazard require separate consideration, but each disease must be carefully defined. Since the presence of pathogenic organisms in the environment usually indicates that the associated diseases are already present in the community, it is important not to blame environmental pollution if in fact the disease has been acquired from other sources, whether directly or indirectly. The acceptability and usefulness of certain species as suitable indicator bacteria thus can be assessed only in relation to specific diseases. All things are relative. There can be no absolute guarantee either of health or freedom from disease, and accurate epidemiological information is essential for evaluation of the risks. Neither unusual diseases nor common, but trivial, infections should be sought in order to justify arbitrary bacterial standards without valid medical or veterinary reasons.

Water Quality Criteria in Britain

The current concept of bacterial indicator organisms as applied to drinking water, swimming pools, bathing beaches, and shellfish waters in Britain are discussed from the viewpoint of a medical and public health bacteriologist concerned with water quality and food hygiene.

Drinking Water

It is axiomatic that no piped supply of adequately treated drinking water should contain any microorganisms known to be virulent. It is inevitable, however, that it will sometimes contain microorganisms which may occasionally cause opportunist disease in vulnerable persons in special situations. The use of indicator bacteria rather than pathogens for monitoring the contamination of potable supplies is universally accepted despite controversy about the choice of organisms or combination of organisms, the best media and methods for the maximum recovery of stressed or damaged organisms, and possible indicator organisms suitable for viral safety. In Britain, emphasis always has been placed on adequate purification and safety of water for drinking rather than on the more elusive target of purifying the original sources. However polluted the raw water, it can in principle be purified, although the best available source is always used. This may indeed be at a price [*25*] but that price may well be less than trying to apply bacterial standards to effluents. British policy has traditionally and effectively monitored the efficiency of treatment and, thus, the safety of the supply rather than the degree of pollution of the raw water. This does not mean that knowledge of the bacterial quality and flora of the untreated water is not desirable—such surveillance would be no more than good practice. But it does imply that extensive monitoring of effluents and water for abstraction merely to comply with numerical and often arbitrary standards may yield expensive information of limited value. Monitoring waters used for abstraction, however, will be required in the future to conform with standards accepted by the European Economic Community (EEC) [*26*].

Although there are no mandatory standards for drinking water, British water supplies are second to none in quality and safety. There is, therefore, no valid reason to change either the views or techniques described in "The Bacteriological Examination of Water Supplies" (see Appendix I). Indeed, before even contemplating any significant change, it would need to be shown by experiment and extensive comparative trials that one area of equivocal results was not simply to be exchanged for another. Since this report has been the mainstay of British practice for nearly half a century, relevant sections from the current edition are cited verbatim in Appendix I. Apart from emphasizing the importance of examining potable waters regularly and frequently by a simple test, they indicate the reasons and work behind the choice of indicator bacteria, media, and the methods recommended. The main emphasis is placed on the absence of any coliform organisms, including *E. coli*, in treated water as it enters the distribution system of piped supplies.

The last edition of Report 71 was published in 1969, and the extent to which its recommendations have been put into practice and some of the difficulties encountered have been collated by Burman [*27*]. It is evident even in Britain that so far there is no universally accepted best method—partly reflecting differences in water sources and partly in laboratory practice. Accuracy consistent with quick results is important, but the possible combinations of published media and methods is so great that a vast amount of work would be required to reach any valid conclusions using routine water samples. In Britain, the use of either natural or experimental chlorine-damaged organisms [*28*] has allowed such evaluations to be made with fewer samples and less time and effort, thus making essential comparison in field trials with actual water samples easier; the unqualified recommendation of Gray's improved formate lactose glutamate (IFLG) medium [29] for the isolation of coliform organisms and *E. coli* in multiple tube tests was made in this way [30]. One of the advantages of this chemically defined medium is its ability to encourage the growth of attenuated *E. coli* which would not grow in other media. The use of simulated samples in water quality control work [31] is being developed in order to achieve greater efficiency and uniformity on an interlaboratory and intercountry basis. Current research includes the evaluation of single tube confirmatory tests for *E. coli*; the identification, enumeration and significance of miscellaneous organisms including actinomycetes and fungi in taste and odor problems, as well as in the maintenance of water quality; and development of materials and testing to ensure that they are inert and do not support the growth of organisms, including pseudomonads in distribution systems [*32,33*]. A suitable substitute will need to be found for Teepol 610, manufacture of which has now ceased. Such changes will be incorporated in the next revision of Report 71. The resistance of viruses to treatment and disinfection and their possible epidemiological significance is important [*34*]. No virus has been isolated yet from a treated piped supply in Britain, and the current position is one of research and surveillance by certain laboratories.

Recreational Waters

This all embracing term covers many different kinds of water used in many different ways. Again, it must be emphasized that customs differ with geography, climate, populations, and their standards of living. It would be difficult, if not impossible, therefore, to lay down meaningful universal criteria, but it is important that health risks should not be extrapolated from particular circumstances to general situations. In Britain, there are no bacterial standards for such waters, although monitoring of bathing waters again will be necessary in the future to comply with the standards agreed by the European Economic Community [35].

Swimming Pools

The current position in Britain is stated in "The Purification of the Waters of Swimming Pools" (see Appendix II). Relevant sections regarding microbial pollution and bacteriological standards are again quoted verbatin (Appendix II) as to do otherwise would be merely playing with words. In an efficiently maintained pool, the water should be normally free from coliform organisms and *E. coli* from 100-ml samples. There are no firm recommendations about the frequency of sampling as individual situations and circumstances vary considerably. It must be emphasized that these bacterial tests check the efficiency of purification and not the health risks. For many infections, such as mycobacterial granulomata, verrucae, and tinea pedis, known to be spread at swimming pools, there are no suitable indicator organisms available. Indeed, regular determinations of pH and free chlorine are probably far more important. This report also contains much useful information on methods of purification, disinfection, and general management.

Bathing Beaches

The potential health hazards of bathing in sewage polluted seawater have been under investigation in Britain for many years. Following extensive outbreaks of poliomyelitis, a working party of the Public Health Laboratory Service (PHLS) was formed in 1953 to investigate the possibility of a causal relationship between this disease and bathing. Despite extensive work over several years, they were unable to find any significant association. The only evidence of health hazards from sea bathing found was four cases of paratyphoid fever all associated with beaches grossly polluted with fecal matter [36]. The international literature on health hazards from bathing has been extensively reviewed by Moore [*21, 37, 38*] and Grunnett [*39*]. It is apparent that the risk of contracting serious disease is minimal, even if morbidity in bathers may be higher than in nonbathers, though Moore [*40*] would dispute the latter.

The case for microbial standards for bathing beaches has been well put by Shuval [*41*]. Since significant epidemiological data are lacking, the case rests on the survival of pathogens, including viruses, compared with coliform organisms discharged in crude sewage to the sea; the volume of water possibly swallowed; and the chances of any individual actually ingesting a pathogen in sufficient numbers to cause illness. These arguments, though logical, are somewhat tenuous. Indeed, Shuval indicates the difficulty of establishing a rational bacterial standard but points out that a dilution of sewage about 10^{-4} would be technically feasible. Assuming further reduction in numbers of coliforms and *E. coli* due to natural factors, he suggests that a standard of 1000 coliforms/100 ml could be applied. He further suggests that, with improved viral isolation techniques [*42*], a microbial standard for bathing beaches based on the absence of enteric viruses in about 100 litres would be more logical because of their longer survival time and their low minimal infectious dose. He states that such criteria should not be regarded as sacred limits but as guidelines towards achieving an optimal environment for man. However, Ingolls [*43*], in considering the role of disinfection, pertinently asks what is an optimum environment, how does one control it, and for whom?

Shuval accepts the difficulty of setting rational standards to eliminate the health risks involved. But what are the health risks involved? Each must be clearly defined and then related to geography, climate, and population and evaluated in the light of the known incidence of the defined infections in the population and their known modes of transmission. These points are discussed fully by Moore [*40*] in presenting the case against setting microbial standards for bathing beaches. It is clearly important to reduce potential hazards to acceptable levels, but no satisfactory significant evidence linking the degree of sewage contamination of coastal or other waters with defined diseases has been described yet, and, to that extent, setting microbial standards for bathing water is irrelevant to public health. Indeed, "by a judicious selection of sampling points and times, a considerable upgrading or downgrading of a given beach could easily be arranged at will" [*36*]. During the past twelve years, surveys by the Water Pollution Research Laboratory workers at a number of beaches around the English coast have reinforced these doubts about the validity of bacteriological standards at least for beaches in Britain [*17*]. Although Grunnett [*39*] considers that Britain is a special case, he nevertheless found that higher morbidity in bathers was independent of the degree of sewage pollution as measured by numbers of *E. coli*, which he regarded as a good indicator organism for this purpose. Much of this morbidity, often trivial, was due to autoinfection for which there is no simple solution. There is, however, no argument against the need for satisfactory aesthetic standards which is compatible with the accepted public health approach that the extent of contamination with fecal microorganisms should be as low as possible consistent with feasibility. For example, in surveillance studies, bathing areas with *E. coli* counts consistently less than 100/100 ml could be regarded as highly satisfactory [*44*].

Vibrios

Vibrios do not represent a waterborne hazard in Britain, and the possibility that cholera might be spread by the water cycle from imported cases is negligible. Any spread is far more likely to occur, and then only to a limited extent, from personal contact with a case—possibly even indirectly via food in the home. In the same way, the potential risk of contracting cholera from recreational waters can only occur if the disease is already present in the communtiy to a significant extent. The possible hazard is thus more theoretical than real. Although more attention is now being paid to noncholera or NAG vibrios—the so-called "water vibrios" as a cause of occasional sporadic and probably ultimately food-borne illness no cases of such infections acquired in Britain have yet been reported. Unless present in large numbers, these organisms have little significance to man in drinking or recreational waters, even allowing for the possiblity that small numbers swallowed in water may pass unaffected through the acid barrier of the stomach, multiply in the small intestine, and then cause infection. However, the role of plasmids in the environment may possibly become more significant in the future.

Apart from recreational uses, many coastal waters are important to the fishery industry. So far as microbial disease in relation to seawater quality is concerned, the main health hazards to man arise from molluscan shellfish such as oysters. These molluscs grow in coastal and estuarine waters, often subject to pollution by sewage effluents. As a large proportion of such shellfish are eaten raw or only lightly processed, infections may be clearly transmitted in this way. These include, but are not limited to, typhoid and paratyphoid fever and infective viral hepatitis. Thus, particular care is required to ensure that molluscs harvested from such areas are suitable for human consumption or are rendered safe by subsequent handling techniques. The suitability of raw shellfish for consumption without further treatment will depend upon the quality of the waters from which the shellfish are derived. Water quality can be assessed either by direct examination or by examination of the shellfish or by a combination of both methods. Baseline studies and surveillance of the growing areas and the shellfish are necessary to determine the degree of sewage contamination. Some countries define what is to be regarded as growing water of acceptable bacteriological quality and thus establish the suitability of harvesting areas. Others define the acceptability of these areas after bacteriological examination of harvested shellfish. The consumption of shellfish from hygienically unacceptable areas should be prohibited, unless they are subjected to approved treatment processes to remove fecal organisms. This can be achieved either by relaying the shellfish in areas where the seawater in unpolluted or holding them in basins or tanks under conditions that allow the shellfish to rid themselves physiologically of polluting bacteria. For this process of purification (depuration), shellfish must be held in tanks of seawater under conditions that allow them to live and function normally. Ideally, tanks should be situated where clean seawater of adequate salinity occurs naturally. In Britain, most purification plants are situated in sewage polluted areas, and the seawater used is usually treated by ultraviolet light for oysters and by chlorination for mussels [*45*].

Although no official standards or methods are laid down in Britain, final control is based not on the numbers in the growing waters but on the *E. coli* content of purified shellfish using recommended methods [*46,47*]. This organism is used to assess the efficiency of purification and thus, the safety of shellfish for human consumption. Although still common in some Mediterranean countries, no cases of enteric fever in Britain have been attributed to shellfish for many years. With regard to enteroviruses in shellfish, on statistical grounds, the absence of *E. coli*, or the presence of only a few, is likely to give adequate protection, and there is no evidence to suggest that this assumption is not true in Britain. It is self-evident that if shellfish, or any foods, are consumed raw, there can be no absolute guarantee of complete safety, and occasional bacterial or viral infections will inevitably occur.

In contrast to contamination with fecal organisms, the marine environment has its own natural flora of microorganisms. These include marine vibrios, among which *V. parahaemolyticus* is a known cause of gastroenteritis in man. Epidemiologically, infection with this organism is almost invariably food-borne, usually from seafoods [*48-50*]. In contrast to many other infections, spread from one person to another is probably not important. In the marine environment, *V. parahaemolyticus* is usually greatly outnumbered by the closely related species *V. alginolyticus*, and, because of this, Barrow and Miller [*50,51*] suggested that marine organisms could be used as additional indicators for assessing the hygienic quality of both raw and processed seafoods. However, as far as seawater itself is concerned, they are normal inhabitants, though *V. parahaemolyticus* presumably represents a potential health hazard if seawater is ingested. However, no case of gastroenteritis caused by *V. parahaemolyticus* in this way has been reported so far among bathers or others at risk. Although a few instances of opportunist infections of skin lesions with closely related organisms have been reported in bathers and fish handlers, the potential recreational health risks are insignificant in relation to its importance as a cause of food poisoning due to poor hygiene practice.

Viruses

Viruses of human or animal origin excreted in feces may be transmitted by water polluted with sewage at any stage of the water cycle. Those which are normally transmitted within communities by the fecal-oral route will presumably also be those most likely to be transmitted by water. These include the enterovirus group (polio, Coxsackie A and B, and ECHO viruses) as well as adeno and reo viruses, hepatitis A [*52*] and, perhaps, the recently identified rota and corona viruses. Others, such as herpes and rhinoviruses also occur but are less resistant to environmental stress, including water treatment. However, those which survive longest may not be the most virulent—quite apart from the question of infectious doses. The enteroviruses are probably best able to survive in water and may remain viable after coliform organisms have succumbed. It follows that the detection of virus particles in any water in the absence of coliforms or other fecal organisms confirms sewage-pollution and reflects current infection either in the human or animal community. Although the presence of *E. coli* indicates sewage pollution, it does not necessarily follow that enteroviruses are also present, although such an assumption would err on the side of safety.

Viruses can be detected readily in sewage treated effluents and receiving waters in Britain [*53*], but there have not been any large outbreaks which in theory might be expected. Although occasional outbreaks of infection have occurred—such as that described by Green et al [7] which involved *Shigella sonnei* and possibly three different enteroviruses—all have been due to failures of chlorination or to subsequent contamination of the water in supply. In contrast, the significance of very small numbers of virus particles surviving normal treatment, including storage, filtration, and disinfection, and passing into a public water supply is unknown but must be assessed both for reassurance and epidemiological reasons. Much useful information on viral inactivation, transmission, and effluent purification in relation to numbers has been collated by Berg [6,54] who takes the approach that any infective virus particle in drinking water is a risk which should be eliminated. This has stimulated development of concentration and isolation techniques [55,56]. Although it may be true that any virus particle may represent a potential hazard, and apart from problems of intermittent pollution, only the virus of hepatitis A has so far been unequivocally associated with water, almost always grossly polluted. In Britain, however, there is no convincing evidence yet of viral infections acquired from adequately treated drinking water, although the epidemiological approach has been criticized on the grounds that very small virus numbers may cause only sporadic infection, thereby allowing the water route to remain unchallenged. It should be remembered, however, that most if not all the common virus infections are spread by means other than water, and, if occasional waterborne infections do occur, they are likely to be obscured by other cases in the community from the more usual modes of transmission. No matter how often viruses may be isolated from water or sewage, it will be very difficult to prove by epidemiology alone that sporadic viral infections are in fact waterborne. If this is so, it follows that the degree of risk is much greater from recreational waters, often obviously polluted. What, if anything, should be done about them? Indeed, taking infective hepatitis as a useful model, the fact is that in Britain the incidence of clinical infections has fallen steadily during the last decade, yet bathing in sewage polluted water is just as popular as ever. From the public health point of view, good hygiene and sanitary facilities are probably much more important for preventing the spread of virus infections than attempting to reduce measurable pollution of the water cycle by creating viral standards.

Future

Conservation of water resources and the safety of public health are inseparable. Coliforms and *E. coli* may be an inadequate index of water quality [25], but, in Britain, they are probably still the best choice for monitoring the safety of water. Do we need or want sterile water for drinking? If cellular immunity has any meaning, perhaps like fluoride, attenuated bacteria or viruses should be deliberately introduced into water supplies in order to prevent clinical infections or even marker organisms into treatment plants to further satisfy ourselves about its safety. When developing and assessing any criteria for the bacterial quality and safety of any water for any purpose, medical and veterinary perspectives must always be taken into account. Only in this way can ideals be reconciled with practicalities.

Let us reach for the sky by all means, but with both feet on the ground and our heads preferably above water.

Acknowledgments

I am grateful to Dr. N. P. Burman, Dr. J. A. Rycroft, and Mr. D. C. Miller for their help and advice in the preparation of this paper.

APPENDIX I—THE BACTERIOLOGICAL EXAMINATION OF WATER SUPPLIES[3]

Reports on Public Health and Medical Subjects No. 71

Her Majesty's Stationery Office, London, 1969

This report has been prepared with the following objects:

1. To outline the principles on which the bacteriological examination of water supplies is based and the way in which the results should be interpreted.
2. To suggest how frequently water should be examined bacteriologically.
3. To explain the precautions necessary in obtaining samples and transmitting them to the laboratory.
4. To describe techniques, the adoption of which will ensure the efficiency of laboratory practice and the comparability of results.

The examination of a water supply properly embodies four lines of investigation—topographical, chemical, biological and bacteriological—each having its uses and indications and each yielding information not otherwise obtainable. This report deals only with the bacteriological aspects.

Bacteriology offers the most delicate test for the detection of recent and therefore potentially dangerous faecal pollution: this is the chief function of the bacteriological examination. Chemical analysis, though lacking the sensitivity of bacteriology in this respect, may nevertheless assist in the hygienic assessment, particularly of proposed new supplies, by revealing past pollution too remote for detection by bacteriological methods.

The information derived from bacteriological tests must be assessed in the light of thorough knowledge of the conditions at the sources of supply, throughout the stages of treatment to which the raw water may be subjected, and in the distribution system.

Contamination by sewage or by human excrement is the greatest danger associated with drinking water. If such contamination has occurred recently, and if among the contributors there are cases or carriers of such infectious diseases as typhoid fever or dysentery, the water may contain the living micro-organisms which cause these diseases and the drinking of such water may result in fresh cases. Sewage-polluted water may also contain the viruses of poliomyelitis, other viruses of the Enterovirus group, or the virus of infectious hepatitis. Animals and birds, particularly seagulls may carry human intestinal organisms pathogenic to man, and the importance of this source of pollution must not be overlooked. Apart from the drinking of contaminated water, its use in the preparation of food, which may allow the multiplication of intestinal pathogens, presents an obvious danger.

The direct search for the presence of specific pathogenic bacteria or viruses in

[3]Reprinted with the permission of the Controller of Her Britannic Majesty's Stationery Office London, England.

water is impracticable for routine control purposes. There are several reasons for this. Pathogens present in water are usually greatly out-numbered by the normal intestinal organisms, and tend to die out more rapidly. Although it is possible to isolate pathogens from water, especially heavily polluted water, large volumes of of water (several litres) may need to be examined, selective media are required for their isolation, and their identification involves biochemical and serological tests on pure cultures. The isolation of viruses requires even more difficult and lengthy procedures. Water bacteriologists have therefore evolved simple and rapid tests for the detection of normal intestinal organisms (e.g. coliform bacteria, faecal streptococci and *Cl. perfringens* (*Cl. welchii*) which are easier to isolate and identify. The presence of normal faecal organisms in a water sample indicates that pathogens *could* be present, but in much smaller numbers. The absence of faecal organisms indicates that pathogens also are probably absent.

Furthermore, if a given supply receives a single contamination from a typhoid carrier, for example, it will probably be a fortnight before a case of typhoid fever develops and another week or longer before it is diagnosed and reported to the Medical Officer of Health. It is improbable that the bacteriologist, after this lapse of time, will succeed in demonstrating the presence of typhoid bacteria. If the contamination is repeated or continuous, then the chance of finding the specific organism is rather greater, but in practice suspicion is usually thrown on the water as a result of epidemiological rather than of bacteriological enquiry. Even if the search for pathogenic organisms were practicable it would not really be suitable as a routine test because what the bacteriologist is concerned with is not so much whether the water does contain pathogenic organisms as whether it could do so. Search for normal faecal organisms thus provides a much wider margin of safety.

A single laboratory examination of any water, whether raw or treated and however favourable the result, does not justify the conclusion that all is well and the supply suitable for drinking purposes. Contamination is often intermittent and may not be revealed by the examination of a single sample. The impression of security given by bacteriological testing of a water at lengthy intervals may, therefore, be quite false. Indeed, the value of bacteriological tests is dependent upon their frequent and regular application. **It is far more important to examine a supply frequently by a simple test than occasionally by a more complicated test or series of tests.** Information gained in the course of years will provide a standard for any particular source of water, a lapse from which must at once arouse suspicion. The most a bacteriological report can prove is that at the time of examination bacteria indicating excretal pollution did or did not grow under laboratory conditions from a sample of the water. It must be emphasized that, when local inspection shows a water as distributed to be obviously subject to pollution, the water should be condemned irrespective of the result of the bacteriological examination.

The organisms most commonly used as indicators of faecal pollution are the coliform group as a whole, and particularly *Escherichia coli.*

> The term "coliform organisms" as used throughout this report refers to Gram-negative, oxidase-negative, non-sporing rods capable of growing aerobically on an agar medium containing bile salts and able to ferment lactose within 48 hours at 37°C with the production of both acid and gas. The group of coliform bacteria fermenting lactose at 30°C is fairly widespread in certain forms of vegetation and is not considered to be of any particular epidemiological importance in the examination of waters. *Escherichia coli* is a coliform organism, as defined above, which is capable of fermenting lactose with the production of acid and gas at both 37°C and 44°C in less than 48 hours; which produces indole in peptone water containing tryptophan; which is incapable of utilizing sodium citrate as its sole source of carbon; which is incapable of producing acetyl methyl carbinol; and which gives a positive result in the methyl red test.

E. coli is undoubtedly of faecal origin, but the precise significance of the presence in water of the other members of the coliform group in water has been much debated (Thresh *et al.*, 1958). All the members of the coliform group (as here defined) *may* be of faecal origin and an explanation of their presence must be sought; unless it can be proved otherwise it should be assumed that they are all of faecal origin.

The Examination for Coliform Organisms

Since coliform bacteria are present in large numbers in faeces and sewage and can be detected in numbers as small as one in 100 ml of water, they are the most sensitive indicators at our disposal for demonstrating the excretal contamination of water. For this reason careful estimation has to be made of the numbers of coliform organisms and of *E. coli* present before deciding whether the pollution is severe enough to render the water potentially dangerous. The whole essence of the test is quantitative.

The coliform examination is usually performed either by adding measured volumes of water to suitable liquid media or by filtering measured volumes of water through membrane filters. The two methods do not give strictly comparable results because, among other reasons, the membrane counts give no indication of gas-production from lactose.

Examination in liquid media consists of several steps, each yielding additional information. In the first stage, the presumptive coliform examination, measured volumes of water are added to a suitable medium in tubes which are then incubated at 37°C for 48 hours. From the number of tubes showing the presence of acid and gas the most probable number of presumed coliform bacteria present in 100 ml of the sample is calculated. This result is known as "the presumptive coliform count", the presumption being that each tube showing fermentation does, in fact, contain coliform organisms.

The next step is to confirm the presence of coliform organisms in each tube showing a presumptive positive reaction. With samples of unchlorinated water such confirmation is not generally necessary, since the presumption that each tube showing fermentation does contain coliform organisms is true in almost every instance. With chlorinated waters false presumptive reactions can be caused by spore-bearing organisms which are more resistant to chlorination than nonsporing organisms; and with samples of chlorinated water it is therefore necessary to confirm the presumption on every occasion.

It is also necessary to ascertain whether any of the coliform bacteria present are *E. coli.* A convenient and rapid test is available which depends on the fact that *E. coli* is almost the only coliform organism which is capable of producing gas from lactose at 44°C.

For examination by membrane filtration techniques, measured volumes of water are filtered through each of two sterile filter membranes. These are placed face upwards on an absorbent pad, saturated with a suitable liquid medium in a petri dish. Both membranes are incubated for a preliminary period at a relatively low temperature, usually four hours at 30°C, and then changed to a higher temperature, one at 35° or 37°C and one at 44°C. Acid-producing colonies are counted after a total incubation time of 18 hours. The results are respectively a presumptive membrane coliform count, and a presumptive membrane *E. coli* count. The presumption in this case is that the colonies are gas-producers as well as acid-producers, and in the case of *E. coli* at 44°C also indole-producers. This is liable to lead to an overestimate of the number of coliform organisms, but the error is on the side of safety and, moreover, the final result is obtained in a much shorter time than by the tube method. If desired, confirmation can be achieved by sub-culturing colonies from membranes to lactose peptone water for incubation at 37° or 44°C. An indole test can also be carried out at 44°C. These confirmatory procedures are not usually necessary, however, except for samples from the distribution system which should not contain any coliform bacteria or *E. coli.*

The Examination for Faecal Streptococci

The faecal streptococci include a number of species found in human and animal intestines, including *Streptococcus faecalis* and *Streptococcus faecium.* Improved techniques for their isolation and differentiation have led to a reassessment of their distribution and significance. The greatest value of the faecal streptococcus test lies in assessing the significance of doubtful results from the coliform test, particularly the occurrence of large numbers of coliform bacteria in the absence of *E. coli.* This is of value both in natural waters and in samples from repaired mains. The finding of faecal streptococci would confirm the faecal origin of the pollution. The greater resistance of faecal streptococci to marginal chlorination also increases their value in assessing pollution of repaired mains.

S. faecium is widespread in both man and animals. Although *S. faecalis* is found in some animals it is not so abundant as in man. These organisms are unlikely to multiply in water and in some circumstances they die out less rapidly than *E. coli.* A differential estimation of streptococcal species may therefore help in determining sources of pollution, the finding of *S. faecalis* probably indicating a human source. Since the greatest danger of direct infection from water comes from human pollution, this differentiation would be of value. The absence of *S. faecalis* in the presence of other faecal streptococci or of coliform bacteria other than *E. coli* should be interpreted with caution.

The Examination for Clostridium perfringens (Cl. welchii)

The test for *Cl. perfringens* has uses similar and additional to the examination for faecal streptococci. *Cl. perfringens* forms spores which survive for a much longer time than the vegetative organisms of the coliform group and usually resist chlorination. The presence of *Cl. perfringens* in a natural water indicates that faecal contamination has occurred, and, in the absence of coliform organisms, that the contamination occurred at some remote time. In the absence of *E. coli,* the occurrence of *Cl. perfringens* in water together with coliform organisms removes any doubt about the faecal nature of the pollution but the combined findings would suggest that this has not been recent. The test is useful in detecting remote or intermittent pollution, especially in shallow well waters supplying a small population where frequent examination by the coliform test is not practicable. Infrequent examination for coliform bacteria may yield misleading results, because these non-sporing organisms may have died out since the last occurrence of pollution. In such instances the demonstration of *Cl. perfringens* will show that the water is subject to faecal pollution.

Colony Counts on Yeast Extract Agar

Colony counts provide an estimate of general bacterial purity, which is of particular value when water is used industrially for the preparation of food and drink. They may also give forewarning of pollution, but the practice in this country is to concentrate on the demonstration of *E. coli* or other members of the coliform group in the water.

Technique of Bacteriological Examination

As the number of coliform organisms in water may be small, large volumes may have to be examined in order to detect a single organism. Direct inoculation techniques on solid agar media are therefore not practicable and other methods must be used. Two procedures available are the multiple tube method, sometimes known as the dilution method or the most probable number (MPN) method, and membrane filtration.

Multiple Tube Method

In the multiple tube method of counting bacteria, measured volumes of the water to be tested, or of one or more dilutions of it, are added to tubes containing a liquid differential medium. It is assumed that, on incubation, each tube which received one or more viable organisms in the inoculum will show growth and the differential reaction appropriate to the organisms sought and the medium used. Provided negative results occur in some tubes, the most probable number of organisms in the original sample may be estimated from the number of tubes giving a positive reaction; statistical tables of probability are normally used for this purpose.

Since coliform organisms ferment lactose with the production of acid and gas, the media used for the presumptive coliform count contain lactose and an indicator of acidity. The tubes of media also contain an inverted inner (Durham) tube for the detection of gas production. In the United Kingdom inhibitory substances are usually added to suppress the growth of other organisms which may be present in water, and thus make the medium more selective for coliform organisms. Inhibitors commonly used include bile salts, as in MacConkey broth, and Teepol 610 as in Jameson and Emberley's (1956) medium. Selective media can also be made by choosing chemically defined nutrients which can be utilized by only a limited number of bacteria as in the various modifications of Folpmers' (1948) glutamic acid medium. In all these media the production of acid and gas is presumed to be due to the growth of coliform organisms.

Choice of Medium

MacConkey broth has the disadvantage that the two main constituents, bile salts and peptone, vary considerably in their inhibitory and nutrient properties. The variability of the bile salts has been overcome by substituting Teepol 610 in Jameson and Emberley's medium, and the variability of all ingredients has been overcome in chemically-defined glutamate media. These media also have the advantage that Teepol 610 is much cheaper than bile salts and the ingredients of glutamate media are the cheapest of all.

A comparison of these media (PHLS Standing Committee on the Bacteriological Examination of Water Supplies, 1968*a*, 1969) showed that Jameson and Emberley's Teepol broth was a good alternative to MacConkey broth but that Gray's (1964) improved formate lactose glutamate medium gave results superior in most respects to MacConkey broth and Teepol broth and was superior to any of the other glutamate media which have been described. This work also extended previous work with MacConkey broth (PHLS Water Sub-committee 1953*a*) and confirmed that, with very few exceptions, the production of acid and gas in these three media when inoculated with unchlorinated water and incubated at 37° C for 48 hours indicates the presence of coliform organisms. Since the presumption that the production of acid and gas is due to the presence of coliform organisms requires further tests to confirm its correctness, this reaction is generally referred to as a "presumptive positive coliform reaction". The presumption is, however, in the United Kingdom, almost always correct for unchlorinated waters incubated at 37° C for 48 hours. Even though some aerobic spore-bearing organisms may cause false presumptive positive reactions in glutamate media, Gray's improved formate lactose glutamate is regarded as the most satisfactory medium for general use.

Incubation and Examination of the Cultures

The inoculated tubes should be incubated at 37° C and examined after 18–24 hours. All those tubes showing acid and a bubble of gas in the inverted inner (Durham) tube, and those in which gas appears on tapping, should be regarded as "presumptive positive" tubes. Each of these should then be subcultured to a tube of confirmatory medium and to a tube of peptone water both for incubation at 44°C in order to give a rapid indication of the presence of *E. coli*. At the same time a tube of confirmatory medium should also be inoculated for incubation at 37°C to confirm that the "presumptive positive tubes" do contain coliform organisms. The remaining tubes should be re-incubated and examined after another 24 hours. Any more tubes in which acid and gas formation become apparent should similarly be regarded as "presumptive positives" and examined for the presence of *E. coli* and for the confirmation of the presence of coliform organisms. With unchlorinated waters it may be sufficient to examine for the presence of *E. coli* without confirming the presence of other coliform organisms in all the presumptive positive tubes. With chlorinated waters, since false presumptive results may be due to the presence of aerobic or anaerobic spore-bearing bacilli, confirmation that coliform organisms are present is essential.

It has been recommended that acid production together with any detectable amount of gas should be regarded as a presumptive positive reaction. This is true of all tubes examined after 24 hours' incubation but when gas is first seen at the 48 hours' examination the acceptance of small amounts of gas may produce, with some water sources, too high a proportion of presumptive positive results which cannot be confirmed as being due to coliform organisms. In such circumstances it may be necessary to use the older rule that sufficient gas to fill the concavity of the Durham tube is the minimum needed before a presumptive positive result after 48 hours' incubation can be recorded. Some true positive results, however, may be missed if small amounts of gas are ignored (PHLS Standing Committee on the Bacteriological Examination of Water Supplies, 1968*a*).

The Sampling Error of the Coliform Count

Various workers, of whom the first was McCrady (1915), have put forward formulae, based on the laws of probability, with which to estimate the number of organisms that are present in 100 ml of water when any given proportion of the tubes inoculated show growth, or acid and gas production, or other characteristic change. The various mathematical approaches have been reviewed by Eisenhart and Wilson (1943); Cochran (1950) has given an excellent introduction to the principles involved in the estimation of bacterial densities by dilution methods; and McCrady (1918), Hoskins (1934) and Swaroop (1938, 1951) have published tables that are particularly suitable for use in water analysis (see Appendix C). By means of these tables it is possible to report the most probable number (MPN) of coliform organisms in 100 ml of water.

The multiple tube method has a very large sampling error. Confidence limits for the MPN are given in tables in International Standards for Drinking Water (WHO, 1963) which should be consulted if required. For the 11-tube (1 × 50 ml, 5 × 10 ml, 5 × 1 ml) method and for the 15-tube (5 × 10 ml, 5 × 1 ml, 5 × 0.1 ml) method the upper limit generally lies between twice and three times the MPN and the lower limit generally lies between a third and a quarter of the MPN. For any given estimation it is possible that the true result lies beyond these limits but this will occur only in an average of 5 percent of all such estimations and therefore the upper limit can, for practical purposes, be regarded as the maximum number of bacteria the sample might contain.

Membrane Filtration

Membranes have the advantage that the conditions of incubation can be easily varied to encourage the growth of attenuated or slow growing organisms. An initial short period of incubation at a low temperature, usually 4 hours at 30°C, can be followed by a further period at a higher temperature, either 35°, 37° or 44°C. If required the membrane may be easily transferred to a different medium for the second incubation. By these techniques it is possible to obtain within a total incubation time of 18 hours direct presumptive coliform and direct *E. coli* counts which do not depend on the use of probability tables.

Counts on membranes are, however, subject to statistical variation and replicate counts of the same water sample will not in general show the same number of organisms. If a count of organisms (C) of greater than twenty is observed approximate 95 percent confidence limits for the true number of organisms can be calculated as follows:

$$\text{The upper limit} \simeq C + 2\,(2 + \sqrt{C})$$

$$\text{The upper limit} \simeq C - 2\,(1 + \sqrt{C})$$

For example if 100 organisms were observed the true number of organisms in the water being tested would lie between 78 and 124. For counts of less than 20 the following values of the limits may be useful:

Membrane Count	*95% Confidence Limits*	
	Upper	*Lower*
1	5·6	0·025
5	11·7	1·6
10	18·4	4·8

Interpretation of Membrane Counts

When interpreting the results of membrane counts it must be remembered that neither gas nor indole production can be detected on membranes. On the other hand false results due to anaerobic spore-forming organisms will not occur and with the medium recommended false results due to aerobic spore-forming organisms will not occur either. The differentiation of gas-producing and non-gas-producing organisms is usually of little significance for treatment control purposes; it becomes of significance only in chlorinated samples and in samples collected in the distribution system which should contain no coliform organisms of any kind. From such samples, yellow colonies from membranes at 35° or 37°C should be subcultured to lactose peptone water to confirm gas production in 48 hours at 37°C, and yellow colonies from membranes at 44°C should be subcultured to lactose peptone water and peptone water to confirm gas and indole production at 44°C in 24 hours.

Choice of Volumes for Filtration

The coliform count and the *E. coli* count are made on separate volumes of water. All samples expected to contain less than 100 coliform organisms per 100 ml require the filtration of 100 ml of sample for each test. The volumes of polluted samples should if possible be so chosen that the number of colonies to be counted on the membranes lies between 10 and 100. In an unexpectedly grossly polluted sample, however, it is possible to give an approximate estimate of numbers even if they greatly exceed 100 per membrane. The use of ruled membranes is recommended to facilitate counting.

Choice of Media and Incubation Procedures

Most of the original work with membrane filters for the coliform group in the United Kingdom was carried out with modifications of MacConkey broth usually preceded by resuscitation on nutrient broth.

Bile salts however have a number of variable properties which influence growth on membranes in addition to the variability of their inhibitory properties. Some samples have a marked effect on colony colour, some encourage colonies to spread and some are apparently more inhibitory on membranes to coliform organisms than to some other organisms. These variabilities have not been entirely overcome even in the standardized bile salts No. 3 sold by some firms. Media containing Teepol as inhibitor are therefore recommended for the enumeration of coliform organisms by membrane filtration. With these media the use of a separate medium for resuscitation is unnecessary.

All the recommendations which follow are based on work carried out at the Metropolitan Water Board (Metropolitan Water Board, 1966a, 1967a). Further simplifications have since been made, so that only one medium is now recommended, namely 0·4 percent enriched Teepol broth (0·4 ET). An additional refinement in the preparation of this medium is to adsorb the basic solution with cellulose triacetate to remove possible traces of toxic ingredients that might become concentrated on the membrane (Burman, 1967*a*). Details of preparation of this medium are given in Appendix B8. In devising these membrane techniques the aim was to obtain a count of coliform bacteria and *E. coli* in 18 hours as high as that from a multiple tube technique in 48 hours at 37°C followed by subculture at 44°C for 24 hours, and which would at the same time restrict the count to the arbitrarily defined coliform group of the water bacteriologists.

In the Metropolitan Water Board Laboratories it was found that in unchlorinated waters or in waters that had become polluted after chlorination, provided no chlorine residue was present, membranes incubated on 0·4 ET for 4 hours at 30°C followed by 14 hours at 35°C generally gave higher counts of coliform organisms than the multiple tube method using either MacConkey broth or Gray's glutamate medium incubated for 48 hours at 37°C. Incubation at 37° instead of 35° gave a slightly lower count.

An *E. coli* count on unchlorinated waters equivalent to multiple tube counts in MacConkey broth was obtained by incubating the membranes on medium 0·4 ET for 4 hours at 30°C followed by 14 hours at 44° ± 0.25°C.

If rapid results are required the membranes may be examined after a total incubation time of 12 hours. If no colonies of any kind are present, a nil count can be assumed. If small colonies of indeterminate colour are present the membranes must be returned to the incubator for the full 18 hours. In a polluted sample a very high proportion of colonies, particularly of *E. coli* can be counted after as little as 10 hours at 44°C.

For the examination of chlorinated samples both for coliform organisms and *E. coli*, a lower initial temperature of incubation and longer total incubation times are required. For coliform organisms membranes should be incubated on 0·4 ET for 6 hours at 25° followed by 18 hours at 35°C. For *E. coli* membranes should be incubated on 0.4 ET for 6 hours at 25° followed by 18 hours at 44° ± 0.25°C.

Delayed Incubation

It has been stressed elsewhere in this report that water samples should be despatched to the laboratory and examined with the minimum of delay and certainly not more than six hours should elapse between sampling and examination. Where this is difficult to achieve, the problem can be overcome by filtering the sample through a membrane on site or in a local laboratory with limited facilities. The membrane is then placed in the normal way on an absorbent pad saturated with a transport medium (Panezai, Macklin and Coles, 1965). This is a very dilute medium on which the organisms survive but do not develop visible colonies in three days at room temperature. If polystyrene petri dishes are used they can be despatched by post to the central laboratory where the membranes should be transferred to Teepol medium and incubated for 4 hours at 30°C followed by 14 hours at 35°C or 44°C. Delays of three days have made little difference to counts of coliform organisms and *E. coli.*

Confirmation and Differentiation of Coliform Organisms

The further investigation of presumptive positive reactions can be considered as two problems. First because of the importance of detecting *E. coli* in water supplies a rapid and simple test is required to show whether or not *E. coli* is present, second it is necessary to confirm that other presumptive positive reactions are due to true coliform organisms. Further differentiation is rarely necessary but may be useful in showing that the water is regularly contaminated with the same organisms and so helping to trace the source of the contamination.

Classification of Coliform Organisms

In previous editions of this report the nomenclature adopted by Wilson and his colleagues (1935) was used to classify the various types of organisms in the coliform group. This classification was based on tests for the production of acid and gas from lactose at 44° C, together with indole-production and methyl-red, Voges-Proskauer and citrate-utilization tests, usually known collectively as the IMViC reactions. More recently changes have been introduced in bacterial nomenclature and, since taxonomists do not attach the same importance as do water bacteriologists to certain reactions including the fermentation of lactose, no direct translation into modern terminology is possible for the organisms listed in previous editions of this report. Complete identification in terms of modern nomenclature would require an extensive series of tests such as those described by Edwards and Ewing (1962) or Cowan and Steel (1965). The relatively simple IMViC tests at present used by water bacteriologists are usually sufficient for identifying any particular coliform organism found without giving it a specific name. As an example *E. coli*, if typed by the IMViC method, would be recorded as+ + − − 44° + and the organism from jute packing, formerly known as Irregular VI, would give the reactions− − + +44°+.

Rapid Detection of E. coli

The detection of *E. coli* depends on the ability of this organism to produce gas from lactose at 44° C. There is ample evidence that in the United Kingdom such gas production is practically specific for *E. coli*. Some other coliform organisms can produce gas at this temperature but few of them are able to produce indole and the indole test, also carried out at 44° C, provides a means of distinguishing most of them from *E. coli*.

With pure cultures, such as can be obtained from membrane counts, 1 percent lactose peptone water is suitable for showing gas production. However, when subculturing from presumptive positive tubes, which may contain spore-bearing organisms, some inhibitory substance which will prevent the growth of these organisms is required. Brilliant-green bile broth has been extensively used to suppress the growth of spore-bearing organisms. In trials carried out by the PHLS Standing Committee on the Bacteriological Examination of Water Supplies (1968*b*) 1 percent lactose ricinoleate broth gave better results than brilliant green bile broth and was less subject to the variability of ox bile or brilliant green.

Each presumptive positive tube should be subcultured to a tube of either 1 percent lactose ricinoleate broth or brilliant-green bile broth and incubated for 24 hours at 44° C. At the same time a tube of peptone water should be inoculated for production of indole after 24 hours' incubation at 44° C. If confirmation is required with membrane counts of *E. coli* grown at 44° C the colonies may be subcultured to 1 percent lactose peptone water and peptone water for incubation at 44° C. Although gas and indole production can frequently be detected after 6 hours' incubation the test should not be regarded as negative until after 24 hours' incubation (Taylor, 1955). For the test to work satisfactorily a temperature in the medium of 44° ±0·25° C is required.

In order to secure this, incubation in a carefully regulated water bath is essential. The variations in temperature in ordinary incubators, even if water jacketed, are too great to permit the maintenance of a constant temperature in the medium. It is useful to include positive and negative control organisms with each batch of tests. Care should be taken in selecting the peptone for the indole tests at 44°C.

Confirmatory Tests for Coliform Organisms

With unchlorinated waters it is usually sufficient to demonstrate the presence of *E. coli* by the method outlined above. With chlorinated waters it is essential to confirm the presence of coliform organisms by subculture to 1 percent lactose ricinoleate broth or brilliant-green bile broth for incubation at 37°C for 48 hours, in order to exclude false positive results due to aerobic or anaerobic gas-producing spore-bearing organisms. Production of gas in these media within 48 hours can be taken as sufficient confirmation that coliform organisms are, in fact, present. Further differentiation should only be necessary in special circumstances.

At the same time as these confirmatory tests are carried out it is advisable to plate out each positive tube derived from chlorinated water on a suitable solid medium such as MacConkey agar. When the tubes are examined for gas production the colonial morphology of the organisms present will then be available for inspection, and pure cultures will be easily obtainable should further differential tests be necessary. If differentiation of coliform organisms from unchlorinated water is required the presumptive positive tube may similarly be plated out on MacConkey agar.

Many organisms found in water differ from the coliform group as defined on page 3 of this report solely in their inability to produce gas from lactose at 37°C, although they may readily be able to do so at lower temperatures. Many organisms naturally found in water, including coliform organisms and *Aeromonas* species, have an optimum growth temperature below 37°C but may produce gas from lactose at 37°C. *Aeromonas* species, which resemble coliform organisms but are of no sanitary significance, may be excluded by carrying out an oxidase test on the colonies by Kovács' (1956) method. Colonies giving a positive oxidase reaction are not coliform organisms.

The Differential Tests

(1) *Fermentation of lactose at 37° C.* One lactose peptone water culture is incubated for 48 hours at 37°C. Acid and gas production within this time is sufficient to show that a coliform organism has been selected.

(2) *Fermentation of lactose at 44° C.* The other tube of lactose peptone water is incubated at 44°C. The presence of any amount of gas in the inverted inner tube after 6 to 24 hours' incubation indicates a positive reaction. Absence of gas production after 24 hours' incubation even though growth or acid production is present is regarded as a negative reaction.

The need for accurate temperature control in water baths for this test has already been mentioned when describing the rapid test for *E. coli* from presumptive positive tubes. The bath should be electrically heated and controlled by a thermostat capable of maintaining the temperature at 44°±0·25°C. Thermostatically controlled heating and circulating units are available which are capable of maintaining the temperature at 44°±0·1°C. The indicator thermometer should be checked against a N.P.L. standard. Two control tubes should be included, one inoculated with a known strain of *E. coli* and the other with a strain of coliform organism such as *Klebsiella aerogenes* which does not produce gas from lactose at 44°C. A recording thermometer provides a useful check.

The Faecal Streptococcus Test

Before describing methods of counting faecal streptococci it is necessary to define the meaning to be given to this term. The species of streptococci normally occurring in human and animal faeces and therefore most likely to be found in polluted water are *Streptococcus faecalis, S. faecium, S. durans, S. bovis* and *S. equinus*. Strains with properties intermediate between these are also common. The term "faecal streptococci" should be used to refer only to these named species and to the intermediate strains which also belong to Lancefield's serological group D. The term "enterococci", often used as a synonym for faecal streptococci, has not been precisely defined, and its use is not recommended. Other streptococci occasionally present in faeces, but not belonging to Lancefield's group D, include *S. mitis* and *S. salivarius*, which originate in the mouth and are swallowed in the saliva. Such strains differ in many of their properties from true faecal streptococci and their presence in water should not necessarily be regarded as evidence of faecal pollution.

The properties of faecal streptococci are fully described in textbooks of bacteriology. Their important characteristics are the ability to grow at 45° C; to grow in the presence of 40 percent bile and in concentrations of sodium azide which are inhibitory to coliform organisms and most other Gram-negative bacteria. Some species resist heating at 60° C for 30 minutes, will grow at pH 9·6 and in media containing 6·5 percent sodium chloride.

Technique for Enumerating Faecal Streptococci

The techniques for enumeration depend on the use of sodium azide as a selective inhibitor, and differentiation by growth at 45° C. A multiple tube method may be used which takes five days for a confirmed result, or a membrane filtration method may be used which gives a result in two days. Although faecal streptococci will normally grow at 45° C, a preliminary resuscitation or growth period at 37° C is recommended since some organisms may temporarily lose this ability outside the body (Allen, Pierce and Smith, 1953).

(a) Multiple Tube Method

Various volumes of water are inoculated into tubes of single or double strength glucose azide broth (Hannay and Norton, 1947) in exactly the same manner as in the multiple tube method for coliform organisms. Inverted inner tubes are not necessary as there is no gas production. The tubes are incubated at 37° C for 72 hours. As soon as acidity is observed a heavy inoculum is subcultured into further tubes of single strength Hannay and Norton's medium, and incubated at 45° C for 48 hours; all tubes showing acidity at this temperature contain faecal streptococci. The most probable number can then be determined from the probability tables. Childs and Allen (1953) avoided the double incubation in azide medium at 37° and 45° C by resuscitating in glucose broth for 2 hours at 37° C and then adding the azide portion of the medium and incubating for 48 hours at 45° C. This method shortens the time and gives significantly higher counts but involves an elaborate procedure after the first 2 hours.

(b) Membrane Filtration Method

The membrane filtration procedure as described for coliform organisms on pp. 311–313 is used except that a different medium and incubation procedure are required. Membranes cannot be washed and re-used. Membranes previously used for coliform organisms must not be used for streptococci. After filtration, the membrane is placed on a well dried plate of Slanetz and Bartley's (1957) glucose azide agar. This is incubated at 37° C for four hours and then at 44° or 45° C for 44 hours. All red or maroon colonies are counted as faecal streptococci (Taylor and Burman, 1964; Mead, 1966). Mead found that initial resuscitation without the inhibitory azide did not increase the recovery of streptococci. Incubation throughout at 37° C will permit the growth of many streptococci which are neither faecal nor of group D.

(c) Confirmatory Tests

The presence of streptococci may be confirmed by direct microscopical examination, which will show typical short-chained streptococci, or by subculture on MacConkey agar, which will show small red colonies after 24 or 48 hours' incubation at 37° C.

Mead (1963 and 1964) has described a single rapid test which will differentiate *S. faecalis* from other faecal streptococci. Cultures are plated on a tyrosine sorbitol thallous acetate agar (see Appendix B16) and incubated at 45° C for three days. Differentiation depends on the ability of *S. faecalis* to reduce 2, 3, 5-triphenyltetrazolium chloride (T.T.C.) at pH 6·2, to ferment sorbitol, to produce tyrosine decarboxylase and to grow at 45° C in the presence of 0·1 percent thallous acetate. Colonies of *S. faecalis* on this medium have a uniform deep maroon colour encircled by clear zones where the tyrosine has been decomposed. As Mead found that *S. faecalis* was particularly associated with man this test when positive suggests that the pollution is of human origin. A negative result does not however exclude human pollution.

The Clostridium perfringens (Cl. welchii) Test

Cl. perfringens is a stout Gram-positive anaerobic spore-forming rod, but spores are not usually formed in laboratory media. This organism is a natural inhabitant of the intestinal canal. The number of *Cl. perfringens* in faeces and sewage is much smaller than that of *E. coli*, and a similar relation between the numbers of *Cl. perfringens* and *E. coli*, is usually found in recently contaminated waters. The *Cl. perfringens* test therefore has little place in the examination of recently contaminated waters, for when *Cl. perfringens* is detectable in 100 ml of such a water *E. coli* is usually present in large numbers. The spores of *Cl. perfringens* can, however, survive in water for a long time and persist when all the other faecal bacteria have disappeared. The chief value of the *Cl. perfringens* test, therefore, is in demonstrating remote or intermittent pollution or in confirming the faecal nature of contamination when only coliform organisms other than *E. coli* are present in the water.

In previous editions of this report, two methods have been recommended for the detection of *Cl. perfringens* in water; one depended on the stormy fermentation of litmus milk and the other on the reduction of sulphite in a glucose sulphite agar medium. Heated samples of the water were used for both methods, so that only spores were detected. Stormy fermentation is a good confirmatory test for *Cl. perfringens* but litmus milk is not the best medium for primary isolation. A much higher isolation rate is given by primary incubation in the differential reinforced clostridial medium (D.R.C.M.) of Gibbs and Freame (1965) followed by subculture of all tubes showing blackening to litmus milk for confirmation by the stormy clot reaction (Burman, 1969). The sulphite reduction test can be improved by reducing the concentration of sulphite in Wilson and Blair's (1925) glucose sulphite iron agar from 2 percent to 0 5 percent, but results have not been so good as with D.R.C.M. followed by subculture to litmus milk.

Technique for Demonstrating Cl. perfringens

The water sample should be heated at 75° C for 10 minutes to destroy nonsporing organisms. One volume of 50 ml should be added to 50 ml of double-strength D.R.C.M. in a 4 oz. (114 ml) screw-capped bottle. Five volumes of 10 ml should be added to separate 10 ml volumes of double-strength D.R.C.M. in 1 oz (28 ml) screw-capped bottles. Separate volumes of 1 ml, and further 10-fold dilutions if necessary, should be added to five 25 ml volumes of single-strength D.R.C.M. in 1 oz. (28 ml) bottles. All the bottles should be topped up if necessary with further single-strength D.R.C.M. to bring the level of liquid up to the neck of the bottle, leaving a small air space. The bottles should then be incubated at 37°C for 48 hours. A positive reaction will be shown by blackening of the medium due to reduction of the sulphite and precipitation of ferrous sulphite.

A positive reaction may be produced by any clostridium. A loopful from each positive culture should be inoculated into a tube of litmus milk which has been freshly steamed and cooled. These tubes are then incubated at 37°C for 48 hours. Those containing *Cl. perfringens* will produce a "stormy clot" in which the milk is acidified and coagulated, the clot is disrupted by gas and often blown to the top of the tube. Growth in the litmus milk is improved by adding to each tube, immediately before subculture, a short length of iron wire, sterilized by heating to redness. A most probable number can then be read from the probability tables in the same way as for coliform organisms.

Standards of Bacterial Quality

Ideally all waters intended for drinking should be free from coliform organisms. Such an ideal standard is not, however, everywhere obtainable at present. When standards are being recommended, a distinction must be made between public supplies of piped drinking water and well or spring supplies used by individual householders or small communities for which the provision of a piped supply is not economically practicable. It is also necessary to consider whether a piped supply is chlorinated or unchlorinated, and whether the sample is taken from the water *entering* the distribution system or from the water *in* the distribution system, either in the authorities' mains or on the consumer's premises. It cannot be stressed too strongly that bacteriological examination has its greatest value when it is frequently repeated. The examination of a single sample can indicate no more than the conditions prevailing at the moment of sampling at that particular point in the supply system. For adequate control of the hygienic quality of the water supply it is necessary to examine bacteriologically samples taken frequently from carefully selected points.

Piped Supplies: Water Entering the Distribution System

(a) Chlorinated Supplies

Efficient treatment, culminating in chlorination, should yield a water free from any coliform organisms, however polluted the original raw water may have been. In practice this means that it should not be possible to demonstrate the presence of coliform organisms in any sample of 100 ml. A sample of water entering the distribution system which shows any deviation from this standard calls for an immediate investigation into both the efficacy of the purification process and the method of sampling. It is important, however, in testing chlorinated waters, that presumptive positive tubes should always be proved to contain coliform organisms by appropriate confirmatory tests (see Part III, Technical).

(b) Unchlorinated Supplies

Piped supplies of unchlorinated water are not recommended. Where small supplies of this sort still exist no water entering the distribution system should be considered satisfactory which yields *E. coli* in 100 ml. If *E. coli* is absent, the presence of not more than 3 coliform bacteria may be tolerated in occasional samples from established non-chlorinated piped supplies, provided they have been regularly and frequently tested, and the topography of the catchment area and the storage conditions are found to be satisfactory. In the routine control of such supplies, if one 100-ml sample shows the presence of any coliform organisms a further sample from the sampling point, or a series of samples, should be examined immediately. If these samples also show the presence of coliform organisms steps should then be taken to discover and if possible remove the source of the pollution. If coliform organisms persist or increase to numbers greater than 3 per 100 ml this indicates that the supply is not suitable for use without chlorination.

Piped Supplies: Samples Taken From the Distribution System

Water which is of excellent quality when it enters the distribution system may undergo some deterioration before it reaches the consumer's tap. It should be borne in mind that just as much deterioration may occur in the distribution system of a chlorinated supply in which there is little or no residual chlorine in the water reaching the consumer as in that of a non-chlorinated supply, and that in this respect the two should be considered on the same footing. Coliform organisms may gain access to the water in the distribution system from booster pumps, from packing used in the jointing of mains, or from washers on service taps. In addition, contamination from outside may gain access to the water in the distribution system, for example, through cross-connections, back-syphonage, defective service reservoirs and water tanks, damaged or defective hydrants or washouts or through inexpert repairs of domestic plumbing systems. Although coliform organisms derived from the tap washers or the jointing material of mains may be of little or no sanitary significance, excretal contamination which gains access to the water in the distribution system is at least as potentially dangerous as the distribution of originally polluted and insufficiently treated water.

Ideally all samples taken from the distribution system, including consumers' premises, should be free from coliform organisms. In practice this standard is not always attainable and the following standard for water collected in the distribution system is therefore recommended:

(1) Throughout any year 95 percent of samples should not contain any coliform organisms or *E.coli* in 100 ml.

(2) No sample should contain more than 10 coliform organisms per 100 ml.

(3) No sample should contain more than 2 *E. coli* per 100 ml.

(4) No sample should contain 1 or 2 *E. coli* per 100 ml in conjunction with a total coliform count of 3 or more per 100 ml.

(5) Coliform organisms should not be detectable in 100 ml of any two consecutive samples.

When any coliform organisms are found the minimum action is immediate resampling. The persistence of 1-10 coliform organisms or 1 or 2 *E. coli*, or the single appearance of higher numbers of either, suggests that undesirable material is gaining access to the water and measures should at once be taken to ascertain and remove the source of pollution.

Small Rural Supplies

In isolated situations where a public water supply may still be economically impracticable and where reliance has to be placed largely on private supplies, the standard outlined above may not be attainable. Such a standard should however be aimed at and everything possible should be done to prevent the access of pollution to the water. By relatively simple measures, such as removal of obvious sources of contamination from the catchment area and by attention to the coping, brick lining and covering, it should be possible to reduce the coliform count of even a shallow well water to lees than 10 per 100 ml. Persistent failure to achieve this, particularly when *E. coli* is repeatedly present, should as a general rule lead to condemnation of the supply.

Frequency of Sampling for Bacteriological Examination

Piped Supplies

The frequency of bacteriological examination for the hygienic control of water supplies and the location of the sampling points should be such as to enable proper control to be kept on the bacterial quality. Sampling points should be chosen to include the raw water, pumping stations, treatment plant, reservoirs, booster pumping stations, and the distribution system. The frequency with which samples should be collected will vary in different circumstances, but it is undoubtedly true that many supplies were in the past sampled far too seldom, and some general guidance is therefore offered on the frequency with which samples from piped supplies should be collected.

Water Entering the Distribution System

(a) Chlorinated Supplies

When water requires chlorination before distribution, a constant check both on the chlorine residuals and the bacterial quality is needed. The importance of the former cannot be overstressed as it ensures that should any inadequately treated, and therefore possibly contaminated, water enter the distribution system as a result of a failure in chlorination, immediate remedial action may be taken. The efficiency of chlorination may be checked most effectively by the use of residual chlorine recorders preferably with automatic control. These however require technical supervision, and for the small supply regular manual testing for chlorine may be all that is practicable. In principle, the bacteriological examination of chlorinated water as it enters the distribution system from each treatment point should be carried out at least once a day, and no doubt with the larger supplies this will be done. With small supplies serving a population of ten thousand or less, daily sampling may be impracticable and reliance should be placed on proper control of residual chlorine dosage with checks on the bacterial quality of the water at, say, weekly intervals.

Some supplies which are of excellent bacterial quality—knowledge of which is itself dependent on the regular bacteriological examination of the raw water—are nonetheless chlorinated as a precautionary measure. Daily bacteriological examination of such water would not appear to be necessary in all instances and it is recommended that in supplies of this kind the raw and treated water should be examined bacteriologically with the same frequency as is proposed in the following paragraph for water which is distributed without chlorination.

(b) Unchlorinated Supplies

All piped supplies should be chlorinated before distribution and a careful check kept on the chlorine residuals, however good the original water may have been. Few public supplies in the United Kingdom are distributed without chlorination and the following recommendations are given largely for the benefit of readers elsewhere.

Where naturally pure waters are distributed without chlorination it would seem reasonable to base the frequency of examination not only on the characteristics of the source but on the population served. The following maximum intervals between successive routine examinations of such water as it enters the distribution system are proposed in International Standards for Drinking Water and European Standards for Drinking Water, both published by the World Health Organization. It is recommended that these be used as a guide in those areas in which such unchlorinated supplies exist.

Population served	*Maximum interval between successive samples*
Less than 20,000	One month
20,000 to 50,000	Two weeks
50,000 to 100,000	Four days
More than 100,000	One day

On each occasion samples should be taken from all points at which water enters the distribution system.

(c) Samples from the Distribution System

As already stated, both chlorinated and unchlorinated supplies may undergo deterioration in the distribution system, and therefore there is no reason to make a distinction between them in considering the frequency of sampling from the distribution system. As with naturally pure water entering the distribution system, frequency of sampling should, again, be determined by the size of the population served by the supply.

As a general guide, one sample per 5,000 of the population served should be examined each month for each supply serving up to 100,000 persons; 1 per 10,000 per month for supplies serving between 100,000 and 500,000 persons; and 1 per 20,000 per month for supplies serving centres of over 500,000 persons. The samples should be spaced out evenly throughout the month. It is considered justifiable to reduce the number of samples to one per 10,000 of population per month when the population served exceeds 100,000, since in systems serving populations of that size the interval between successive samples will be very short. Further reduction of the frequency is considered justifiable in distribution systems serving centres of more than 500,000 persons. In such supplies the number of persons served per pipe-mile would be considerably higher than in supplies serving smaller or more scattered populations. These figures are similar to those given in International Standards for Drinking Water (WHO 1963) and European Standards for Drinking Water (WHO 1961).

It is the responsibility of water undertakings to arrange for the collection and examination of samples from the distribution system and to keep medical officers of health of the local authority districts served by them informed of the quality of the water distributed in their areas. The custom by which medical officers of health also arrange for the independent collection and testing of samples is a valuable check. Water undertakings may find it useful to discuss the matter with medical officers of health in order to arrange for the interchange of information and to avoid unnecessary duplication and the collection of excessive numbers of samples.

Small Rural Supplies

In order to assess the quality of an established supply bacteriological examinations should be carried out at intervals of one month or less, but in many areas this is not practicable. If samples are examined less frequently than once a month it is impossible to say that a supply is satisfactory. If there has been any change in use of the land in the vicinity, or quarrying, or gravel digging, additional samples should be examined to see if these have had any effect on the bacterial quality of the water.

When a new well or spring is being brought into use samples should be collected relatively frequently and under various climatic conditions so that any variation in the quality of the water can be observed.

References

Allen, L. A., Pierce, M. A. F. and Smith, H. M., 1953. *J. Hyg., Camb.*, 51, 458.
Bardsley, D. A., 1934, *J. Hyg., Camb.*, 34, 38.
Barritt, M. M. 1936. *J. Path. Bact.*, 42, 441.
Burman, N. P., 1955. *Proc. Soc. Wat. Treat. Exam.*, 4, 10.
Burman, N. P., 1961. *J. appl. Bact.*, 24, 368.
Burman, N. P., 1967*a*. *Proc. Soc. Wat. Treat. Exam.*, 16, 40.
Burman, N. P., 1967*b*. *Recent advances in bacteriological examination of water.* In: *Progress in microbiological techniques,* edited by C. H. Collins. London, Butterworth, p. 185.
Burman, N. P., 1969. In the press.
Childs, E. and Allen, L. A., 1953. *J. Hyg., Camb.*, 51, 468.
Cochran, W. G., 1950, *Biometrics,* 6, 105.
Cook, G. T., 1952. *J. Path. Bact.*, 64, 559.
Cowan, S. T. and Steel, K. J., 1965. *Manual for the identification of medical bacteria.* London, Cambridge University Press.
Edwards, P. R. and Ewing, W. H., 1962. *Identification of enterobacteriaceae. 2nd edition.* Minneapolis, Burgess.
Eisenhart, C. and Wilson, P. W., 1943. *Bact. Rev.*, 7, 57.
Folpmers, T., 1948. *Antonie v Leeuwenhoek,* 14, 58.
Gibbs, B. M. and Freame, B., 1965. *J. appl. Bact.*, 28, 95.
Gray, R. D., 1964. *J. Hyg., Camb.*, 62, 495.
Hammarström, E. and Liutov, V., 1954. *Acta. path. microbiol. scand.*, 35, 365.
Hannay, C. L. and Norton, I. L., 1947. *Proc. Soc. appl. Bact.*, No. 1, 39.
Harvey, R. W. S., 1956. *Mon. Bull. Minist. Hlth,* 15, 118.
Harvey, R. W. S. and Thomson, S., 1953. *Mon. Bull. Minist. Hlth,* 12, 149.
Hobbs, Betty C. and Allison, V. D., 1945*a*. *Mon. Bull. Minist. hlth,* 4, 12.
Hobbs, Betty C. and Allison, V. D., 1945*b*. *Mon. Bull. Minist. Hlth,* 4, 63.
Hoskins, J. K., 1934. *Publ. Hlth. Rep., Wash.*, 49, 393.
Hynes, M., 1942. *J. Path. Bact.*, 54, 193.
Iveson, J. B., Kovács, N. and Laurie, W., 1964. *J. clin. Path.*, 17, 75.
Jameson, J. E. and Emberley, N. W., 1956. *J. gen. Microbiol.*, 15, 198.
Koser, S. A., 1923. *J. Bact.*, 8, 493.
Kovács, N., 1928. *Z. ImmunForsch. exp. Ther.*, 55, 311.
Kovács, N., 1956. *Nature, Lond.*, 178, 703.
Leifson, E., 1935. *J. Path. Bact., 40, 581.*
Leifson, E., 1936. *Am. J. Hyg.*, 24, 423.
McCartney, J. E., 1933. *Lancet,* 11, 433.
McCoy, J. H., 1962. *J. appl. Bact.*, 25, 213.
McCrady, M. H., 1915. *J. infect. Dis.*, 17, 183.
McCrady, M. H., 1918. *Can. J. publ. Hlth,* 9, 201.
Mead, G. C., 1963. *Nature, Lond.*, 197, 1323.
Mead, G. C., 1964. *Nature, Lond.*, 204, 1224.
Mead, G. C., 1966. *Proc. Soc. Wat. Treat. Exam.*, 15, 207.
Metropolitan Water Board, [1966*a*]. *Rep. Results chem. bact. Exam. Lond. Wat.*, 1963-1964, 41, 17.
Metropolitan Water Board, [1966*b*]. *Rep. Results chem. bact. Exam. Lond. Wat.*, 1963-1964, 41, 31.
Metropolitan Water Board, [1967*a*]. *Rep. Results chem. bact. Exam. Lond. Wat.*, 1965-1966, 42, 15.
Metropolitan Water Board, [1967*b*]. *Rep. Results chem. bact. Exam. Lond. Wat.*, 1965-1966, 42, 18.

Ministry of Housing and Local Government, 1967. *Safeguards to be adopted in the operation and management of waterworks.* London, HMSO.
Moore, B., 1948. *Mon. Bull. Minist. Hlth,* 7, 241.
Moore, B., 1950. *Mon. Bull. Minist. Hlth,* 9, 72.
Moore, B., Perry, E. L. and Chard, S. T., 1952. *J. Hyg., Camb.,* 50, 137.
O'Meara, R. A. Q., 1931. *J. Path. Bact.,* 34, 401.
Panezai, A. K., Macklin, T. J. and Coles, H. G., 1965. *Proc. Soc. Wat. Treat. Exam.,* 14, 179.
Public Health Laboratory Service Water Sub-Committee, 1952. *J. Hyg., Camb.,* 50, 107.
Public Health Laboratory Service Water Sub-Committee, 1953*a*. *J. Hyg., Camb.,* 51, 268.
Public Health Laboratory Service Water Sub-Committee, 1953*b*. *J. Hyg., Camb.,* 51, 559.
Public Health Laboratory Service Water Sub-Committee, 1953*c*. *J. Hyg., Camb.,* 51, 572.
Public Health Laboratory Service Standing Committee on the Bacteriological Examination of Water Supplies, 1968*a*. *J. Hyg., Camb.,* 66, 67.
Public Health Laboratory Service Standing Committee on the Bacteriological Examination of Water Supplies, 1968*b*. *J. Hyg., Camb.,* 66, 641.
Public Health Laboratory Service Standing Committee on the Bacteriological Examination of Water Supplies, 1969. *J. Hyg., Camb.,* In the press.
Rappaport, F., Konforti, N. and Navon, B., 1956. *J. clin. Path.,* 9, 261.
Ross, A. I. and Gillespie, E. H., 1952. *Mon. Bull. Minist. Hlth,* 11, 36.
Simmons, J. S., 1926. *J. infect. Dis.,* 39, 209.
Slanetz, L. W. and Bartley, C. H., 1957. *J. Bact.,* 74, 591.
Swaroop, S., 1938. *Indian J. med. Res.,* 26, 353.
Swaroop, S., 1951. *Indian J. med. Res.,* 39, 107.
Taylor, C. B., 1942. *J. Hyg., Camb.,* 42, 17.
Taylor, C. B., 1951. *J. Hyg., Camb.,* 49, 162.
Taylor, E. W., 1955. *J. Hyg., Camb.,* 53, 50.
Taylor, E. W., and Burman, N. P., 1964. *J. appl. Bact.,* 27, 294.
Taylor, E. W. and Whiskin, L. C., 1951. *J. Instn Wat. Engrs,* 5, 219.
Thresh, J. C., Beale, J. F. and Suckling, E. V., 1958. *Examination of waters and water supplies.* 7th edition by E. Windle Taylor. London, Churchill.
Wilson, G. S., Twigg, R. S., Wright, R. C., Hendry, C. B., Cowell, M. P., and Maier, I., 1935. *Spec. Rep. Ser. med. Res. Coun., Lond.,* 206.
Wilson, W. J. and Blair, E. M. McV., 1925. *J. Hyg., Camb.,* 24, 111.
Wilson, W. J. and Blair, E. M. McV., 1927. *J. Hyg., Camb.,* 26, 374.
World Health Organization, 1961. *European standards for drinking-water.* Geneva, World Health Organization.
World Health Organization, 1963. *International standards for drinking-water.* 2nd edition. Geneva, World Health Organization.

APPENDIX II—THE PURIFICATION OF THE WATER OF SWIMMING POOLS[4]

Her Majesty's Stationery Office, London, 1975

Introduction

1.1 The first edition of this booklet was published in 1929. It was revised in 1951 but that second edition is now out of date in many respects. We were appointed to revise it again. This third edition is a good deal longer than either of its predecessors, partly because new developments have had to be discussed and partly in the hope that readers will secure a better understanding of the fundamentals of the processes involved. A summary is, however, included for quick reference.

1.2 The prime reason for maintaining pool water in an adequate state of purity is so that bathing shall be without risk to life and health, but another most im-

[4]Reprinted with the permission of the Controller of Her Britannic Majesty's Stationery Office, London, England.

portant reason is to make the pool so attractive and inviting that people want to use it. The aim of this publication is to give assistance in maintaining not only safe but attractive pool conditions.

1.3 Much of what was said in the 1951 booklet is said again, because it is still sound and far from out of date, but there have been some developments which have needed detailed explanation and careful assessment. It is inevitable that these will appear to have been given prominence. But well-known and tried precautions and procedures are probably more valuable than new ones, and the importance of a matter should not be judged by the amount of space devoted to it in this booklet.

1.4 One major development in the last 20 years has been the growth in the number of pools on school premises, which we believe may now number some 6,000 or 7,000. They vary greatly both in size and facilities provided, and in most cases their type and pattern of use is very different from pools provided for the general public. These variations are reflected to some degree in the practices relating to purification.

1.5 We have had to keep these matters in mind when dealing with purification procedures, and it will be for the pool operator to apply our recommendations in the light of the circumstances of his case.

1.6 It is necessary to emphasise that this publication is concerned only with swimming pools which have been built and equipped for the purpose. Uncontrolled bathing places in rivers, ponds, or canals, or bathing in the sea, present quite a different problem, and the standards of quality we recommend here are not applicable to them.

The Need for Purification

Pollution

2.1 Pool water is regarded as polluted when it is unsuitable or unacceptable (or less suitable or less acceptable than it might reasonably be) for swimming, having regard to health, safety, amenity, freedom from causing irritation, and maybe other factors. This is a strictly limited definition, but is adequate for the circumstances. If the water is unpleasant to drink this may not matter (sea water is unpleasant to drink but generally suitable for bathing). If it is toxic to algae this is generally a desirable property. But if it is unsafe to drink this would be important, for some of it might be drunk. And if it irritated the skin it would be unacceptable even if it could be demonstrated that no actual harm could be caused. These are merely examples of what would not be, and what would be, regarded as pollution in this particular case, and the significance of possible contaminants should be assessed accordingly.

Pollution Not Derived From the Persons of Bathers

2.2 Atmospheric pollution may cause some nuisance though not, we believe, on the scale it could before the Clean Air Acts came into operation. A film of dust may collect on the surface of some open air pools and a slight deposit may form on the bottom. Though not dangerous these detract from the appearance of the water and should accordingly be removed. Similarly, at certain times of the year open-air pools may be contaminated by leaves, seeds blown about in the wind, and even grass cuttings from nearby lawns, and these, too, have an adverse visual effect and the normal purification treatment should be such as to remove them.

2.3 Pathways round some pools are accessible to people entering from outside, and filth from roads may be deposited from shoes. This should not enter the pool directly, for drainage ought to be away from the pool, but can be picked up and carried in on the feet of bathers. There may be other ways by which soil of various kinds can find its way into a pool. This pollution is dealt with in the purification process but it is something to be discouraged as much as possible.

2.4 From time to time new chemicals for use for specialised purposes in swimming pools are advocated, and in the past few years this has taken place at an ever increasing rate. The use of some of these is discussed later on. A chemical of value in one direction may actually be a pollutant in relation to some other factor, and it is not always easy to decide whether a new chemical clearly of value for one specific purpose might pose a risk to the health of bathers. Most pool owners and operators are not in a position to reach an informed opinion on the matter. We are now able to report that the Department's Committee on New Chemicals for Water Treatment, set up initially to consider chemicals for treating supply waters, has now extended its terms of reference to cover new chemicals for swimming pool water treatment. The Committee does not comment on the value of new chemicals but merely states whether, in its opinion, the proposed use could be in any way objectionable on health grounds.

Pollution Derived From the Persons of Bathers

2.5 A wide variety of substances can, in theory, be transmitted to pool water from the bodies of those using it. None of them is wanted but, given reasonable cleanliness and hygiene, the amounts are such that they are unlikely to do harm. Nevertheless, it is part of the purpose of purification to prevent any accumulation of any impurity to such an extent as to be unsightly, or noticeable in any other way, or to constitute any possible risk to health. The substances may be divided into three classes.

2.6 The first comprises substances originating in the bodies of the bathers. They include mucus from the nose, saliva from the mouth, sweat and dead scaly matter from the skin, urine and traces of faecal matter. These can never be avoided entirely, but good hygiene on the part of the bathers, encouraged by the provision of pre-cleansing facilities, can reduce them to very small proportions.

2.7 The second class comprises substances collected on the skin in the course of pre-bathing activities and not washed off before entering the water. They include general 'soil' from streets, offices, shops etc which some bathers may not see the necessity of washing off before bathing, and in some cases contaminants deriving from the place of work of the bathers. Again the quantities will be small, but could accumulate if purification did not remove them.

2.8 The third class comprises substances previously applied to the bodies of the bathers—various powders, creams, lotions, oils and so on—whether for the promotion of health, beauty or attractiveness. We do not know that any detailed investigation has been made of how much of what kind of applications comes off during bathing, but it is unlikely that many bathers will feel the necessity for removing them before entering a pool. There does not appear to be any real public health necessity for this, but there is no doubt that some foreign matter enters the water in this way and it should not be allowed to accumulate unduly.

Pollution by Bacteria and Viruses

2.9 This subject is dealt with in a separate section (following the precedent of previous editions) because the general public are conscious of the fact that some diseases can be water borne and naturally they require assurance that the water in which they bathe is free from risk to health.

2.10 There are two general points about the possibility of spread of infection at swimming pools. First, although the number of bacteria and other organisms which enter the pool water from bathers is quite considerable the vast majority are non pathogenic. Most of these organisms live in vast numbers as harmless parasites in or on healthy people and only in exceptional cases could they cause disease. A few are known causes of disease, but in a well ordered swimming pool their numbers in the water are so small that the risk of disease occurring in bathers is almost non existent. Secondly, disease may be spread at swimming pools by ways other than water as an intermediary. Wherever people congregate together, for instance in theatres, public transport, and sports centres, there are opportunities for disease to spread by personal contact or air borne infection, and swimming pool establishments are no exception. This is a sound reason for avoiding overcrowding of the pool and changing areas and for maintaining pool surrounds and facilities in a hygienic condition.

Transmission of Infection by Swimming Pool Water

2.11 Various diseases can be transmitted by swimming pool water if adequate standards of hygiene are not maintained.

a. Gastro-intestinal infections may be spread by water, but in a well-ordered pool, if any faecal organisms responsible for these diseases are introduced into the water they are rapidly dispersed, diluted and inactivated by the residual chlorine. Outbreaks of infection attributable to swimming have been reported in the past, but in these instances the pools and other waters concerned were shown to be polluted with sewage. We know of no authenticated cases of gastro-intestinal infection associated with swimming in a well-ordered pool.

b. Skin infections such as furunculosis, scabies and pediculosis may be associated with visits to a swimming pool, but if they do occur they are more likely to result from direct contact with an infected person than from any organisms or parasites shed by bathers into the water. Inflammation of the ear after bathing is generally caused by the swelling of wax and skin casts rather than by infection.

Tinea pedis (athlete's foot) may be spread by contact with infected footwear or floor surfaces and there is evidence that the disease can be acquired during visits to swimming pools. The fungus which causes the disease has been isolated from floor surfaces and it is probable that the floors of changing cubicles and of walkways are infected by skin fragments carrying the fungus which are shed by bare foot infected bathers.

Verrucae (plantar warts)—there is some evidence that warts are infectious and may be caused by a virus. It is possible therefore that verrucae are spread by infected skin fragments in a similar manner to the tinea pedis which cause athlete's foot. If such skin fragments are shed into the water of the pool they will be removed by the filters.

Swimming pool granuloma is characterised by a small nodule or ulcer which develops at the site of an abrasion. The causative organism is a mycobacterium, and outbreaks of infection have always been associated with swimming pools where the pool sides or bottom were sufficiently irregular or rough to provide crevices which harboured the organism. Infection should not arise in a well maintained pool.

Erythema and rashes due to chemicals used for water purification may sometimes occur in hypersensitive persons.

c. Conjunctivitis may be caused by infection or by physical or chemical irritation. Owing to the difference in composition between the fluid of the eye and pool water, any prolonged swim, particularly if it includes much diving or underwater swimming, may cause a slight mechanical or osmotic conjunctivitis. A few individuals who are particularly sensitive to traces of chemicals present in the water may also develop some temporary eye irritation. But conjunctivitis caused by bacterial infection is far more likely to be spread by direct contact or via infected articles such as towels than by the water itself, even supposing that the bacteria could survive the disinfectant treatment given in a well conducted pool, which is unlikely. If spectators complain of eye trouble then better ventilation is an obvious remedy (see Sports Council Technical Unit for Sport Design Bulletin No.1).

Several outbreaks of pharyngo conjunctival fever, a virus infection, have been associated with swimming pools. The disease is characterised by a sore throat and conjunctivitis and is accompanied by fever. Although a direct connection with pool water has not been proven beyond doubt, the circumstances suggest that water was probably the source of infection in reported incidents. The virus is readily inactivated by chlorine and risk of spread of the disease by pool water should not arise in pools where a satisfactory level of free chlorine residual is maintained.

d. Nasopharyngeal and respiratory infections are commonly spread by airborne infection and bathers are more likely to contract these diseases in crowded areas than via the medium of swimming pool water. Outbreaks of otitis media (middle ear disease) are unlikely to pass unnoticed, particularly in the many schools which have their own pools, yet few instances of an association between the disease and swimming are recorded. If the disease occurs following swimming it may be the result of infected mucus from the nasopharynx being forced into the eustachian tubes while swimming.

e. Poliomyelitis. The incidence of poliomyelitis in the community is now so very small that it no longer calls for any special discussion.

2.12 The conclusions to be drawn from the available evidence are that risk of contracting disease while swimming should not arise in a properly controlled swimming pool. Where epidemics associated with swimming pools have occurred, investigations have usually revealed clear evidence of unsatisfactory standards of hygiene and water purification, creating conditions conducive to the transmission of infection.

2.13 However, the fact that the number of recorded incidents is small is no argument for relaxing standards. This satisfactory record can only be maintained by the use of disinfection processes for pool water which have been shown to be effective and by frequent checks to ensure that a high standard of treatment and water quality is achieved.

2.14 The congregating together of numbers of people particularly in changing areas is inevitable, but the chances of air borne spread of infection taking place will be minimised if overcrowding is prevented, and good ventilation is provided.

2.15 The prevention of foot infections merits special mention. Although epidemiological evidence of the spread of these conditions is incomplete, it is advisable that sufferers from plantar warts and athlete's foot should not attend swimming pools until adequate medical treatment has been received. Good foot hygiene is an unqualified advantage because it both reduces the risk of infected skin fragments being disseminated and protects the skin against infection.

2.16 Infected skin fragments rubbed off on to walkways are readily swept away by high pressure hoses, but the organisms inside them may not be quickly reached by disinfectants. Floor surfaces should be cleansed mechanically once or twice a day preferably by high pressure hosing, followed by flooding with a disinfectant such as one containing a 1% solution of available chlorine, which should be left in contact with the floor for as long as possible. Skin fragments are less likely to be rubbed off in the pool itself; if they are, the filters will remove them as a matter of course and any free microorganisms will be dealt with by the disinfectant.

2.17 Attractive pre-cleansing facilities with warm water and soap always available should be provided on a generous scale and the importance of their use canvassed widely. Where it is confidently felt that such facilities would be conscientiously used, then, in the case of indoor pools, they could be held to make unnecessary the difficult provision of attractive disinfectant footbaths. Otherwise, and particularly in the case of open air pools, the latter should be provided since there is evidence that these footbaths are of some value, though they cannot be relied upon to cleanse the feet entirely and prevent contamination of floor surfaces, pool surrounds and walkways.

Urea in Pool Water

2.18 It has long been known that urination in swimming pools takes place. Although no threat to health in well ordered pools, it is grossly unhygienic. The organic matter introduced into the pool in this way no doubt consumes chlorine and adds to the cost of maintaining purity. In particular, its content of nitrogenous matter (eg creatinine) not only consumes relatively large quantities of chlorine, but might well produce chlorinous compounds which could be confused with, or interfere with the action of, compounds of chlorine and ammonia (eg dichloramine) (see para 4.28). It is not improbable, either, that the chlorinous compounds so produced are malodorous. So there is every reason for attempting to reduce urination to the absolute minimum even if, in some cases, it may be quite involuntary.

2.19 Until recently there was no method of estimating the amount of urination which takes place. Some work by Lomas,[5] however, suggests that at least one constituent of urine (mostly probably creatinine) produces a chlorinated derivative which gives a reaction for dichloramine but which does not (as dichloramine does) decompose in the presence of excess free chlorine. From the information given in the paper, and text book data on the composition of urine, it appears that the quantity involved might be surprisingly large. So we asked the Water Research Centre to carry out a brief investigation, with special reference in this case to urea, which is the principal nitrogenous constituent of urine and which has recently become of some importance in the river pollution field.

2.20 The work of the Research Centre revealed another complicated chemical situation involving chlorine. This cannot be unravelled here, but two important facts emerged. The first is that urea is broken down by free residual chlorine but only slowly, so that several milligrammes per litre could be present in a well-used pool. The second is that methods became available for estimating the approximate volume of urine which enters the pool in unit time (one day) and consequently of the average contribution from each bather.

2.21 Experience with these methods has been very limited so far and, for various technical reasons, is not easy to interpret. We think that it would be a good thing if at least a few pool authorities would attempt to estimate what happens at pools under their control. All we can say at the moment is that it looks as though the average urine contribution per bather is in the region of 20-50 millilitres (1-2 fluid ounces).

2.22 This figure appears to us to be much higher than it should be in a well behaved community, even allowing for the possibility that in a few cases urination is involuntary due to the shock of immersion. The contribution from this source can hardly be much if the bladder is emptied immediately prior to precleansing and bathing, as it should be.

2.23 We think there should be greater efforts made to minimise urine contamination. The appropriate facilities should always be provided, where possible in a situation where it is almost easier to use them than not. Printed reminders should be exhibited in places where they would be most effective. The education of children in the practice of hygiene should include this point, and they should be reminded of it when using school pools or visiting others in an organised body.

Continous Filtration and Disinfection

3.10 These are now generally accepted as the essential components of a swimming pool water purification system. We repeat that no pool should be built without providing them at the outset. Any purpose-built pool still lacking either of them

[5]P. D. R. Lomas, *Journal of the Association of Public Analysts*, 1967, Vol. 5, p. 27.

should be supplied with them, the urgency of doing so being dependent upon existing conditions, amount of use, results of routine analyses and so on.

3.11 The processes of filtration and disinfection should both be 'continuous', but this does not necessarily mean that the application of the disinfectant must be continuous, even though in the larger pools it usually is, and should be. For smaller pools, however, it may be taken to mean that the disinfectant must be continuously present in effective concentration, even though it may be added only intermittently.

3.12 It is convenient to distinguish sharply between the functions of filtration and disinfection. The purpose of the former is to produce a clear and attractive water; that of the latter, a water free from any health hazard from bacteria, viruses and other infective organisms. The processes are, however, inter-related to some degree. A clear water is easier to disinfect than one containing particulate matter, and it is quite likely (and generally accepted though not rigidly proved) that the choice of disinfectant and its method of use will affect clarity. Nevertheless, for the present purposes, filtration and disinfection are considered quite separately, and any interrelationship is ignored. Disinfection is considered first, since the health aspect should manifestly have priority, and because the practice of disinfection has more nearly approached a standard one.

Standards for Swimming Pool Water

12.1 We do not lay down rigid or inflexible standards for any property of swimming pool water; we think it would be a mistake to do so. Our use of the word 'standard' is more akin to 'guideline' or 'criterion of good quality'. To us a judgement of whether a pool water is of good, satisfactory, fair or poor standard is more than consultation of a printed table of limits. It is a matter of judgement by one who knows the principles on which the guidelines have been laid down and can relate them to particular circumstances, and who knows when a departure from them requires an investigation or demands an immediate closure of the pool on health or safety grounds. We do not believe that a table of figures can replace knowledge, experience and understanding. In passing it might be remarked with profit, that this country has at least as safe and satisfactory a public water supply as any other country in the world but no mandatory standards of quality. It should be emphasised that the lack of absolute standards for pool water is no excuse for slackness or complacency. Rather the reverse; it should stimulate the attitude 'how much better can I get than the standard' rather than 'if I manage to reach the standard what more can they expect'.

Bacteriological Standards

12.2 In this country the quality of public water supplies is judged on the results of the tests for coliform organisms and particularly Escherichia coli. Large numbers of these organisms are present in human excreta and sewage and, although the majority of these organisms are not capable of causing disease, their presence in water is an indication that harmful intestinal organisms may also be present. It is impracticable to search for intestinal pathogens (harmful organisms) in routine water examination, as when present they are usually much smaller in number than faecal coli, and they also tend to die off more quickly than coliforms. If no faecal organisms are found in the water, intestinal pathogens are not likely to be present.

12.3 It has been argued that coliform organisms are not necessarily the best indicators for judging the bacterial quality of swimming pool water, since the predominant organisms may not be from human excreta but from the nose, mouth and skin. Some of these bacteria, such as the staphylococcus, are more resistant than coliforms to the disinfectant action of chlorine, and small numbers may therefore be present in pool water from which coliforms have been eliminated.

12.4 All the evidence we have, however, indicates that when coliforms are absent and a satisfactory level of free chlorine residual is maintained throughout the pool, the risk of infection to bathers from the small number of organisms remaining in the pool water is minimal or non-existent. Consequently we consider the routine bacteriological examination should continue to be that for coliform organisms and E. coli. These tests are described in detail in the Bacteriological Examination of Water Supplies (1969).[6] Colony counts are also used as an additional test to monitor the efficiency of the water treatment processes.

12.5 Water as sampled from a pool should contain disinfectant, usually chlorine. After sampling, *unless the chlorine is immediately destroyed*, disinfectant action will continue, and during the time elapsing until the analysis takes place the bacteriological condition may improve a great deal, leading to falsely satisfactory results. What needs to be known is the quality of the water at the moment of sampling. Accordingly, a quantity of sodium thiosulphate (hypo) must be added to sample bottles to neutralise the chlorine immediately. It is vitally important that this should *not* be washed out by rinsing the bottle before filling it with the sample. Current instructions for sampling for water testing (Report No. 71) prescribe an amount of hypo (18 mg/1 $Na_2S_2O_3$, $5H_2O$) which will neutralise 5 mg/l of chlorine, which is adequate with a large factor of safety for public water supplies. But the factor of safety is much smaller for pool waters, which may contain 2 mg/1 or more of free chlorine and at the same time appreciable quantities of chloramines, particularly dichloramine. It is in fact by no means impossible that on occasions the total chlorine may exceed 5 mg/1 and that some may remain therefore after all the hypo in the sampling bottle has been destroyed. If there is the slightest possibility of the *total* residual chlorine in a pool exceeding 5 mg/1 the bacteriologist should be consulted with a view to doubling the amount of hypo in the sample bottle. For small pools, only chlorinated at intervals, it is always wise to do this.

12.6 In sampling water supplies from distribution systems conditions are usually such that any sample is typical of the water being sampled, and therefore the exact place, and time of sampling are not of preeminent importance. But in a swimming pool, water near the bottom of the deep end may not have received any pollution during the previous hour or two and, because it has been in the presence of disinfectant all that time, should invariably be uncontaminated. A sample from the shallow end during an interval when there are no bathers, eg lunch time at a school pool, should always be similarly satifactory. But one taken at the shallow end with a number of bathers present could contain contamination which had only entered the pool a few seconds before. The bacterial state of that water would depend a great deal upon the concentration of disinfectant and its speed of action. And of course it is just that water which is most likely to reach other bathers. The speed of action of the disinfectant is therefore most important. This of course is a major reason for the preference for free residual chlorine over chloramines—its action is one or two orders of magnitude more rapid. It is also apparent that routine sampling at the deep end or in quiet periods would tell us little or nothing about speed of reaction. It would merely tell us that after a time, which may be an hour or two, the disinfectant was effective.

12.7 It is of course necessary to do frequent residual determinations of samples at the deep end. But sampling for bacteriological analysis at the deep end need

[6]The Bacteriological Examination of Water Supplies, Reports on Public Health and Medical Subjects, No. 71. Department of Health and Social Security, Welsh Office, and Ministry of Housing and Local Government, Her Majesty's Stationery Office, London, England, 1969.

not be carried out frequently. Instructions as to the proper place and time of sampling cannot be laid down with any precision. Normally however it should be done at the shallow end when bathers are present and active. Whoever takes a sample should record whether this was the case or not, together with the time and place of sampling, and the amount of chlorine residual in the water, and information should be included with the certificate of analysis.

12.8 For special pools (diving pools, paddling pools, remedial pools etc) the necessary variation in sampling technique from this norm is a matter of common sense.

12.9 When filtration is adequate and disinfection properly operated coliforms and E. coli will not normally be detectable in 100 ml samples of water. This should be achieved when a free chlorine residual is present in a concentration of not less than 0·5 mg/l. For reasons which will be discussed later we consider that a free chlorine residual of more than 0.5 mg/l should be maintained. Experience with other disinfectants is not nearly so extensive and a corresponding statement about suitable concentrations of alternatives to chlorine cannot be made with confidence. Each method has to be subjected to detailed investigation in practice to establish the concentrations of disinfectant necessary to maintain satisfactory water quality under varying conditions of pool usage.

12.10 A well equipped and well run pool tested bacteriologically should show the water normally to be free from coliform organisms and E. coli in 100 ml. This is as near a standard as we are prepared to recommend. To ask for more would be unrealistic—it would always be possible in a crowded pool for a random sample to include water which has been contaminated a few seconds before and no disinfection process could be expected to deal with bacteria as quickly as that. But to demand less would be a failure to require the best practicable. It is to be noted that we do not say that one or a few organisms of the coliform type constitutes a potential threat to health—they would constitute evidence of shortcomings in operating the pool and these shortcomings should be put right. The proper reaction to an unsatisfactory sample is therefore an investigation and where necesary an institution of remedies. If those responsible for the pool in question were unwilling or unable to institute remedies then doubtless the Environmental Health Officer or the Medical Adviser would feel bound to recommend its closure if they considered there was a risk to public health.

12.11 The frequency of bacteriological testing depends upon the type of pool, the equipment available, its reputation and so on. No rule can be laid down. Clearly a children's pool which is abnormally crowded, a pool in which changes have been recently made in the method of operation or in the chemicals used, a pool which has recently given some unsatisfactory results, are all cases requiring more frequent bacteriological checks than others. At the other extreme no pool should go without its occasional bacteriological check, the samples being taken on a day and at a time which should come as a surprise to whoever is immediately responsible for operation, but almost always when the pool is being well used.

12.12 The technique of sampling is important and particular care must be taken to avoid contamination of the sample and to neutralise the chlorine residual in samples taken for the bacteriological examination. The part of the pool from which a sample is taken should always be recorded since results obtained at the shallow end when the bathing loads are heavy may differ from those at other parts of the pool.

12.13 Whenever a sample is taken for a bacteriological analysis a chlorine determination should be made on another sample taken as nearly as possible at the same time, and in the same place, as the first. The result of this determination should be taken into account in interpreting the bacteriological findings. If for instance there were less than 0·5 mg/l free residual present this might explain the presence of any coliforms found. The remedy would be plain.

12.14 The results of bacteriological examination are not however immediately available, and regular and frequent determinations of free residual, if chlorine is used, are far more important as a check on the condition of the water, and as a guide to immediate action. Sometimes a chlorine recorder is a justifiable expense. Otherwise the frequency of determination depends upon the pool and is a matter of experience. For no pool should it be less frequent than 2 or 3 times a day, and for many it should be at least once an hour. On large pools with continuous chlorination, samples taken near both inlet and outlet should always be tested; on small pools, where complete mixing is relied upon and chlorination is intermittent, then single samples taken before and after the morning chlorination, and once or twice during the rest of the day, again depending on experience and bathing load, may be adequate. Normally the pool attendant can be trained to do the test, but supervising officers should do a test whenever they pay a visit. A record should be kept of all determinations, not necessarily for retention permanently, but as evidence of the state of affairs over the previous few weeks.

12.15 The figure of 0·5 mg/l free residual chlorine has been mentioned as being adequate to ensure almost always an absence of coliforms from the water. But it may not always be adequate to ensure the continuance of the chlorine in the free residual form. Too much ammonia, from urea, may be introduced into the water for this, and once the free residual has been converted to combined residual, or decomposed to hydrochloric acid, it is not easy to re-establish it, and in the meantime the disinfectant is principally in the chloramine form and has much less efficacy. The free residual must therefore be large enough to maintain the residual in the free form, and the figure of at least 1 mg/l was recommended in the previous edition. In many cases this certainly is adequate, but when the bathing load is very heavy or chlorination is intermittent, and therefore some hours may elapse between additions, a still higher residual of up to 2 mg/l may be essential. At many pools it may be desirable to work at such a figure.

12.16 At the same time too much chlorine is not good. Even if it does not give rise to complaints of smells, eye smarting etc, people wish to bathe in water, not chlorine, and increased residual should not be used merely because it appears to be acceptable. The exact value adopted must suit the circumstances. It must be effective, and be known to be somewhat greater than the minimum necessary, but the upper limit should not be fixed with the idea of 'the higher the safer'. Anything over 2 mg/l needs a good deal of justifying.

12.17 If water for a pool, including make-up water, is not already of bacteriological quality equal to that of potable water, it should be suitably disinfected before it enters the pool. Backwash water, if not taken from the pool, should be of similar bacteriological quality.

Chemical Standards

12.18 Mains water is generally of chemical composition suitable for pools, and, except for sea water pools, is and should be used. Exceptionally, a private source, eg a deep borehole, may be available and subject to its bacteriological condition, may be used if desired. In that event the possibility of its containing dissolved iron or manganese which would form a precipitate on aeration should be examined; preliminary treatment could be necessary. River water should not be used. If, exceptionally, a case occurred where there was no other source it should first be given full treatment to potable standards. If sea water is used the point and times of abstraction should be chosen with care. Preliminary settling might well be necessary, and its essential bacteriological quality suitably assured.

12.19 The *p*H of pool water should be maintained as near as possible to 7·5 or 7·6 and always within the range 7·2 to 8·0. If dolomitic materials are relied on to correct the *p*H automatically the 'equilibrium' *p*H may well be somewhat different from the recommended value of 7·5 to 7·6, but it should normally be within the acceptable range and if this is so a departure from the optimum need cause no concern.

12.20 The purification system of pools is designed to kill bacteria and other organisms, to remove suspended impurities and to destroy as far as possible dissolved organic impurities. Inorganic impurities added to the water, and the remaining organic impurities or their oxidation products, are not destroyed and accordingly accumulate progressively. If no pool water were ever discarded this accumulation would continue indefinitely and a time would come when, on health or other grounds, the water would no longer be acceptable as bathing water. If pool water is gradually discarded and replaced by fresh water then the concentration of each impurity would eventually reach a constant value, the magnitude of which would depend upon the rapidity of water replacement. It might or might not reach an unacceptable level.

12.21 Very often backwash water is taken from the pool and replaced by mains water. The rate of replacement is variable depending on type of filter, frequency of backwashing and other factors. But in our view when pool water is used for backwashing the concentration of the impurities mentioned will never reach a figure which justifies serious consideration as to whether it is acceptable or not.

12.22 When the pool water is not used for backwashing, or when backwashing is not practised, as with pad filters which are discarded or laundered separately after use, there is little automatic replacement of pool water to counteract the indefinite accumulation of impurities, and the question how far this can be allowed to go on must be faced. To tackle the problem properly would necessitate much investigation and development work. The impurities would first of all have to be identified, and this could be difficult. Then methods for the estimation of the concentration of each would need to be worked out, and to be of practical use they would have to be such as could be performed simply and cheaply. Then the toxicological and other properties of the compounds would have to be determined and finally permissable figures derived. It seems to us that all this is quite unjustified merely in order to save a pool full of water now and again. We would prefer, from this angle, that pool water should be used for backwashing and then discarded. If this is not done the pool should be emptied and refilled once a year at the very least. With normal usage and proper and conventional treatment then neither health considerations nor amenity compel more frequent changing.

Physical Standards

Colour

12.23 No standard can be given for colour. Any colour except an apparent light blue may give an unfavourable impression. Deep colour and particularly colloidal colour, interferes with clarity and is therefore taken account of in the standard expressed as distance seen through. Chlorine and the use of coagulants both help to remove or prevent certain types of colour. Colour due to algae can be avoided by preventing their growth.

Clarity

12.24 This has been fully discussed (paras 6.7 to 6.22). In no circumstances should a pool be allowed to become so turbid that the botton in the deepest part cannot readily be seen. This is the very minimum requirement but much more is necessary in order to have attractive water. We believe that at a normal size pool the filtration system and its operation should be such that at times of normal high bathing loads, but not short period peak loads, a clarity of at least 12·2 m (40 feet) is maintained. This means that for much of the time this figure should be considerably exceeded. Any reduction to below 12.2 m (40 feet) due to temporary peak loading should be corrected within one turnover period of the end of the peak. The figure quoted would also be the minimum applicable to diving pools, but here there should be no question of abnormal peak loads of such a size as to bring the clarity below the figure quoted.

12.25 For small shallow pools such clarity is not necessary on safety grounds but they should be maintained at all normal loads in a condition that the average spectator would describe as 'clear' without any qualification. What this would mean in terms of distance seen through is not yet certain, but in many cases it would probably mean that the clarity instrument (the semi-periscope) could not be used because the clarity would be greater than the length of the pool. To fix a standard in these circumstances would be useless, nevertheless, a check that the clarity does exceed the length of the pool would be useful as evidence of the performance of the filtration plant.

12.26 No clarity need be laid down for genuine paddling pools. The chief concern should be to maintain as good a bacteriological condition as the nature of the case makes possible.

12.27 In due course it may be possible to relate clarity, as defined above to turbidity given by a specific type of instrument. In that event the determination of clarity would be greatly simplified and it would become possible to measure it in a small sample. This would make possible a standard filter performance, which at present cannot either be measured or even defined. A stage might then be reached when a requirement could be laid down to the effect that 'At a given rate of treatment per unit of filter area and with a liquid of clarity X the filtrate shall not be below a clarity Y'.

12.28 Pending that stage it seems still to be necessary to design filtration systems on the basis of a rate of filtration per unit area of filter, and this matter has already been discussed in some detail. Unfortunately, it is at present quite impossible to lay down any figure which is based either on experiment or on experience backed by quantitative evidence. Research is urgently required and it must be of a very rigidly controlled type if its results are to be authoritative. The essays we have ourselves been able to make in this field could not be nearly rigidly enough controlled and any conclusions from them are extremely dubious. They have done little more than underline the urgent necessity of special investigations. Meanwhile we can offer no more on the question of standardisation of filtration rates than has already been given (para 8.13). We do not repeat that here because it is necessary that it should be read in conjunction with the whole discussion.

References

[1] Davies, J. G. in *Microbial Aspects of Pollution*, G. Sykes and F. A. Skinner, Eds., Academic Press, London, England, 1971, pp. 1–9.

[2] Holden, W. S., *Water Treatment and Examination*, Churchill, London, England, 1970.

[3] Downing, A. L. in *Microbial Aspects of Pollution*, G. Sykes and F. A. Skinner, Eds., Academic Press, London, England, 1971, pp. 51–69.

[4] Majesty's Stationery Office, London, England, 1939, revised 1967.

[5] "Safeguards to be Adopted in the Operation and Management of Waterworks," Her Majesty's Stationery Office, London, England, 1939, revision 1967.

[6] Berg, G., *Transmission of Viruses by the Water Route*, Interscience, New York, 1967.

[7] Green, S. M., Scott, S. S., Mowat, D. A. E., Shearer, E. J. M., and Thomson, J. M., *Journal of Hygiene, Cambridge*, Vol. 66, 1968, p. 383.

[*8*] George, J. T. A.,Wallace, J. G., Morrison, H. R., and Harbourne, J. G., *British Medical Journal*, Vol. 3, 1972, p. 208.
[*9*] "8th Report," Royal Commission on Sewage Disposal, His Majesty's Stationery Office, London, England, 1912.
[*10*] Gameson, A. L. H., Munro, D., and Pike, E. B., in *Institute of Water Pollution Control, Symposium on Water Pollution in Coastal Areas*, 1970, p. 34.
[*11*] Harvey, R. W. S., Martin, D. R., Foster, D. W., and Griffiths, W. C., *Journal of Hygiene, Cambridge*, Vol. 67, 1969, p. 517.
[*12*] Fennell, H., *Journal of the Association of Water Officers*, Vol. 2, 1975, p. 19.
[*13*] McCoy, J. H. in *Microbial Aspects of Pollution*, G. Sykes and F. A. Skinner, Eds., Academic Press, London, England, 1971, pp. 33–50.
[*14*] "44th Report of the Results of Bacteriological Examination of the London Waters for the Years 1969–1970," Metropolitan Water Board, London, England, 1970.
[*15*] Fennell, H., James, D. B., and Morris, J., *Proceedings*, Society for Water Treatment and Examination, Vol. 23, 1974, p. 5.
[*16*] Gameson, A. L. H., Bufton, A. W. J., and Gould, D. J., *Water Pollution Control*, Vol. 66, 1970, p. 501.
[*17*] Gameson, A. L. H. in *Marine Pollution and Marine Waste Disposal*, D. A. Pearson and E. Frangipane, Eds., Pergammon Press, London, England, 1975, pp. 387–399.
[*18*] Savage, W. G., *The Bacteriological Examination of Water Supplies*, H. K. Lewis, London, England, 1906.
[*19*] Bonde, J. G., *Health Laboratory Science*, Vol. 3, 1966, p. 124.
[*20*] *Working Party on Sewage Disposal*, Her Majesty's Stationery Office, London, England, 1970.
[*21*] Moore, B. in *Microbial Aspects of Pollution*, G. Sykes and F. A. Skinner, Eds., Academic Press, London, England, 1971, pp. 11–32.
[*22*] Bradford Hill, A., *Proceedings of the Royal Society of Medicine*, Vol 58, 1965, p.295.
[*23*] Perry, W. D., Siegel, A. C., and Rammelkamp, C. H., *American Journal of Hygiene*, Vol. 66, 1957, p. 96.
[*24*] Brisou, J., *Bulletin of the World Health Organization*, Vol. 38, 1968, p. 79.
[*25*] Dutka, B. J., *Journal of Environmental Health*, Vol. 36, 1973, p. 39.
[*26*] "Council Directive Concerning the Quality Required of Surface Water Intended for Abstraction of Drinking Water in Member States (75/440), " *Official Journal of the European Communities*, No. L194, 25 July 1975, p. 26.
[*27*] Burman, N. P., *Proceedings*, Society for Water Treatment and Examination, Vol. 23, 1974, p. 355.
[*28*] "Report by the Public Health Laboratory Service Standing Committee on the Bacteriological Examination of Water Supplies," *Journal of Hygiene, Cambridge*, Vol. 67, 1969, p. 367.
[*29*] Gray, R. D., *Journal of Hygiene, Cambridge*, Vol. 62, 1964, pp. 495–508.
[*30*] "Report by the Public Health Laboratory Service Standing Committee on the Bacteriological Examination of Water Supplies," *Journal of Hygiene, Cambridge*, Vol. 67, 1969, p. 367.
[*31*] Gray, R. D. and Lowe, G. H., *Journal of Hygiene, Cambridge*, Vol. 76, 1976, p. 49.
[*32*] Hutchinson, M., *Proceedings*, Society for Water Treatment and Examination, Vol. 23, pp. 174–189.
[*33*] Burman, N. P. and Colbourne, J. S., *Journal of the Institute of Plumbing*, Vol. 3, No. 2, 1976, pp. 12–13.
[*34*] Pike, E. B. in *Ecological Aspects of Used-Water Treatment*, Vol. I, C. R. Curds and H. A. Hawkes, Eds., Acacemic Press, London, England, 1975.
[*35*] "Council Directive Concerning the Quality of Bathing Water (76/160)," *Official Journal of the European Communities*, No. L31, 2 May 1976, p. 1.
[*36*] "Report by a Working Party of the Public Health Laboratory Service," *Journal of Hygiene, Cambridge*, Vol. 57, 1959, p. 435.
[*37*] Moore, B., *Bulletin of Hygiene, London*, Vol. 29, 1954, p. 689.
[*38*] Moore, B. in *Water Pollution Control in Coastal Areas*, Institute of Water Pollution Control, London, England, 1970.
[*39*] Grunnet,K., Salmonella *in Sewage and Receiving Waters—Assessment of Health Hazards due to Microbially Polluted Waters*, FADL's Forlag, Copenhagen, 1975.

[40] Moore, B. in *International Symposium on Discharge of Sewage from Sea Outfalls*, London, 28 Aug. 1974.
[41] Shuval, H. I. in *International Symposium on Discharge of Sewage from Sea Outfalls*, London, 28 Aug. 1974.
[42] Shuval, H. I. in *Developments in Water Quality Research*, H. I. Shuval, Ed., Ann-Arbor Humphrey Science Press, Mich., 1970, pp. 47–71.
[43] Ingolls, R. S., *Proceedings*, Society for Water Treatment and Examination, Vol. 22, 1974, p. 147.
[44] "Guides and Criteria for Recreational Quality of Beaches and Coastal Waters," WHO Report Euro 3125 (1), World Health Organization, Geneva, 1975
[45] Wood, P. C., "The Production of Clean Shellfish," Ministry of Agriculture, Fisheries and Food, Fisheries Laboratory, Burnham-on-Crouch, England, 1969.
[46] Sherwood, H. P. and Thomson, S., *Monthly Bulletin of the Ministry of Health and the Public Health Laboratory Service*, Vol. 12, 1953, p. 103.
[47] Ayres, P. A., *Journal of Applied Bacteriology*, Vol. 39, 1975, pp. 353–356.
[48] Sakazaki, R. in *The Microbiological Safety of Food,* B. C. Hobbs and J. H. B. Christian, Eds., Academic Press, London, England, 1973, pp. 19–30.
[49] Sakazaki, R. in *The Microbiological Safety of Food*, B. C. Hobbs and J. H. B. Christian, Eds., Academic Press, London, England, 1973, pp. 375–385.
[50] Barrow, G. I. and Miller, D. C. in *Microbiology in Agriculture, Fisheries and Food*, F. A. Skinner and J. G. Carr, Eds., Academic Press, London, England, 1976, pp. 181–195.
[51] Barrow, G. I. and Miller, D. C., *Lancet*, Vol. 2, 1969, pp. 421–423.
[52] Taylor, F. B., *Journal of the American Waterworks Association*, Vol. 66, 1974, p. 306.
[53] Poynter, S. F. B., Slade, J. S. and Jones, H. H., *Proceedings,* Society for Water Treatment and Examination, Vol. 22, 1973, p. 194.
[54] Berg, G., *Bulletin of the World Health Organization*, Vol. 49, 1973, pp. 451, 461.
[55] Shuval, H. I. and Katzenelson, E., *International Symposium on Discharge of Sewage from Sea Outfalls*, London, 28 Aug. 1974.
[56] Jakubowski, W., Clarke, N. A., Hill, V. F., and Akin, E. W., *Water Research Centre Colloquium on Drinking Water Quality and Public Health,* Marlow, England, 4-6 Nov. 1975, Paper No. 9.

D. A. Hunt[1]

Indicators of Quality for Shellfish Waters

REFERENCE: Hunt, D. A., "**Indicators of Quality for Shellfish Waters,**" *Bacterial Indicators/Health Hazards Associated With Water, ASTM STP 635*, A. W. Hoadley and B. J. Dutka, Eds., American Society for Testing and Materials, 1977, pp. 337–345.

ABSTRACT: A chronological review of the history of significant research contributing to the development of current bacteriological criteria and standards for shellfish waters is described beginning with Prescott, Winslow, and Eijkman in 1904 and culminating in a proposal for a fecal coliform standard in 1974. Coliform and fecal coliform standards are defined, and the significance of the presence of these organisms in shellfish waters is discussed. The National Shellfish Sanitation Program presumes that both coliform and fecal coliform standards represent equivalent degrees of pollution and resultant health hazard when measuring levels of diluted sewage related to point sources of pollution in shellfish growing area waters. These two indicator groups are used in conjunction with sanitary surveys for the classification of shellfish growing areas and interpreted according to the limitations of the groups.

KEY WORDS: bacteria, water, coliform bacteria, coliform group, fecal coliform group, shellfish, treatment plant effluents, sanitary survey

In the United States, the development and use of indicator bacteria to assist in the determination of the sanitary quality of shellfish growing area waters have to a large degree paralleled and been influenced by the test procedures for potable water. Historically, state medical officers or state sanitary engineers responsible for the sanitary control of water supplies also have been responsible, either directly or indirectly, for the classification and control of shellfish growing area waters, and state laboratory directors have provided the analytical support for control programs of both types of water. Under these conditions, it might be predicted that the methodology, indicator systems, and technical developments used to monitor potable waters also would be evaluated by microbiologists concerned with pollution in marine or estuarine waters. The history of indicator systems development for shellfish waters is also a part of the history of sanitary microbiology of potable waters. The purpose of this paper is to describe the development of

[1]Assistant Chief, Shellfish Sanitation Branch, Division of Food Technology, Bureau of Foods, Food and Drug Administration, Washington, D. C. 20204.

bacteriological standards for shellfish growing area waters and the bacteriological indicator groups upon which the standards are based.

Historical Developments

The multiple tube fermentation method for the recovery of *Bacillus coli* and the relationship between the presence of *B. coli* and domestic sewage in receiving waters had been well established when Prescott and Winslow published *Elements of Water Bacteriology* in 1904. The authors [*1*][2] concluded that the "finding of a few colon bacilli in large samples of water, or its occasional discovery in small samples, does not necessarily have any special significance. The detection of *B. coli* in a large proportion of small samples (1 cc or less) examined is imperatively required as an indication of recent sewage pollution." The term "large proportion" was not defined [*2*], but the concept was established that the presence of coliform bacteria in a large proportion of 1-ml portions of a water sample indicated a potential health hazard. By these statements, the authors described an indicator criterion and standard for the detection of fecal wastes in water which has continued to influence the promulgation of microbiological standards for environmental waters. Another paper published in 1904 also influenced microbiological criteria and standards for shellfish growing area waters. Eijkman [*3*], noting that some organisms recovered in the test for *B. coli* were ubiquitous in the environment and not necessarily indicative of fecal contamination, recommended a high incubation temperature of 46°C to inhibit growth of nonfecal types.

These two publications provided the basic concepts which led to the development of the National Shellfish Sanitation Program's "approved" growing area microbiological criteria and standards [*4*] which are used in the United States, Canada, Japan, Korea, Mexico, and perhaps other countries.

National Shellfish Sanitation Program Developed

Following the shellfish borne outbreak of typhoid fever during the 1924-1925 oyster harvest season [*5*], the Report of the Committee on Sanitary Control of the Shellfish Industry in the United States [*6*] was submitted to the Surgeon General, U.S. Public Health Service, giving rise to what is presently known as the National Shellfish Sanitation Program (NSSP). The report also might be considered to be the first edition of the NSSP's *Manual of Operations* [*4*], as the basic concepts for sanitary control of the shellfish industry outlined in the report form the principal sections of the 1965 edition of the manual.

Concerning a growing area microbiological standard, the report states,

[2]The italic numbers in brackets refer to the list of references appended to this paper.

"the committee is not prepared to recommend any precise bacterial standards for water ... until additional data ... have been assembled and considered. In light of present knowledge, it would probably be unfair and unnecessary to apply to such water the rigid standards which are applied to the drinking water supplied in interstate commerce. It is considered, however, that the waters should ordinarily not show the presence of *B. coli* in 1 cc amounts, tests for *B. coli* being made in 10 cc, 1 cc and 0.1 cc amounts, according to the Standard Methods of the American Public Health Association." Three significant steps in the development of an "approved" growing area microbiological standard were established: the method was standardized, the indicator group defined, and a standard recommended.

Early Studies on Shellfish Growing Areas

Perry [*7, 8*] published a comprehensive study on the significance of indicator organisms in oysters and overlying waters in 1928 and 1929. This was followed by a series of studies by Hajna and Perry [*9–11*] on a modification of the Eijkman test for the recovery of *Escherichia coli*. On the basis of these studies, Perry [*12*] made the following recommendations to the American Public Health Association's (APHA) Committee on the Examination of Shellfish for Fecal Pollution in 1936. (1) The whole oyster rather than just the shell liquor should be examined; (2) the new procedure should include such edible molluscs as oysters, clams, and mussels; and (3) *E. coli* rather than the colon group should be the index of pollution for both shellfish and shellfish waters. In another publication in 1939, he stated [*13*], "many coliform bacteria, particularly of the *cloacae* types, are present in shucked market oysters or shell oysters when the temperature exceeds 15.6°, but are without significance as indicating pollution. *Escherichia coli* would seem to be the logical bacterium to use as an index of fecal pollution in shellfish and shellfish growing waters."

Six revisions were made in the APHA methods for Bacteriological Examination of Shellfish and Shellfish Waters between 1935 and 1943. The 1943 Report of Committee [*14*] listed the modified Eijkman Lactose Medium incubated at 45.5°C as an alternative method to the eosin methylene blue (E.M.B.)–Koser Citrate test for confirming the presence of *E. coli*.

It was also in 1943 that Hajna and Perry [*15*] published the results of studies on the EC Medium which was a modification of the Modified Eijkman test. The formula was modified by increasing lactose from 3 to 5 g and adding 1.5 g of bile salts. Results of tests were reported as *E. coli* most probable number (MPN). The following year the Recommended Procedure for the Bacteriological Examination of Shellfish and Shellfish Waters [*16*] stated, ". . . there is some evidence not fully confirmed that some coliforms found in salt water may be of little sanitary significance . . . there is every justification for the use of recognized bacteriological procedures for the determination of the coliform group of bacteria, for in the great majority of cases the sanitary engineer can demonstrate beyond doubt that the presence of coliform organisms is directly related to pollution of human or animal origin. In the revised procedure, the short methods previously suggested for estimating *Escherichia coli* were eliminated as unsatisfactory." Thus, the committee disregarded 16 years of work on an elevated temperature test and Dr. Perry resigned from the committee. The formula for the EC or fecal coliform test has remained unchanged, but the incubation temperature was decreased by later researchers, and water bath incubators replaced air incubators for stricter temperature control.

In a study to develop a wholesale market standard for oysters conducted by state and Public Health Service Laboratories in 1958 and reported by Kelly in 1960 [*17*], data from two of the Maryland State Department of Health Laboratories using water jacketed incubators at 45.5°C were statistically indistinguishable from data from other laboratories using water bath incubators at 44.5°C. Therefore, the third edition of *Recommended Procedures for the Bacteriological Examination of Sea Water and Shellfish,* published in 1962, listed both procedures and incubation temperatures. The research of Geldreich and associates at the Taft Center in Cincinnati [*18, 19*], Tennant and associates in Canada [*20, 21*], Hosty and Perry at the Alabama and Maryland State Health Laboratories, Stevens, Grasso, and Delaney [*22*] at the Lawrence Experiment Station, Fishbein and associates in the Food and Drug Administration's Washington laboratories [*23*] Bidwell and Kelly [*24*], and Andrews, Presnell, and others in the Public Health Service's Shellfish Sanitation Laboratories [*25*] made substantial contributions toward the understanding of the relationships between the coliform and fecal coliform groups in determining the sanitary quality of growing area waters and shellfish meats.

During the 1958 National Shellfish Sanitation Workshop, Dr. Arthur Novak [*26*], Louisiana State University, suggested amending the coliform definition as follows: "Bacteria of this group (coliforms) which will produce gas from EC Medium within 48 hours at 44.5°C in a water bath will be referred to as 'fecal coliforms'." This definition was accepted by the workshop, with the understanding that the EC test recovered bacteria other than *E. coli*, and that bacteria recovered by the EC test were presumed to be of fecal origin and therefore of greater sanitary significance than the wider spectrum of organisms recovered by the standard coliform test.

The first official growing area microbiological standard was published in 1946 in the Manual of Recommended Practice for Sanitary Control of the Shellfish Industry [*27*]. It was defined as follows: "The median bacteriological content of samples of water . . . shall not show the presence of organisms of the coliform group in excess of 70/100 ml of water . . . ." By 1965, [*4*] the standard had been modified, and stated, "a) The coliform median MPN of the water does not exceed 70/100 ml; b) Not more than 10% of the samples ordinarily exceed an MPN of 230/100 ml for a 5-tube decimal dilution test or 330/100 ml where the 3-tube decimal dilation test is used) in those portions of the area most probably exposed to fecal contamination during the most unfavorable hydrographic and pollution conditions; and c) These foregoing limits need not be applied if it can be shown by detailed study that the coliforms are not of fecal origin and do not indicate a public health hazard."

Studies Leading to a Fecal Coliform Standard

Studies comparing EC and coliform data from shellfish growing area waters in Massachusetts, Maryland, Virginia, Alabama, Canada, and Oregan were reported in the Proceedings of the 1961 National Shellfish Sanitation Workshop [*28*]. The Maryland report stated that a 70 coliform MPN value was equivalent to an *E. coli* value of 14. The summary of the Massachusetts report stated "it does not appear that the overall classification of any of the areas would be changed materially if the *E. coli* (fecal coliform) data were used instead of the coliform data . . . ." The Canadian report states, "We regard the E.C. confirmation test (fecal coliform) as a useful, effective method for the determination of *E. coli* densities, but we do not believe that it should replace the standard coliform test in the bacteriological control of Canadian shellfish growing areas." The Alabama report stated, "A more significant association with pollution conditions was revealed by the EC test (fecal coliform) since the proportion of coliforms that responded positively to the test decreased with remoteness from pollution." Thus, we see a wide interest in the test and a variety of opinions of its value.

The data collected from individual studies reported at the 1961 workshop were tabulated and analyzed by Beck [*29*] and presented at the 1964 workshop with the following recommendations for an approved growing area standard: "The fecal coliform test enumerated domestic pollution potential in each area on a more consistent basis than the coliform test . . . . In an approved shellfish growing area, a median fecal coliform MPN of 7.8 shall not be exceeded and not more than 10% of all samples tested shall exceed an MPN/100 ml in excess of 33 (46 per 100 ml where the 3-tube decimal dilution test is used)."

The workshop took no affirmative action on the recommendation. One weakness of the study was that the data were not correlated with specific classified NSSP growing areas according to NSSP standards and criteria. The 1968 workshop concluded that the fecal coliform criterion showed promise and recommended another study. In 1968, Strobel [*30*] reported that "equivalent fecal coliform values for a median coliform MPN of 70/100 ml ranged from 15 to 67 per 100 ml" for the areas he studied in New York. He also stated "the present coliform standard used as a component part of the sanitary survey in the certification of shellfish waters, was derived from a study of Raritan Bay by Kerr in 1941" [*31*]. This statement is misleading as the Raritan Bay study, in general, reaffirmed standards of safety for shellfish water quality which had been used by public health officials in the United States since the early part of the twentieth century. Cook [*32*] reported in 1969 that fecal coliform equivalents to a 70 coliform value in waters in Mississippi Sound ranged from 13 to 20.

During the 1971 workshop, Slanetz et al [*33*] recommended "a standard of not over 12 fecal coliforms per 100 ml of water, with a tolerance of that not more than 10 percent of the samples shall exceed 15 per 100 ml . . . ." Salinger [*34*] arrived at a 7.8 MPN standard using Maryland data with upper 90 percentiles of 33 and 43 for 5-tube and 3-tube decimal dilution tests. Presnell [*35*], working with Gulf coast waters, recommended "a median fecal coliform MPN of 23/100 with not more than 10 percent of samples in excess of 79/100 ml and no sample with an MPN in excess of 130/100 ml samples . . ." for the 5-tube test. Presnell's proposal is the only one presented with a maximum limit.

An NSSP Microbiology Task Force met in Washington, D.C., in 1973 to review shellfish growing area standards, criteria, and methodology. This task force included professional microbiologists from Environmental Protection Agency, National Marine Fisheries Service, Food and Drug Administration (FDA), Environment Canada, industry organizations, the academic community, and state shellfish control program representatives. The task force concluded that the fecal coliform group is scientifically and logistically superior to the coliform or fecal streptococci indicator groups as a microbiological indicator of fecal pollution in estuarine waters. It recommended that FDA and a select group from the task force meet and develop a plan to collect fecal coliform data on selected estuaries for a nationwide study, including Canada, to be presented to the 1974 workshop. The study protocol was developed and participating control agencies were requested to collect both coliform and fecal coliform data from routine monitoring stations on both sides of the closure lines in representative growing areas. Sixteen shellfish producing states and two Canadian provinces participated in this study.

It is recognized that coliform/fecal coliform ratios will vary according to distance from pollution source, dilution, degree of treatment, and possibly other factors. For this reason, the study concentrated on the bacteriological quality of the water at that point in the estuaries where coliform and sanitary survey data indicated that restriction of harvesting was necessary. The fecal coliform equivalent value to the 70 MPN was 14 for data on both sides of the closure line; overall, there was good correlation between coliform and fecal coliform data. On the basis of this study, supported by the conclusions of studies previously referred to, FDA proposed [*36*] a fecal coliform microbiological standard as follows: "The median fecal coliform MPN value for a sampling station shall not exceed 14 per 100 ml of sample and not more than 10 percent of the samples shall exceed 43 for a 5-tube, 3 dilution test or 49 for a 3-tube, 3 dilution test." The proposal was accepted by the Microbiology Task Force which recommended it to the workshop. At the present time, the NSSP accepts the total coliform or the fecal coliform criterion. It should be noted that Canadian data from the Atlantic region indicate that fecal coliform organisms constitute 25 to 60 percent of the total coliform flora.

At a meeting between the FDA and Canadian shellfish control personnel in September 1974, Menon, a microbiologist with the Canadian Environmental Protection Service, working with Tennant's data, indicated that a fecal coliform MPN of 23 is equivalent to a 70 coliform value in Canadian Atlantic waters.

Utilization of Indicator Groups

A bacterial indicator used as an adjunct to sanitary surveys for the classification of shellfish growing area waters should detect the presence of viable fecal contamination at relatively high dilutions, distinguish between bacterial flora of sanitary significance and bacterial flora of little or no sanitary significance, be relatively simple and easy to enumerate, be legally acceptable, and correlate with point sources of pollution as determined by shoreline survey and hydrographic and meteorological data. No single indicator group contains all of the desired characteristics; therefore, each group should be interpreted according to its specific limitations.

The NSSP *Manual of Operations* [*4*] states "... a water MPN of 70/100 ml is equivalent to a dilution ratio of about 8 million cubic feet of coliform free water per day for the fecal material from each person contributing sewage to the area ...." This degree of dilution of fecal wastes in the estuary is presumed by the FDA to represent an adequate margin of safety to prevent enteric disease transmission by shellfish. Although both coliform and fecal coliform standards theoretically represent equivalent levels of fecal dilution in growing area waters affected by point sources of pollution, the coliform criterion permits a greater range in MPN values than the fecal coliform standard when waters contain less than 70 coliforms per 100 ml of sample. On the other hand, the fecal coliform criterion restricts growth of nonfecal types, resulting in a more specific but less sensitive test, when used in conjunction with the 14 MPN standard.

Conclusion

It is recognized that there is no constant pathogen/indicator ratio in sewage or in waters receiving sewage treatment plant effluents. The ratio changes according to the level of disease in the population, type or degree of sewage treatment, sewage dilution by storm drains or other types of wastewater, dilution or distance from the point source in the receiving waters, die off characteristics, and other factors. Examination of large numbers of samples of water for enteric pathogens or viruses is technically feasible but economically prohibitive for monitoring hundreds or thousands of sampling stations.

At the present stage of technology, shellfish control agencies in the United States rely upon the sanitary survey and bacteriological monitoring for coliform or fecal coliform or both indicator organisms to determine the incidence and concentrations of fecal wastes in shellfish growing area waters. The establishment of closure lines in shellfish areas affected by point sources of pollution is based upon the detection and enumeration of viable indicator organisms of fecal origin, as the presence of detectable levels of feces in shellfish waters establishes a potential for shellfish borne illness.

If closure lines were determined on the basis of the detection of enteric microbial pathogens and viruses only, the safety factor provided by the coliform indicator groups would be eliminated and a real health hazard to shellfish consumers would be established. Specific types or degrees of wastewater treatment may result in bacterial destruction and virus survival. Under these conditions, the indicator role of the bacterial group is negated and the surviving viruses may be utilized as indicators of pollution. However, under conditions of continuous pollution from waste treatment outfalls, as exist in shellfish growing areas in the United States, bacterial indicator groups usually can be traced from point sources of pollution to greater distances in the estuary, in higher dilutions and smaller sample volumes, and at a fraction of the time and cost than is the case with viruses. The coliform and fecal coliform standards and criteria, when applied as adjuncts to sanitary surveys according to NSSP guidelines, are believed to be of equal value in determining the degree of pollution and corresponding health hazard potential of shellfish growing waters.

References

[*1*] Prescott, S. C. and Winslow, C. E. A., *Elements of Water Bacteriology*, Wiley, New York, 1904, p. 81.

[*2*] Furfari, S. A., "History of the 70/100 ml MPN Standard," Memorandum to Chief, Water Resources Program, National Center for Urban and Industrial Health, 1968.

[*3*] Eijkman, C., *Centralblatt f, Bakt. Ab 1*, Vol. 0, No. 37, p. 742, 1904.

[*4*] *Manual of Operations*, Part I, National Shellfish Sanitation Program, 1965.

[*5*] Lumsden, L. L., Hasseltine, H. E., Leake, J. P. and Veldee, M. V., "A Typhoid Fever Epidemic Caused by Oysterborne Infection," *U.S. Public Health Reports*, Supplement No. 50, 1924–25.

[*6*] Report of Committee on Sanitary Control of the Shellfish Industry in the United States, *Public Health Reports*, Supplement No. 53, 1925.

[*7*] Perry, C. A., *American Journal of Hygiene*, Vol. 8, No. 5, 1928, pp. 694–722.

[*8*] Perry, C. A., *American Journal of Hygiene*, Vol. 8, 1929, pp. 580–613.

[*9*] Perry, C. A. and Hajna, A. A., *Journal of Bacteriology*, Vol. 26, 1933, pp. 419–429.

[*10*] Hajna, A. A. and Perry, C. A., *Journal of Bacteriology*, Vol. 30, 1935, pp. 479–484.

[*11*] Perry, C. A. and Hajna, A. A., *American Journal of Public Health*, Vol. 25, 1935, pp. 720–724.

[*12*] Examination of Shellfish for Fecal Pollution Committee Report, American Public Health Association Yearbook, 1935–36, pp. 111–117.

[*13*] Perry, C. A., *Food Research*, Vol. 4, No. 4, 1939, p. 394.

[*14*] "Recommended Methods of Procedure for Bacteriological Examination of Shellfish and Shellfish Waters," *American Journal of Public Health*, Vol. 33, No. 5, May 1943, pp. 582–591.

[*15*] Hajna, M. S. and Perry, C. A., *American Journal of Public Health*, May 1943, pp. 550–556.

[*16*] "Recommended Procedure for the Bacteriological Examination of Shellfish and Shellfish Waters," *American Journal of Public Health*, Sept. 1947, pp. 1122–1129.

[*17*] Kelly, C. B., "Bacteriological Criteria for Market Oysters," DHEW/PHS, Robert A. Taft Sanitary Engineering Center Technical Report F60-2, 1960, pp. 1–16.

[*18*] Geldreich, E. E., Clark, H. F., Kabler, P. W., Huff, C. B., and Bordner, R. H., *Applied Microbiology*, Vol. 6, 1958, p. 347.

[19] Geldreich, E. E., Bordner, R. H., Huff, C. B., Clark, H. F., and Kabler, P. W., *Journal of the Water Pollution Control Federation*, Vol. 34, No. 3, March 1962, pp. 295–301.

[20] Tennant, A. D. and Reid, J. E., *Canadian Journal of Microbiology*, Vol. 7, 1961, pp. 728–739.

[21] Tennant, A. D., Reid, J. E., and Bastien, J. A. P., "A Comparison of the Coliform and Fecal Coliform Indices of Water Pollution with Special Reference to Canadian Atlantic Shellfish Growing Areas," Laboratory of Hygiene M. S. Report, No. 64-6, 1964.

[22] Stevens, A. P., Grasso, R. J., and Delaney, J. E., *Proceedings*, Eighth National Shellfish Sanitation Workshop, DHEW/PHS/FDA, No. (FDA) 75-2025, 1974, pp. 132–136.

[23] Fishbein, M., Surkiewicz, B. F., Brown, E. F., Oxley, H. M., Padron, A. P., and Groomes, R. J., *Applied Microbiology*, March 1967, pp. 233–238.

[24] Bidwell, M. H. and Kelly, C. B., *American Journal of Public Health*, Vol. 40, pp. 923–928.

[25] Andrews, W. H., Diggs, C. C., Presnell, M. W., Miescier, J. J., Wilson, C. R., Goodwin, C. P., Adams, W. N., Furfari, S. A., and Musselman, J. F., *Journal of Milk and Food Technology*, Vol. 38, No. 8, Aug. 1975, pp. 453–456.

[26] *Proceedings*, Shellfish Sanitation Workshop, Department of Health, Education and Welfare/Public Health Service, Aug. 26–27, 1958, p. 6.

[27] Manual of Recommended Practice for Sanitary Control of the Shellfish Industry, Public Health Bulletin No. 295, Federal Security Agency, U.S. Public Health Service, Washington, D.C., 1946.

[28] Jensen, E., *Proceedings*, Shellfish Sanitation Workshop, DHEW/PHS, November 28–30, 1961.

[29] Beck, W. J., *Proceedings*, Fifth National Shellfish Sanitation Workshop, Leroy Hauser, Ed. DHEW/PHS, 17–19 Nov. 1964, pp. 143–154.

[30] Strobel, G. A., *Journal of the Sanitary Engineering Division*, Proceedings of the American Society of Civil Engineers, Vol. 94, No. SA4, Aug. 1968, pp. 641–655.

[31] Kerr, R. W., Levine, B. S., Butterfield, C. T. and Miller, A. P. "A Report on the Public Health Aspects of Clamming in Raritan Bay," Public Health Service Report, 1941. Reissued by Division of Sanitary Engineering Services, PHS, DHEW in June 1954.

[32] Cook, D. W., "A Study of Coliform Bacteria and *Escherichia coli* in Polluted and Unpolluted Oyster Bottoms of Mississippi and a Study of Depuration by Rebedding," Gulf Coast Research Laboratory, Ocean Springs, Mississippi.

[33] Slanetz, L. W., Bartley, C. H., and Stanley, K. W., *Proceedings*, Seventh National Shellfish Sanitation Workshop, DHEW/PHS, 20-22 Oct. 1971, pp. 197–205.

[34] Salinger, A. C., *Proceedings*, Seventh National Shellfish Sanitation Workshop, DHEW/PHS, 20–22 Oct. 1971, pp. 206–207.

[35] Presnell, M., *Proceedings*, Seventh National Shellfish Sanitation Workshop, DHEW/PHS, 20–22, Oct. 1971, pp. 210–213.

[36] Hunt, D. A. and Springer, J., *Proceedings*, Eighth National Shellfish Sanitation Workshop, DHEW/PHS/FDA, No. (FDA) 75-2025, 1974, pp. 97–104.

Summary

We noted in our introduction to this publication that it was our objective to provide a review of concepts of indicators and the classification, significance, and ecology of bacterial indicators and pathogens of current and emerging interest as well as their application as indicators of water quality. It was our intent to challenge current concepts, put different indicator systems into perspective, and provide insight into alternative approaches to ensure the safety of waters for various uses. It was our wish in particular to serve the individual or public agency called upon to establish or review water quality objectives and standards, select bacterial indicators of public health hazards, or interpret water quality data.

In this publication, we have developed the historical basis for existing bacterial standards of water quality, not only in North America but in Europe and South Africa as well. Barrow has examined the British attitude towards water quality criteria in relation to microbial health hazards. His paper provides a comprehensive review of concepts of indicators of potential health hazards which should be of particular value to the reader concerned with standards. The logic behind the use of the coliform group, and *Escherichia coli* in particular, as indicators of fecal pollution and their past success justify their continued use. The value of this paper is enhanced by the appended Report No. 71 on the Bacteriological Examination of Water Supplies and the report on the Purification of the Water of Swimming Pools. In Germany, total plate counts at 20°C have long been used, and the history of their use has been reviewed by Müller. Coliforms and *E. coli* in drinking water are also limited by the German Drinking Water Law. Confirmation of coliforms and *E.coli* includes the cytochrome oxidase test to differentiate these from *Aeromonas* which may be present in water samples causing gas formation in lactose broth. As in Britain, other indicators may be employed under special circumstances. No regulations exist for surface swimming waters, but a test for *E. coli* is commonly employed. A test for salmonellae may be employed, and these organisms are regarded as a potential health hazard when present. The standard for swimming pools requires that water quality be suitable for drinking. However, other indicators, such as fecal streptococci and *Pseudomonas aeruginosa*, may be of use in the evaluation of swimming water quality. Guidelines prepared by the World Health Organization and discussed by Suess likewise call for assays of pathogens under certain circumstances.

It is important, in selecting an indicator, to know what it really indicates. That this is not always evident was emphasized by Barrow and McCabe.

Pseudomonas aeruginosa, staphylococci, vibrios, *Yersinia*, clostridia, *Bifidobacterium*, and *Candida albicans*. Those aspects of significance, classification, methodology, and behavior of each in water which might influence their use as indicators are examined. Grabow also concluded on the basis of his studies of R plasmids associated with coliforms and recommended that when water quality standards permit the presence of coliforms, they should specify that these organisms not carry R plasmids bearing resistance determinants for antibiotics commonly employed for therapy. Since drinking waters should be free of coliforms, such a standard might apply primarily to recreational, irrigation, or shellfish waters.

Farmer and Brenner examined taxonomic questions which have begun to receive attention, particularly as they relate to classification of coliforms and to distinctions among strains of indicators and pathogens of environmental or fecal origin. Classification may not be simple. New tools have made it possible to clarify relationships and to establish the bounds of some species. A better understanding of the distinctions among environmental strains which may not be pathogenic or indicate fecal pollution and strains capable of causing disease or of fecal origin may be important. These questions appear also in many of the discussions of individual indicators and pathogens.

In five papers, the application of indicators in the assessment of water quality for shellfish, recreation, swimming pools, and drinking is examined. Kraus in a review and updating of the 1970 report of the Committee on Environmental Quality Management on "Engineering Evaluation of Virus Hazard in Water," presents a strong case for the use of cyanophage as an ideal indicator of polluted aquatic environments. Hunt reviewed the history of research contribution to the development of coliform and fecal coliform standards for shellfish waters. Cabelli examined the relationship of potential indicators of health hazards associated with recreational waters to disease. Such indicators need not be indicators of fecal pollution. Cabelli stressed, as he and other authors already had done, the need to relate rates of disease to bacterial densities in water. In studies at marine beaches, a statistically significant correlation was demonstrated between *E. coli* and enterococci and gastrointestinal symptoms among swimmers. Mood reviewed bacterial contaminants of swimming pool waters, their significance, value as indicators, and susceptibilities to chlorine, concluding that, when a bacteriological test was desired, a total plate count at 35°C normally was the most useful. Finally, Ptak and Ginsburg reviewed comprehensively indicators in drinking waters, including the standard plate count.

This publication, devoted primarily to an examination of bacteriological standards and indicator systems, would not fulfill its purpose if it did not include discussion of alternatives to bacteriological standards. In 1972, Geldreich[1] proposed that satisfactory plate counts could be achieved in

[1]Geldreich, E. E., Nash, H. D., Reasoner, D. J., and Taylor, R. H. *Journal of the American Water Works Association*, Vol. 64, 1972, pp. 596–602.

drinking waters if chlorine residuals of 0.3 mg/litre and turbidities of less than 1 turbidity units could be maintained. The rationale for the turbidity standard has been developed more thoroughly by Symons and Hoff,[2] who cited reduced efficacy of chlorination at higher turbidities, the breakthrough of viruses as turbidities exceed 0.5 turbidity units, reduced taste and odor problems, and reduced nutrients for bacterial regrowth in distribution systems. The limit on turbidity has been incorporated into the National Interim Primary Drinking Water Regulations in the United States. These regulations include the provision that a supplier of drinking water may, with the approval of the state and based upon a sanitary survey, substitute the use of free chlorine residuals for not more than 75 percent of the required bacteriological samples. McCabe in his paper has challenged the use of indicator bacteria for quality control in community water systems and has developed the rationale for residual chlorine standards. This paper contains much to provoke thoughtful consideration of the objectives of bacteriological examination of waters. However, Ptak and Ginsburg concluded that retention of the coliform test, with restrictions on turbidity and chlorine residual, supplementation of total coliform tests with standard plate counts, and the institution of multiple parameter tests when high counts are encountered seems more logical.

A. W. Hoadley
School of Civil Engineering, Georgia Institute of Technology, Atlanta, Ga., editor.

B. J. Dutka
Microbiology Laboratories, Canada Centre for Inland Waters, Ontario, Canada; editor.

[2]Symons, J. M. and Hoff, J. C. in American Water Works Association 1975 Water Quality Technology Conference, Atlanta, Ga., Paper No. 2.

Index

S

T

V